THE VALUES OF PRECISION

THE VALUES OF PRECISION

Edited by M. Norton Wise

PRINCETON UNIVERSITY PRESS

PRINCETON, NEW JERSEY

PUBLISHED BY PRINCETON UNIVERSITY PRESS, 41 WILLIAM STREET,
PRINCETON, NEW JERSEY 08540
IN THE UNITED KINGDOM: PRINCETON UNIVERSITY PRESS,
CHICHESTER, WEST SUSSEX

LIBRARY OF CONGRESS CATALOGING-IN-PUBLICATION DATA

THE VALUES OF PRECISION / EDITED BY M. NORTON WISE
P. CM.
INCLUDES INDEX.
ISBN 0-691-03759-0
1. PHYSICAL MEASUREMENTS—HISTORY.
I. WISE, M. NORTON.
QC39.V26 1994 389'.09—DC20 94-19562 CIP

THIS BOOK HAS BEEN COMPOSED IN ADOBE SABON

PRINCETON UNIVERSITY PRESS BOOKS ARE PRINTED
ON ACID-FREE PAPER AND MEET THE GUIDELINES FOR
PERMANENCE AND DURABILITY OF THE COMMITTEE ON
PRODUCTION GUIDELINES FOR BOOK LONGEVITY
OF THE COUNCIL ON LIBRARY RESOURCES

PRINTED IN THE UNITED STATES OF AMERICA

1 3 5 7 9 10 8 6 4 2

CONTENTS

PREFACE AND ACKNOWLEDGMENTS

THIS VOLUME launches the Princeton Workshop in the History of Science, a series of volumes stemming from workshops hosted by the Program in History of Science at Princeton University. Every year or two the workshop takes up a theme of current interest to historians of science and to the larger intellectual community. For 1991–92 the topic was "Values of Precision," which this volume represents. Subsequent topics will be "Visualization in the Sciences" and "Materials." The sessions of each workshop meet on Saturdays three or four times per year. Each session normally brings three precirculated papers, whose authors discuss their intentions and what they think they have accomplished. A prepared commentary by one of the participants provides constructive criticism. Discussion, however, is the primary goal and occupies most of each session. Authors then rework their papers to respond to these discussions which, as they continue through the year, open new dimensions and develop recurring themes, uniting the papers in unexpected ways.

Like the workshops themselves, the published results should maintain the sense of an ongoing discussion. We intend to open topics rather than to close them, to suggest interesting directions of research and to pursue some of them, conscious always of what has been left unexplored and what has evaded explanation. We do, however, have a loose agenda which represents the present direction of the Program. Wishing to avoid what have become sterile arguments over internal versus external causation in the history of science, of realism versus relativism, and of social versus cognitive factors, we are attempting to elaborate a broadly cultural history which precludes simple dichotomies. Some have dubbed it "cultural internalism." Our aim is to thoroughly embed the technical practices of the sciences in the cultural settings in which they find their resources. Typically those settings include both very local ones like laboratories and more global ones like nations. And they are temporal as well as spatial, whether at the level of traditions of instrumentation or of political traditions. We want to encourage a variety of studies which span the entire spectrum of such "cultural" interpretation and which merge without abrupt shifts of ideology or vain attempts to isolate "nature" from "nurture."

In pursuit of this aim, we seek workshop topics which seem to preclude the dichotomies from the outset. "Precision" is such a topic. It has forced us to look simultaneously at the wide variety of means that people have invented for attaining precision and at the wide variety of purposes it has served. Thus, taking our cue from a pun, we have moved constantly along the axis from "precision values" to "values of precision." The result of all this flexibility, we hope, has enriched the pun, showing how numerical values attach to things that are valued, and how the attachment informs both the means of valuing and the valued thing.

The stimulation which this workshop provided is owing to the energetic participation of authors, commentators, and audience alike, numbers of whom came regularly from as far as Washington, D.C., and New Haven. I thank them all for making it happen. That it went off with few hitches is owing to the constant attention to organization of Ms. Peggy Reilly, whom many of the participants have particular reasons to thank, whether for sending papers, arranging accommodations, rapid reimbursement, or just lunch. For all of this, and for her professional skill in assisting the editorial process, I am most grateful. My colleagues Gerry Geison, Charles Gillispie, Mike Mahoney, and Nancy Nersessian constantly supplied ideas, comments, and enthusiastic support, which made the process most enjoyable. Elaine Wise supplied hard work, hospitality, and as always, her good judgment and sharp wit.

M. Norton Wise
June 1993

THE VALUES OF PRECISION

INTRODUCTION

M. Norton Wise

PRECISION," as we use it here, concerns primarily precision instruments and precision measurements. Linguistic and conceptual precision are secondary. But especially in its quantitative sense, precision carries immense weight in the twentieth century. A glance at the yellow pages of any phone book will show that the term validates popularly not only luxury automobiles, air conditioning, electric wiring, and satellite communications; but haircuts, mortgages, and paint jobs. It connotes trustworthiness and elegance in the actions or products of humans and machines. Precision is everything that ambiguity, uncertainty, messiness, and unreliability are not. It is responsible, nonemotional, objective, and scientific. It shows quality. These "values of precision" have become part of our heritage. We tend to transfer them automatically to sharply delimited statements and objects, especially numerical ones, like a 40 percent chance of rain, even when the numbers mean little. But how did numbers come to represent the cutting edge of modernity?

Genesis

It was not always so, either popularly or scientifically. This volume explores some of the ways in which quantitative precision (and to some degree the qualitative precision associated with it) came to be a value of such import in Western culture. Most scholars now agree that exact measurement emerged only in the late eighteenth century as a general characteristic of the physical sciences. In originally launching this claim, Thomas Kuhn referred to the movement toward precision in all things as a "second scientific revolution," in which the classical mathematical sciences—mechanics, optics, astronomy—extended their reach into the so-called "Baconian sciences" of the seventeenth century—heat, electricity, magnetism, chemistry. John Heilbron has importantly revised Kuhn's thesis to argue that it was not mathematicians turning to natural philosophy ("physics" in the Aristotelian sense), but the reverse. Natural philosophers began to assimilate parts of so-called mixed mathematics (e.g., mechanics) to their own relatively new efforts at quantification in heat, magnetism, electricity, and associated fields like chemistry. Ian Hacking too revises Kuhn's "revolution" to argue that it was the Baconians who did the work of quantifying. He is interested primarily in the "avalanche of numbers" produced in the collection and publication of statistical data in the early nineteenth century, but his Baconian thesis is general.[1]

The unproblematic core of these differing studies is their recognition that it was only in the late eighteenth century that enthusiasm for precision instruments and numbers developed over a broad spectrum, extending beyond astronomical and optical instruments. The marine chronometer and the pocket watch symbolize this development for, respectively, the world and the individual, for navigating with precision to locate ships of war and commerce and for regulating with precision the work schedules of a self-consciously responsible middle class. Control of materials and machining and an associated division of labor in the burgeoning watchmaking industry led the way for the factory system in other areas.[2] More important for the sciences were new instruments that measured quantities of matter, both ponderable and subtle: perfected chemical balances, electric and magnetic torsion balances, thermometers and calorimeters, and other instruments for reifying quantities of an even more nebulous physical nature, such as eudiometers for measuring the quality of the air. Equally significant were new instruments of calculation and counting, from the grand schemes of the variational calculus and probability calculus to mundane techniques for reducing to reliable numbers the amount of wood in a forest or the average consumption of meat for a citizen of the new French republic.[3]

So far we are speaking of instruments of precision in a narrow sense, that of the technical means that people use to quantify, the means for attaining precision values. Nearly all studies of quantification in the sciences have focused on these instruments. Most never raise explicitly the question of the values of the people who have pursued precision, which people, and to what ends. The pursuit itself has often been taken as a self-explanatory good, the evident motor of progress. Recent interpreters of measurement, however, are as concerned to know why something has been valued as to know how its value has been determined. That is the interest we are following here. "Values of Precision" calls attention to the fact that precision values always have another face, often hidden, the face that reveals the culture in which instruments of particular kinds are important, because the quantities they determine are valued.

Extending Control

Seen from within the sciences, the most obvious value of precision measurement has been the discovery of new phenomena and new regularities. Tycho Brahe's planetary observations, grounding Johannes Kepler's laws of elliptical planetary motion, are a classic example, as are James Joule's measurements to confirm the interconvertibility of heat and work, and Rosiland Franklin's x-ray diffraction pictures for James Watson's and Francis Crick's discovery of the double helix of DNA. Discovery itself, however, while often present, is not our primary focus. It is rather the role of precision in bringing wide domains of experience under systematic order. Here "domain" may refer to the natural or the social world, extending over different things or different communities. This perspective highlights a general theme of the workshop, that the generalized drive for precision has regularly been linked to attempts to extend uniform order and control over larger territories.

A second shift from commonly held views of science is also important. Mathematization of nature, the search for unifying mathematical laws like Newton's laws of motion, will not appear as the primary motivation for quantification. Certainly mathematization and quantification have been closely related and have often depended on one another, but as Heilbron and Hacking have insisted, quantification did not emerge historically as a campaign of mathematicians, or not in general, even though mathematicians were sometimes among the primary quantifiers. In large part it derived from the need of administrators for reliable information about particular aspects of the world in order to be able to make reasonable plans: availability of human and material resources, cost estimates, tax revenues, life and annuity tables, maps, location of ships, etc. In other words, when we ask about the most general source of the desire to quantify, we find it more nearly in the requirements for regulating society and its activities than in the search for mathematical laws of nature, even though such laws may both rely on and drastically improve quantification (by rationalizing design of instruments, establishing theoretical expectations for response, and revealing sources of error).

This recognition shows immediately that in looking for the sustaining features of precision measurement we will want to move smoothly from concern with controlling domains of nature to controlling economic and political domains. When Sir William Thomson, as one of the great promoters of quantification in the second half of the nineteenth century, uttered his oft-quoted lines, "when you can measure what you are speaking about and express it in numbers you know something about it; but when you cannot measure it . . . you have scarcely, in your thoughts, advanced to the stage of *science,*" he was talking about electrical units of measurement and their standardization.[4] The demand for these units and standards emerged earliest and most forcefully from the telegraph industry, which increasingly served as the medium of imperial government and international commerce. Thomson played a leading role in this enterprise as inventor and marketer of telegraphic instruments, as well as theorist of electromagnetism. Such relations between science, industry, and government have been ubiquitous in the history of precision, beginning at least with the crusade for the metric system in late eighteenth-century France (Alder). They were institutionalized first at the Cavendish Laboratory under Maxwell and Rayleigh (Schaffer), then at the *Physikalisch-Technische Reichsanstalt* in Berlin, and soon at government standards laboratories in other countries.

A recurring theme, therefore, will be precision in the centralized bureaucratic state: "Exactitude; État," as represented on our cover in a modernist celebration of the values of precision by Pierre Fix-Masseau. The trains must run on time. Rusnock's paper on French attempts to construct a census in the late eighteenth century as well as Alder's discussion of the metric system illustrate the point that the need for centralizing states and expanding commercial enterprises to regulate extensive human and material resources supplied one of the most important grounds for the explosive growth in precision measurement in that period. Several subsequent papers, by Schaffer, Olesko, Porter, Gooday, and Warwick show how these needs continued to operate through the nineteenth century, in ever more nuanced forms: international standards (Schaffer), error analysis (Olesko), uniform rating of insurance companies (Porter), self-registering meters (Gooday), and calculating machines (Warwick). Military needs have been another major source of the coupling between measurement, industry, and bureaucracy. Merritt Roe Smith and David Hounshell have both discussed this coupling elsewhere for United States armories in the mid-nineteenth century, stressing that interchangeable parts manufacture entered the famous "American System" through precision manufacture of armaments and that bureaucratic organization was the hallmark of the system thus refined. Precision machining and manufacturing will appear at many points in the papers of this volume, but especially in Part Three, characterized by mass distribution of instruments of precision. Gooday's account of electrical meters, Sweetnam's discussion of diffraction gratings, and Warwick's story of calculators would all be unthinkable without precision in the production process. Despite their obvious importance, however, we will not dwell on production techniques themselves nor on their military significance. These would be highly desirable topics for another volume focusing on the technologies of precision.[5]

Traveling

If controlling territory in the natural and social worlds is seen as one of the most important motives for precision measurement, then mobility will appear as one of its great virtues. While qualities do not travel well beyond the local communities where they are culturally valued, quantities seem to be more easily transportable, and the more precise the better. But there are problems. This is especially apparent in the papers of Part One. Can population be estimated in the same way for differently structured communities (Rusnock)? Will the same measure of agricultural land seem reasonable for the North and South of France (Alder)? Will Lavoisier's measured weights of oxygen and hydrogen in the decomposition of water be recognized in England (Golinski)? For the moment, it may be useful simply to observe that when qualities are successfully quantified, they also become objectified, and it is as objects that they travel. Five ohms of electrical resistance (say one of the standard resistors distributed by the British Association) travels like five pounds of potatoes. Both depend on reifying a quality as a physical object. Their range is limited, however, by the extent to which agreement can be reached both on the fact that five ohms or five pounds is an appropriate measure of the quality in question and on the measuring instrument that establishes the numbers. The relevant audience must agree that any proposed measure registers something valuable *for them* about quality, and not some artifact of arbitrary or interested opinion. For the numbers will travel just so far as their objectivity is trusted.

To repeat, precision measurements require agreement about what is valued and how it is to be valued. Higher precision requires more nuanced agreements as well as more refined instruments. Thus precision, along with the objectivity it reifies, is a joint product of material and social work. Numbers do travel well, but the apparent ease is a product of difficult adjustments between different parties. This is an ever-present theme of the volume. Schaffer takes it up explicitly for Maxwell's work on the British "ohm" and I will echo his analysis in my own commentaries. The problem of agreement raises two further themes: standardization and dissent.

Standardization, Precision, and Accuracy

A distinction between precision and accuracy has become common in the twentieth century and at every session of our workshop the question arose of when this distinction emerged and what its significance was. We reached no consensus. Traces of the discussions remain in several of the papers in the somewhat different meanings that different authors assign to the present distinction and in their accounts of similar concerns voiced earlier. These differences are part of the history and should not be effaced. Nevertheless, it will be useful for comparative purposes to specify at the outset one common version of the precision-accuracy distinction. This version will also show why precision and standardization are so closely linked in several of the papers, particularly those of Part Two.

Imagine a rifle fixed to a rigid support and aimed at a target. If successive firing yields a tightly grouped set of holes in the target, the rifle might be said to be very precise, even if its accuracy is poor, that is, even if the group lies some distance from the center of the target. A similar distinction is often applied to measurements. They are precise if they show little spread on repeated trials, and they are accurate if they yield values close to a "true" value. The error bars seen on the data points of a graph are intended to give the degree of precision of the measurements. They do not include systematic errors in the measurement, which will affect its accuracy (like the aim of the rifle). Such errors may require adjustments to the instrument, to the method of using it, or to the theory of its operation (like adjusting the sight of the rifle, firing without exerting pressure to the right, or accounting for a downhill aim).

The distinction thus drawn between precision and accuracy may seem staightforward enough, but to see that it is not an obvious one to make we need only look at the 1971 edition of the *Oxford English Dictionary.* Under "precision" comes "*Arms of precision*: a firearm fitted with mechanical aids, such as rifling, graded sights, etc., which make it more accurate of aim than weapons without these." Here it is assumed that precision implies accuracy, or that the firearm is not precise unless it is also accurate. This assumption seems to have been typical in measuring practices generally until at least the late nineteenth century, when the width of statistical distributions acquired a commonly cited measure, the standard deviation (or width at half-maximum for the bell-shaped curve of errors), and when machine-shop practice distinguished between the capacity to machine parts to match a pattern (precision) and the newly available capacity to machine parts to a precisely specified dimension (accuracy).[6] Consequently, except where otherwise discussed (Rusnock, Olesko), the papers in this volume assume that, as a historical matter, precision has normally implied accuracy.

Occasionally precision has been taken to refer simply to a large number of digits stated in a measurement. This is the practice that Lavoisier sometimes followed in claiming precision, and for which he was roundly criticized (Golinski). He was vulnerable because his critics took precision to imply in addition a high degree of reliability in the numbers stated. Here and in nearly all of the other papers a central question concerns the means of establishing the reliability of a measurement. Problems of establishing precision thereby become simultaneously questions of establishing agreement among a community. Precision requires standardization.

To see why this is so in general terms, we need only observe that claims to precision have minimally required that successive measurements should yield very nearly the same value. An individual, therefore, such as Maxwell in the paper by Schaffer, might claim precision in a measurement of the ohm carried out with a single instrument if he was able to attain consistency among the results of several trials. But as this example shows, such claims have not been widely accepted unless the measurement was consistent with results obtained by others, typically under different circumstances and with different concerns. Precision thus requires agreement about standards of comparison. This point is important because it shows that precision is never the product simply of an individual using a carefully constructed instrument. It is always the accomplishment of an extended network of people. Precision within the laboratory requires an extensive set of agreements about materials, instruments, methods, and values that reach out into the larger culture.

Seen from a somewhat different perspective, reliability is never a matter simply of a reliable instrument. It is also a matter of people judged to be reliable using methods that display their reliability. Establishing precision is also a matter of establishing credibility and trust, an issue that Golinski, Olesko, Porter, and Gooday take up explicitly.

Local and National Dissent

If establishing uniformity by agreement is a prerequisite for precision, then dissent is its nemesis. Although present at all levels, dissent within a national context has been most prominent. It appears, for France, in the papers by Rusnock and Alder; and for Britain, in those by Porter, Gooday, and Warwick. All of these papers reveal how intensely local are the cultural meanings attached to any given project for attaining numerical precision and setting standards, whether by promotors or detractors. Their values may be located by method, discipline, and ideology, as well as economic interest and geographical region. These are the first boundaries that any claim to precision must cross, and they are as prominent today as they were two centuries ago.

In a recent special section of the *New York Times* on "Education Life," for instance, the lead article discusses "The Fight over National Standards" under the headline, "Fears mount that national tests and standards will impose rigidity in the classroom." A featureless face (fig. 1) with an eyeball calibrated on several scales depicts a standardized student, the product of uniform measuring instruments tuned to standards that "would wring local and regional flavor out of curriculums." There were old standardized tests, of course, the multiple-choice exams that gave the name "objective" to numerical scores. But their objectivity had been constructed by agreements that did not properly credit the strengths of females and minorities, and they failed to assess writing and thinking ability. New instruments will have to satisfy a wider range of interested parties. They face intense opposition too from parents' groups represent-

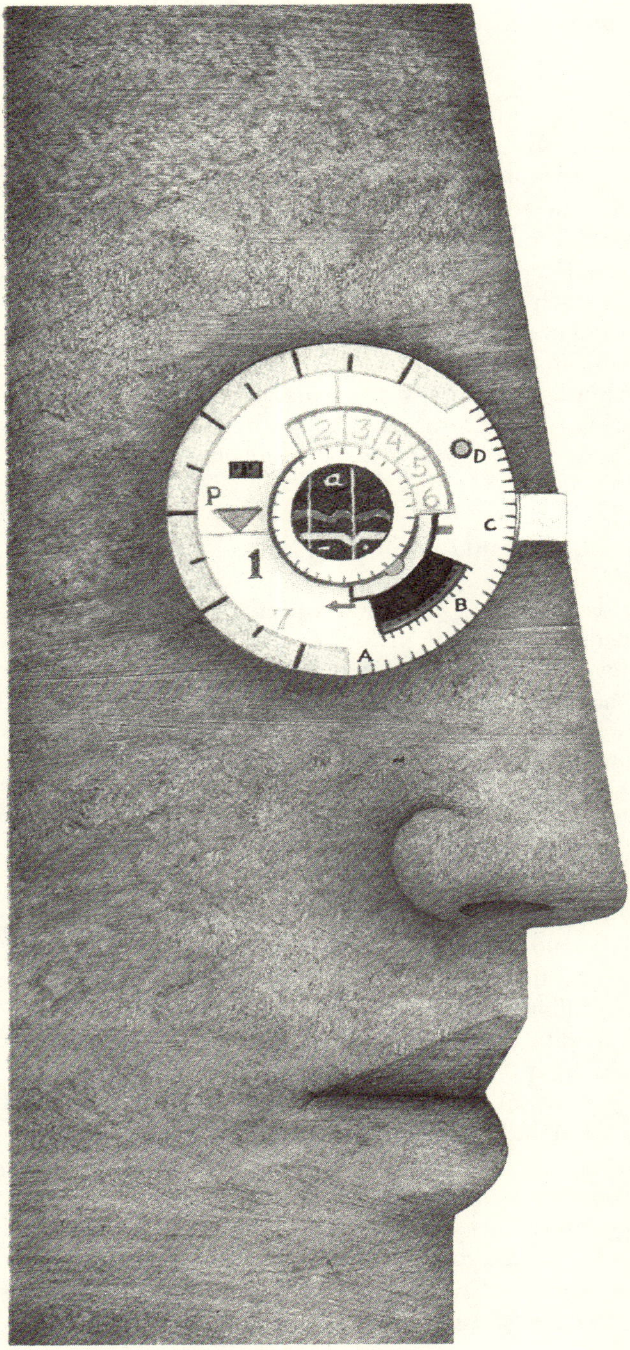

FIG. 1. Illustration by James Endicott for "The Fight over National Standards." Reproduced by permission of the artist.

ing local options and local control, education associations representing the academic freedom of teachers, and poorer districts who believe that national standards will benefit the wealthy. Uniformity has a high price. It will "require the expensive retraining of teachers and the wider use of computers and calculators." To obtain consensus about whether, what, and how to measure, therefore, the proponents of national standards have mounted a broad public relations campaign. Their strength lies in the support of large sectors of the business community and in federal government sponsorship of the National Standards Project. They derive their fervor from an apparent loss of international competitiveness among American students, workers, scientists, and industries. Any standards ultimately established, therefore, will represent a complex set of negotiations among all of these parties before they can move freely through the nation as objective tests.[7] The problems are remarkably like those discussed by Alder in the contest over the metric system in France.

Again, if precision travels only with difficulty across such local divisions, it crosses international boundaries with even more obvious breakdowns. Golinski shows the difficulty Lavoisier had in convincing contemporary British chemists of either the precision or the demonstrative power of his chemical measurements, however often he repeated them. Olesko observes in a distantly related case that in Britain statistical techniques for attaining precision did not acquire throughout the nineteenth century the validity they carried in France and Germany. Or as Ernest Rutherford famously remarked, "If your experiment needs statistics, you ought to have done a better experiment."[8] The difference in attitude toward error analysis contributed to important conflicts about electrical standards, as Schaffer shows. There may be something quite general here, and in several of the other papers, that could be ascribed to national styles. Is there a British style in establishing standards, for example, rooted in a social order based on consensus and class (agreement among responsible "gentlemen") rather than a society of legal uniformity; or are the observed differences between nations more the product of particular networks of people than of national characteristics? This is a tortured issue. My own remarks will stress networks, but the workshop has not decided the question of how cultural differences should be located. It has, nevertheless, pointed to a variety of fruitful avenues of research.

These then are some of the themes which will emerge in the essays that follow: the bureaucratic state, controlling territory, traveling, standardization, and dissent. Their interrelations will suggest why and how precision has become the sine qua non of modernity. In the interest of maintaining the character of the workshop as a continuing discussion, I have avoided giving introductions to the completed papers and have instead placed a running series of comments at the end of each section, intended as representations of what came out of a highly participatory activity and of its recurrent themes. These remarks will draw on the commentaries and discussions as they actually occured, albeit without attribution of individual contributors. I thank them for their indulgence in this selective collation of their many insights, and even moreso for their original participation. The formal participants, in addition to those in this volume, were Charles Gillispie, John Heilbron, Richard Sorrenson, Harold James, Bruno Latour, Anson Rabinbach, Timothy Lenoir, Keith Neir, John Carson, Joseph O'Connell, Deborah Warner, and Theodore Arabatzis.

Notes to Introduction

1. Thomas S. Kuhn, "The Function of Measurement in Modern Physical Science" (1961), in *The Essential Tension: Selected Studies in Scientific Tradition and Change* (Chicago, 1977), pp. 178–224 ("second scientific revolution," 220); and in the same volume, "Mathematical versus Experimental Traditions in the Development of Physical Science," 31–65. John Heilbron, Electricity in the 17th and 18th Centuries: A Study of Early Modern Physics (Berkeley, 1979), chap. 1; reprinted in *Elements of Early Modern Physics* (Berkeley, 1982), chap. 1; id., "A Mathematicians' Mutiny, with Morals," in *World Changes: Thomas Kuhn and the Nature of Science*, ed. Paul Horwich (Cambridge, Mass., 1993), 81–129. Ian Hacking, *The Taming of Chance* (Cambridge, Eng., 1990), 60–63. J. A. Bennett, "The Mechanics Philosophy and the Mechanical Philosophy," *History of Science* 24 (1986), 1–28, makes the point that in the seventeenth-century mechanical instruments were the province of mathematicians, not natural philosophers, and that it was the mechanical philosophy, with its close ties to real and metaphorical mechanics (both the people and the subject), that brought mathematics into natural philosophy. This would suggest an important role for the makers of mathematical instruments, who became the makers of philosophical instruments, in the transfer of mathematics into natural philosophy.

2. The best general history of time is David S. Landes, *Revolution in Time: Clocks and the Making of the Modern World* (Cambridge, Mass., 1983), chaps. 7–8 and 14–16 on eighteenth-century watches and chaps. 9–11 on marine chronometers. Samuel Guye and Henri Michel, *Time & Space: Measuring Instruments from the 15th to the 19th Century*, trans. Diana Dolan (New York, 1971), 109–50, give a useful survey of chronometers and watches with numerous photographs. The classic for the significance of pocket watches and factory clocks on daily life is E. P. Thompson, "Time, Work-Discipline, and Industrial Capitalism," *Past and Present* 38 (1967), 56–97. See also Ulla Merle, "Tempo! Tempo!—Die Industrialisierung der Zeit im 19. Jahrhundert," in *Uhrzeiten: Die Geschichte der Uhr und ihres Gebrauches* (Frankfurt, 1989), 161–217.

3. A wide spectrum of examples is collected in Tore Frängsmyr, J. L. Heilbron,

and R. E. Rider, eds., *The Quantifying Spirit in the 18th Century* (Berkeley, 1990); also Simon Schaffer, "Measuring Virtue: Eudiometry, Enlightenment, and Pneumatic Medicine," in *The Medical Enlightenment of the Eighteenth Century* (Cambridge, 1990), 281–318; M. Norton Wise, "Mediations: Enlightenment Balancing Acts, or the Technologies of Rationalism," in Horwich, *World Changes*, 207–56.

4. William Thomson, "Electrical Units of Measurement" (delivered to the Institution of Civil Engineers, 1883), in *Popular Lectures and Addresses*, 3 vols. (London, 1891–94), vol. 1, 73–136, on 73.

5. David Hounshell, "The *System*: Theory and Practice," in *Yankee Enterprise: The Rise of the American System of Manufactures*, ed. Otto Mayr and Robert C. Post (Washington, D.C., 1981), 127–52. More generally on "uniformity" as an ideal of the Ordnance Department of the U.S. Army, see Merritt Roe Smith's essay in the same volume, "Military Entrepreneurship," 63–102, and his *Harpers Ferry Armory and the New Technology* (Ithaca, N.Y., 1977). Ken Alder discusses early visions of interchangeable parts manufacture in the French army's involvement in the metric system in "The Origins of French Mass Production and the Language of the Machine Age, 1763–1815" (unpublished Ph.D. dissertation, Harvard University, 1991). For the production of military-industrial precision today see Donald MacKenzie, *Inventing Accuracy: A Historical Sociology of Nuclear Missile Guidance* (Cambridge, Mass., 1990).

6. Frederick A. Halsey, *Methods of Machine Shop Work, for Apprentices and Students in Technology and Trade Schools* (New York, 1914), 82–83, 90–92. Even Halsey, however, does not distinguish precision from accuracy in the way indicated here but only the capacity to repeat given sizes using a plain caliper from the capacity to measure those sizes using a micrometer caliper.

7. William Cellis III, "The Fight over National Standards," in "Education Life," Section 4A, *The New York Times*, August 1, 1993, 14–16. A corresponding debate in Britain is even more acrimonious, as reported in the following article by William E. Schmidt, "Britain Flunks a Test of its National Curriculum," 17–19. For discussion of related issues, see Theodore Porter, "Objectivity as Standardization: The Rhetoric of Impersonality in Measurement, Statistics, and Cost-Benefit Analysis, *Annals of Scholarship* 9 (1992), 19–59; and his forthcoming book, *Trust in Numbers: The Pursuit of Objectivity in Science and Public Life* (Princeton: Princeton University Press, 1995).

8. Quoted by Sir Brian Pippard in a review for *Nature* 362 (1993), 216, of Charles Ruhla, *The Physics of Chance: From Blaise Pascal to Niels Bohr* (Oxford, 1992). Pippard agrees with Rutherford and criticizes Ruhla, whom he guesses to be, "like all too many of his colleagues [French?], in thrall to a tradition that considers no measurement complete unless its probable error is stated."

PART ONE

ENLIGHTENMENT ORIGINS

ONE

QUANTIFICATION, PRECISION, AND ACCURACY:

DETERMINATIONS OF POPULATION IN THE ANCIEN RÉGIME

Andrea Rusnock

THE IMPULSE to count, measure, and calculate quickened dramatically in eighteenth-century Europe. Recently baptized the "quantifying spirit" by historians of science, this movement has been most closely associated with the development of new scientific instruments such as the barometer and electrometer[1] But the quantifying spirit was not confined to scientific instruments: it spread to other domains as well. In this essay, I examine the quantification of population in eighteenth-century France and argue that the methods of measuring population, like other scientific instruments, became increasingly accurate and precise, especially after 1760.[2]

The concept of population gained saliency in early modern Europe, in large part due to the consolidation of centralized monarchies during the sixteenth and seventeenth centuries. Historians locate the origins of the modern nation-state precisely during this period, when definitions of the royal state came to include population as well as territorial extent. The emphasis on population by the monarchies of the sixteenth century had a clear basis in taxation: monarchs such as Francis I in France and Henry VIII in England sought to devise new ways to finance increasingly costly and sizable wars.

By the seventeenth century, discussions of population had become commonplace in political and economic writings and notably new voices called for quantified accounts of population. Foremost among these new voices was Sir William Petty who made the numerical determination of population an integral part of his political arithmetic. Petty's French contemporary, Marshal Vauban, the leading military engineer during the reign of Louis XIV, also underlined the importance of measuring population. In a memoir written in 1693 and addressed to the King, Vauban noted that ". . . the king and his ministers do not seem clearly to have seen that the greatness of kings is measured by the number of their subjects.

The obvious proof of this truth is that where there are no subjects there is neither prince nor state, nor any domination at all."[3] Throughout the eighteenth century, discussions of population abounded, in part because population came to be defined as a measure of government—not just of a king's strength as the mercantilists had advocated, but of the goodness of government. Rousseau articulated this new definition when he asked, "And what is the surest evidence that they [i.e., members of a political association] are so protected and prosperous? The numbers of their population."[4]

In the first section of this essay, I review the attempts made by the monarchy to monitor and count the population of France and I argue that the administrative structure of government in the old regime made such a task impossible. In the seventeenth and eighteenth centuries, France did not possess the necessary bureaucracies for collecting statistical information and when reform-minded officials sought to collect data about the population they frequently encountered strong opposition. Nonetheless, it is significant that the state instigated the first attempts to quantify population.

Alongside these calls for quantified accounts of population flourished an extensive literature on population which contained no numerical figures at all. Early in the eighteenth century, the political philosopher Montesquieu had linked depopulation with despotism, which triggered an extended debate about whether the French population had increased or decreased. Writers who contributed to this literature debated the causes of population growth and decline: they made no attempt to establish the number of persons at a given time. Their writings, however, spurred others to formulate quantified accounts of population.

The quantification of population was not an easy task, nor did it necessarily produce accurate knowledge of the population. In fact, many of the earliest figures given for the French population were not based on any empirical evidence whatsoever. Over the course of the seventeenth and eighteenth centuries, providing more accurate accounts of population became a matter of concern among administrators and the enlightened public. In the seventeenth century, there was widespread agreement that a census was the best method to secure an accurate count of population. A century later, any hope of an accurate census was abandoned and government officials turned to calculation instead. This shift in opinion is in part related to the changing fortunes of the French monarchy during this period. Whether accurate or not, the figures given for populations in both the seventeenth and eighteenth centuries were usually precise. Writers customarily gave the number of a population to the last individual, even when that number was a product of calculation involving an admitted approximate figure.[5]

In the second section of this essay, I provide an account of the depopulation debates provoked by the writings of Montesquieu and others, who, it must be emphasized, did not quantify population. Those who opposed the arguments of depopulationists, however, did turn to quantification. I provide two examples—stimulated both by administrative needs of the French state and by the depopulation debates—to ascertain the population by calculation using a universal multiplier. In the final section, I examine two analyses of this calculation technique; one made by the intendant Auget de Montyon, the other by the natural philosopher Pierre Simon de Laplace. Although motivated by different interests, both men introduced concerns of accuracy and precision into their discussions of the universal multiplier.

Parochial Registers and Royal Censuses

The various administrative reforms and initiatives aimed at collecting numerical accounts of the French population during the seventeenth and eighteenth centuries fell largely into two categories: parochial registers and government censuses. The relative frequency of these initiatives illustrates the acute concern on the part of the monarchy for its population and suggests the failure on the part of the royal bureaucracy to secure the information they sought.

With the rise of the centralized state in early modern Europe came attempts by monarchs to monitor the population of their realms. In 1539, at roughly the same time that Henry VIII of England announced his royal edict concerning vital registration, the royal ordinance of Villers-Cotterêts was issued which mandated the registration of baptisms and deaths in all parishes throughout France. As in England, members of the clergy were to be responsible for maintaining these registers; however, a striking difference in the French specification was the requirement that the registers be signed by a notary as well as by the curé. In theory the clergy were to be enlisted by the French state, and clerical documents were to be monitored by royal officials. In practice, this was not the case. Historians and demographers who have used these registers have never found an example of a register that was in fact signed by a notary.[6]

Louis XIV, who reigned from 1643 to 1715, came closer than any other French monarch to putting in place a sufficient bureaucracy to collect reliable information about the population. In April, 1667, a new edict, formally the ordinance of Saint-Germain-en Laye (commonly called the Code Louis), was put into effect. Among other specifications, this code called for the registration of births, marriages, and deaths, and required the curés to take the registers to a royal judge once a year. Although not required to do so by the code, many curés began systematically recording the age of death in these registers.

In 1691, the government introduced the sale of offices for "clerks, keepers, and conservators of the registers of baptisms, marriages, and burials"; in 1705, offices for "controllers of registers and extracts of baptisms, marriages, and burials"; and in March 1709, offices for "alternative secretary clerks." These actions can be interpreted as an attempt by the monarchy to transfer record keeping from the Church to the royal bureaucracy. By recording information about the French population itself, the royal bureaucracy sought to increase its control over the accuracy and distribution of that information. Upon Louis XIV's death, however, the offices created during his reign concerning vital registration came under attack from both the nobility and clergy and were eventually suppressed in December 1716.[7]

In addition to these changes which specifically concerned vital registration, Louis XIV introduced numerous reforms in royal administration, all of which tended to strengthen the royal bureaucracy at the expense of the church, nobility, and provincial estates. France was divided into *généralités*, regions that were administered by royal intendants, commissioners sent by the royal council.[8] The intendants provided information about their *généralités* directly to the King and his council, and the men who held these offices proved to be the most active in collecting quantitative information about the population during the eighteenth century.[9]

In contrast to his grandfather, Louis XV introduced relatively minor changes concerning vital registration during his reign (1715–74). In 1736, for instance, following suggestions made by the King's advisor Joly de Fleury in 1729, a *Déclaration royale* was issued which reformed the Code Louis. Curés were now required to keep duplicate registers, one on official paper and one on regular paper, and they were instructed to record the death of infants in the registers. The *Déclaration* formed the bureaucratic basis by which information about the population was collected in the French state, and there were no further attempts at reform until the Revolution.[10] Thus, throughout most of the eighteenth century, registration of births, deaths, and marriages remained in the hands of the clergy.

In addition to parochial registers, the civil government did conduct periodic censuses which provided quantitative information about the population.[11] A series of censuses for different regions of France dated back to the Middle Ages, but they were infrequently conducted, and usually for only one specific area.[12] During Louis XIV's reign one motivation for the enumerations undertaken came from taxation. Louis's penchant for extravagance, coupled with the expenses of war, placed France in a financially precarious position and made taxation a critical issue. The only direct tax in place was the *taille*, a feudal tax levied against those who did not serve in the military. Many towns were exempt, and thus the brunt of the tax fell on the countryside. In the latter part of Louis XIV's reign, new poll taxes were proposed and adopted. The *capitation* (1695) and the *dixième* (1710) were levied in an attempt to tax everyone with no exceptions.

The idea of the *capitation* had been proposed by Sebastien le Prestre de Vauban (1633–1707), one of Louis XIV's chief advisors.[13] A reformer as well as an engineer, he advocated the unification of weights and measures, the abolition of internal customs, and the suppression of abuses associated with the collection of taxes—reforms that were not achieved until the Revolution.[14] Vauban expressed his interest in population and in finding ways to measure it in many of his writings, including the memoir cited at the beginning of this paper. In "Note sur le recensement des peuples," for example, Vauban lamented: "Without a census repeated every year, or at least every two or three years, one cannot know precisely [*précisément*] the number of subjects, the true state of their wealth and poverty, what they do, how they live, their commerce and employment, if they are well or ill. . . ."[15] Vauban consistently argued that a census was the only method to arrive at a precise figure for the population.

In 1686, Vauban privately printed a twelve-page pamphlet entitled *Méthode générale et facile pour faire le dénombrement des peuples*, in which he incorporated many of the ideas expressed in his "Note sur le recensement des peuples."[16] Ten years later, he published an example of the type of census he advocated in "Description géographique de l'élection de Vézelay" (1696). This work contained a thorough description of the geography of Vézelay, its crops and products ("The wines are mediocre there," he thought), and an enumeration of the population.[17] He included a table which concisely summarized all the quantitative information that he had collected. In this table, Vauban provided precise figures for the number of men, women, houses, cows, and so forth for each parish in Vézelay. He was careful to point out that his numbers were based on a *census*, not on an estimate. Vauban asserted: "Here is a true [*véritable*] and sincere description of this small and poor land, made after a very exact [*très-exacte*] inquiry, founded, not on simple estimations which are nearly always faulty; but on a good, formal census, well rectified."[18] In this statement, Vauban used the words *très-exacte* in a sense which combined both the modern meanings of precise and accurate.

Vauban's call for a yearly census was ignored by Louis XIV, and censuses for specific taxes were undertaken only intermittently during Louis's reign. In 1694, for example, the minister of finance ordered the intendants to conduct a national census prior to the establishment of the *capitation*. Three years later another census was undertaken as part of a set of memoirs entitled *Mémoires pour l'Instruction du duc de Bour-*

gogne, which were in response to a questionnaire sent by the Duke of Beauvillier to all the intendants who, with the exception of one who died, responded with evaluations of their *généralités*. According to modern demographers who have used these documents, however, "the majority [of intendants] were content with earlier figures, taken from enumerations of hearths or from the inquiry for the *capitation*."[19] Orders for a new census, thus, did not necessarily result in new enumerations, and in this instance, the limitations of the French royal bureaucracy are clearly revealed.

During the reign of Louis XV, no general censuses were taken, although numerical information about trade and population was gathered in certain locales.[20] Near the end of Louis XV's reign, the Abbé Joseph Marie Terray (1715–78), the Controller-General of France, ordered the intendants to establish an annual account of the changes of the population and made explicit his concerns with accuracy. "*Monsieur l'intendant*, it is very important for the administration to know exactly the state of the population of the kingdom," he wrote in 1772. "I beg of you, consequently, to have made each year an exact résumé of the population of your *généralité* conforming to the model list which you will find herewith." Importantly, Terray did *not* request a census. "It is not an enumeration by persons, dwellings [*ménages*] or households that I ask of you, that enumeration, although easy, would demand too much time and trouble to be renewed each year; what I ask is that you have sent to you each year by the clerks of the royal jurisdictions a résumé of births, marriages, deaths in all parishes. . . ."[21]

This is a curious statement. On the one hand, Terray stated that undertaking a census is a relatively easy task, and yet his contemporaries openly acknowledged the difficulties involved, as we will see below. On the other hand, Terray requested a quantitative account of births, deaths, and marriages, figures which provided the basis for calculating population according to widely accepted methods at the time, which will be discussed in the next section. In other words, Terray's request tacitly endorsed the latter method of calculating population, rather than relying upon census figures.

Replies to Terray's request are preserved in the Archives Nationales, many entered on printed forms entitled "État de la population de la Généralité ____, Année ____." The introduction of a printed form is itself an important innovation, and insured a more consistent collection of information from *généralité* to *généralité*.[22] Existing replies to Terray's request reveal that the numbers of births, marriages, and deaths were listed down to the last individual. A handwritten memoir from the Revolutionary period testified to the relative success of Terray's plan: "The old regime opened the route and traced the way for us. Every year the curés sent to the intendants, and they to the government, an extract of the marriages, births and deaths in their *généralité* and the government in publishing these tables offered to both savants and men of state precious mines . . . of facts that could serve as the basis for their theories and their conceptions. Why not employ the same methods. . . ?"[23]

It is clear from the above sketch that the French monarchy was interested in ascertaining the population of France. Government actions to secure figures for the population concentrated on improving record-keeping. From Louis XIV on, new regulations were promulgated in order to monitor the population. Numerous obstacles, however, prevented the king from having an accurate account of population. Foremost among these was the very structure of the administration itself. Servants of the crown were remunerated in a variety of ways. Many of these positions were venal offices; once bought, the owner of the office had little obligation to the royal bureaucracy. Other royal officials owed their position to personal ties to the monarch, and still others received payment from the public they served. The variety of these arrangements underscores the highly individualistic character of the royal bureaucracy. It is very difficult for a bureaucracy to operate "efficiently" given these financial and personal arrangements.[24] A further obstacle came from what we would today consider the highly irrational organization of royal offices in the old regime. The various offices in eighteenth-century France were products of historical circumstances, and as a result there was considerable overlap in jurisdiction and duties. Add to this the unclear boundary between ecclesiastical functions and royal functions, and the administrative picture looks muddier still. Terray's solution to these difficulties was perhaps the best: rely upon the local clergy to record births, deaths, and marriages within each parish and then leave it to the intendants—royal administrators directly appointed by the crown—to tally these individual reports.

The Debates over Depopulation

The French government's interest in population was reflected more generally in the debates over depopulation. During the eighteenth century, it became a matter of common assent that France—indeed, Europe—had become progressively depopulated since antiquity and especially since the time of Louis XIV; however, there were a few who contested this view. Montesquieu raised the alarm over depopulation early in the century in his *Lettres persanes* (1721), in which he asserted that the ancient world had been more populated than the modern one. "After a calculation as exact as may be in the circumstances," he wrote, "I have found that there are upon the earth hardly one-tenth part of the people which there were in ancient times. And the astonishing thing is, that the depopulation goes on daily: if it continues, in ten centuries the earth will be a desert."[25] Although he provided no evidence of his calculations, he did appeal to their accuracy, their *exactitude*.

Montesquieu developed this theme in detail in *Esprit des Lois* (1748). Considering the relationship between form of government and population, he argued that republics were the most densely populated countries, and that despotic governments caused depopulation.[26] Subsequently, depopulation became a sign of despotism, and the philosophes were quick to take up this point, which ensured the debate on depopulation a prominent role in discussions on government.[27] Aside from his allusion to the calculation above, Montesquieu did not present any quantitative evidence in support of his contention that France had suffered depopulation; he only provided reasons for why it had.[28]

Second only to Montesquieu, the Marquis of Mirabeau was perhaps the most influential eighteenth-century depopulationist. In *L'Ami des Hommes ou Traité de la population* (1756), Mirabeau put forth a number of reasons for why France was suffering depopulation and, like Montesquieu, provided no numerical evidence for his claims. Instead, he asked, "Why have the majority of the States in Europe been visibly depopulated?"[29] Disagreeing with others who cited that clerical celibacy, emigration to the New World, and war were the chief causes of depopulation, Mirabeau argued that it was the result of "the decline of agriculture on one hand, and on the other hand, luxury and too much consumption by a small number of inhabitants. . . ."[30] His central point was that "the measure of subsistence is that of the population," and, in colorful language he commented: "People multiply like rats in a barn if they have the means to subsist."[31] Thus, Mirabeau like Montesquieu provided reasons for why France had suffered depopulation and allegedly continued to do so. Both writers asserted the depopulation of France without providing quantitative evidence for their claims.

Many other works that addressed depopulation in the 1760s and 1770s were also qualitative. In 1767, for instance, the Abbé Pierre Jaubert provided a lengthy account of the causes of depopulation (corruption of morals, war, emigration), and the means to remedy these problems. But he made only one reference to a census: one ordered by the Parlement of Bordeaux *to be* taken.[32]

An exception to the form of most of the French depopulationist works is found in the writings of the physiocrat François Quesnay, who provided numerical figures purportedly documenting the depopulation of France. He stated that the population of France had been twenty-four million in 1650, nineteen million in 1701, and had declined by 1750 to only sixteen million.[33] In general, however, Quesnay was neither concerned with increasing population, nor with numerical estimates of population. In his most famous work, *Tableau Économique* (1759), Quesnay remarked that he paid "less attention. . . . to increasing the population than to increasing the revenue; for the greater well-being that a high revenue brings about is preferable to the greater pressure of subsistence needs which a population in excess of the revenue entails; and when the people are in a state of well-being there are more resources to meet the needs of the state, and also more means to make agriculture prosper."[34] Quesnay did not accept the general premise that depopulation was a sign of despotic government; rather, he was concerned with population movement, with urbanization and migration from the countryside. He did, however, cite figures regarding revenue in support of his argument that France had suffered depopulation: "It is proved by the registers of baptisms, marriages, and burials, and by the consumption of wheat in Paris, that this city has not increased in inhabitants for a long time." He concluded "that depopulation of the countryside is not compensated by the population of this capital."[35] Quesnay reaffirmed the physiocratic argument that the wealth of the country lay in agriculture and the rural population.

Thus the depopulationist position was well sketched by the mid-1760s. The writings of Montesquieu and Mirabeau had provided ample reasons to explain why the French population was less than it had been under Louis XIV. With the exception of the population figures presented by Quesnay, those who asserted depopulation did not turn to quantification for proof. This stands in marked contrast to those who disputed the depopulationists' claims and typically provided a quantified account of population in their works.

In light of the depopulation debates, ascertaining the population of France had become a matter of political as well as administrative importance. Abbé Terray's request to the intendants in 1772 for annual population figures takes on added significance in this context. Providing a quantitative account of the French population was no easy task as we saw in the discussion above on government enumerations. Because no regular census existed in France, those who sought to quantify the French population turned to alternate methods. The two most common methods of estimating population involved a universal multiplier. In the first, the number of hearths taken from tax records was multiplied by the number of people per hearth—usually 4 or 5. In the second method, the number of annual births taken from parish registers was multiplied by a number, anywhere from 25 to 28, which yielded a figure for total population.[36] This number was typically attained by taking a census of several small parishes and establishing the proportion of births to total population.

The idea of a universal multiplier is itself noteworthy. Eighteenth-century writers contested the actual number but rarely did they criticize the technique of employing a universal multiplier. Instead, objections focused on the empirical figures used in both methods of calculation. Some individuals dismissed the use of tax records by arguing that they were never accurate because there were always people who managed to avoid paying the tax. Others criticized the use of parish registers and characterized the curés as unreliable. Accordingly, Louis XIV's attempt to create royal offices for keeping vital records could be read as distrust of the curés' activities.

An example of the first method (the hearth method) can be found in the writings of the Abbé Jean-Joseph Expilly (1719–93). Expilly held a variety of governmental posts throughout his career and he was a member of the Société Royale des Sciences et Belles-Lettres de Nancy.[37] In 1762, he undertook to correct and update Saugrain's *Dictionnaire Universel de la France* (1726), in large part to refute the depopulationists' claims. Almost fifty years earlier, the publisher Claude Saugrain had provided population information about every town in France listed alphabetically.[38] A typical entry read as follows:

> Dimancheville, in Orleannois, Diocese of Orleans, Parlement of Paris, Intendance of Orleans, Election of Pithiviers, has 126 inhabitants.[39]

Saugrain gave no indication of how he arrived at the number of inhabitants.[40] Expilly drew on this earlier work of Saugrain and took issue with the commonly employed multiplier of 4 persons per hearth. "It is true that it is common enough that one understands by a hearth one family; but this meaning will not work in Provence, or Dauphiné, or in Brittany," he charged.[41] Instead Expilly proposed a multiplier of 5, and argued that cities usually had closer to 6 persons per hearth, while the countryside had 4, so that 5 was an average. Using the number of hearths found in Saugrain's *Nouveau dénombrement du Royaume*, and experimenting with a coefficient of 4.5 and 5, Expilly calculated the population of France as 20,905,413.

Others who sought to calculate the population of France and refute depopulation claims relied upon parish registers, which provided the annual number of births. A good example of this type of calculation is found in the work published under the name of Louis Messance: *Recherches sur la population des généralités d'Auvergne, de Lyon, de Rouen, et de quelques provinces et villes du Royaume* (1766); and *Nouvelles Recherches sur la Population de la France* (1788). Messance was the secretary to the intendant A. M. de La Michodière, but scholars have argued that La Michodière was in fact the author.[42] La Michodière had a long and distinguished career in the French government which provided one source for his interest in population. In his second book, he revealed his other motivations for writing: "In 1756, the book of *l'Ami des Hommes* appeared, and nearly everyone believed in the depopulation of France based on the words of the author."[43]

In the opening sentence of his first book La Michodière proclaimed: "The ordinary year of births [*l'année commune des naissances*] must be a sure rule [*règle sure*] for determining the number of inhabitants living in a province. . . ."[44] His advocacy of this technique rested on pragmatic considerations. "All the Politicians and Administrators of States have always thought that an exact census [*dénombrement exact*] of the inhabitants of a realm was a necessary operation for the different parts of Government." But, contrary to Terray's assertion, he admitted, "It proved difficult to carry out each century, so long as time and care required, because the fears that it could inspire in the people always prevented the inquiries ordered by the government." Censuses were historically associated with taxation, and as a result the French were not at all anxious to cooperate with the royal government.

Births, then, would provide the basis for calculating the population and the information was already available.[45] Certain parishes and small towns could undertake a census and these figures would be used to calculate the multiplier for the entire province. To complete his argument, La Michodière explicitly equated a census with calculation: "It seems, however, possible to form an exact census or at least one very nearly approaching the truth [*un dénombrement exact, ou du moins très-approchant de la vérité*] of the inhabitants of an entire province. . . ." by using this method.[46] La Michodière deemed the results of his technique accurate.

The efforts of La Michodière most clearly represent the administrator's aim to document population figures empirically. He focused on providing figures from the clerical registers kept throughout France—figures which would serve as a basis for calculation. La Michodière was typical of many administrators of this period: his concern with accuracy focused on record-keeping, not on the methods of calculation.

Administrative and political considerations, then, combined to encourage the assessment of population. Royal bureaucrats, such as La Michodière, referred explicitly to depopulation as a motivation for their work. By the latter half of the eighteenth century, any hopes of a regular census—advocated by Vauban during Louis XIV's reign—were dashed. Numbers of population were to be calculated by using a universal multiplier and there was widespread agreement that calculation based on a universal multiplier provided an exact—or very nearly exact—account of the population.

Refining the Universal Multiplier

Two of La Michodière's contemporaries addressed the limitations of calculation for determining population. The first, a fellow intendant, Antoine Auget, baron de Montyon, published *Recherches et Considérations sur la Population de la France* in 1778, which today is considered a masterpiece in early demography.[47] The stated author of this book was J. B. Moheau, secretary to Montyon, but most scholars agree that the book was in fact written (or at least directed) by Montyon. With this in mind, I will refer to the author as Montyon.[48]

The book was divided into two parts, the first entitled "State of the Population"; the second, "On the causes of the increase or decline of the population." This division reflected the need to demonstrate that France had not lost population, and to examine the putative causes of depopulation. Montyon's approach was that of a naturalist, and it came through most clearly in his discussion of fecundity and mortality. "The production of flowers, plants, grains, fruits, and animals, although liable to variations and irregularities, is however subject to certain rules," he noted. "What is this order of production for the most precious being of all that covers the surface of the earth?"[49] Taking his cue from the animal world, he investigated the most fertile months for humans, the effects of city versus country on fecundity, and the variations of fecundity from year to year.[50] He viewed mortality in a similar way by comparing mortality rates across different geographical areas, sexes, ages, and classes.[51] Montyon adopted the common position that population was the base of a nation's strength and wealth. "Moreover, putting aside military and financial considerations," he observed, "it is an established fact that the intrinsic and relative force of States consists principally in the population, and singularly in the number of individuals who can handle a spade, manage a plough, work at a trade, carry arms, finally reproduce: that is the basis of the strength of nations. . . ."[52] Montyon evaluated the various ways of calculating population and determined that the best method was the one based on the annual number of births. Montyon, along with many others, acknowledged that the number could range anywhere from 23 to 31 depending on the region in France.[53] Nonetheless, he advocated the use of a single multiplier. From his investigations, he concluded "that the term of proportion that seems to suit best [*mieux convenir*] the estimation of the population of the realm is that of 25 1/2."[54]

Montyon distinguished himself from his contemporaries, however, by arguing that a census would *never* produce an accurate count of population. "A head-by-head enumeration of the inhabitants of a realm would not make known exactly [*exactement*] the number, the less because it cannot be done at the same time in all places; further the moment when the census is complete, it is no longer true, and the unexpected birth or disappearance of some individuals changes the state of things."[55] He acknowledged that a census would provide the most precise account of population; however, he noted that "in seeking only the degree of necessary truth [*le degré de vérité nécessaire*], the census is certainly the operation that gives the most precise notion [*la notion la plus précise*] of the number of inhabitants; but it can become, and often becomes very faulty by the inapplication or default of orders by the clerks. . . ."[56] Montyon went on to explain his notion of "the degree of necessary truth." He insisted that an administrator had no need of a complete census and hence an accurate account of population. Assessing the size of the population did not admit the degree of certainty other subjects did. Montyon asserted: "There are many truths for which it is not given to man to attain with precision [*précision*]; but there are also many for which mathematical accuracy [*l'exactitude mathématique*] is nothing more than a perfection without an object or with little utility. The man of State who wants to know the strength of the population of a country, only needs an approximation [approximation]. . . ."[57] Here Montyon equated precision and accuracy as similar epistemological goals, and dismissed both as unnecessary to the determination of population.

Despite Montyon's insistence that approximations were all that were necessary, he carried out his mathematical calculations very rigorously and thoroughly. He computed figures for the universal multiplier from records of births and censuses for several towns, and presented them in a well-designed table. The precision of Montyon's figures is readily apparent in the following figures:[58]

Paris	25 1/114
Riom	24 137/255
Limoges	23 1439/5717
Lyon	23 3079/4152

Censuses were the only method which would provide an exact number of the population; yet a complete census could never be carried out due to the fallibility and changeability of humans. Hence, all figures for a population must necessarily be approximate, which, Montyon argued, were sufficient for the "man of State." Significantly, Montyon did not carry over his argument concerning approximation to the methods and results of using a universal multiplier to calculate population. Instead he employed very precise figures in his calculations. This restriction of the use of approximation points to a very different understanding from our own of precision and accuracy.

The second individual to analyze the limitations of the universal multiplier was the natural philosopher Pierre-Simon de Laplace. In 1783, Laplace examined the determination of population in an article published in the *Mémoires* of the Paris Royal Academy of Sciences.[59] This memoir is one of several written by Laplace on the calculus of probabilities between 1772 and 1786.[60] In this particular paper, he presented what later became known as "Laplace's method" of calculating population.[61]

Focusing on the method of using the annual number of births, Laplace stated: "The ratio of the population to births, determined by the preceding method [of partial censuses], can never be rigorously exact [*rigoureusement exact*]. . . ." Further, Laplace reasoned, "The population of France, deduced from the annual births, is therefore only a probable result, and consequently susceptible to errors."[62] Thus Laplace introduced notions of probability into the determination of population: that is, Laplace specified that the calculations based on a universal multiplier yielded only *probable* figures—not exact ones—and sought to limit the degree of inaccuracy. Whereas Montyon had argued that only approximate figures were necessary to the administrator, Laplace stated that only probable figures could be calculated.

Using the calculus of probabilities, Laplace determined the extent of the census necessary to limit the error of the result from calculating the total population from the annual number of births. "Now I find by my analysis, that to have a probability of 1000 to one, to not be mistaken by half a million in this evaluation of the population of France, it is necessary for the census which serves to limit the factor 26 [i.e., the universal multiplier] to have been 771,469 inhabitants [i.e., the minimum size of the partial census]."[63] "It follows," he concluded, "if one wants to have the precision [*précision*] that is demanded by the importance of this subject, one must carry out a census of 1,000,000 to 1,200,000 inhabitants." In short, Laplace argued that one needed sufficiently large and accurate partial censuses in order to calculate a precise universal multiplier.[64]

In his later work, *Essai philosophique sur les probabilités* (1814), Laplace explained his method in lay terms. First, echoing Montyon, he posited that undertaking a census was "a laborious operation and one difficult to make with exactitude."[65] Advocating instead the method of calculation based on the annual number of births, Laplace then described "the most precise [*le plus précis*] means" for ascertaining the universal multiplier—the ratio of the population to the annual births. He asserted that "this number, divided by that of the inhabitants, will give the ratio of the annual births to the population in a manner more and more accurate [*sûre*] as the enumerations become more considerable."[66] Laplace's solution was thus one of compromise: partial census combined with precise calculation.

Laplace's work in this area did not remain confined to the scientific community. By the end of the eighteenth century administrative officials and natural philosophers were working together to count and calculate various measures of the population. In the 1783 *Mémoires* of the Royal Academy of Sciences, a report drawn up by A. P. Dionis de Séjour, marquis de Condorcet, and Laplace concerning the population of France was published, which presented tables of figures for the number of towns, villages, and inhabitants.[67] Large portions of this report depended upon the efforts of the intendant La Michodière, whose papers were announced and deposited in the Academy of Sciences.[68] Entitled "Table alphabétique des villes, bourgs, villages dans différentes régions de France, Moyenne des naissances, mariages, morts sur les années 1781, 1782, 1783," La Michodière's table provided the annual number of births, deaths, and marriages of each parish for each election.[69] It formed the basis for the academicians' studies, and in this way the Academy supported these early demographic inquiries. This cooperation flourished during the revolutionary period, when many natural philosophers served as bureaucrats in the new governments.

Laplace's later work epitomized this cooperation. In his *Essai philosophique sur les probabilités*, Laplace provided an account of a partial census undertaken by the French government in 1802 upon his recommendation. According to Laplace, "In thirty districts spread out equally over the whole of France, communities have been chosen which would be able to furnish the most exact information. Their enumerations have given 2,037,615 individuals as the total number of their inhabitants on the 23rd of September, 1802." Thus Laplace shared the administrator's concern for providing accurate numbers for calculating the universal multiplier. He then gave figures for the annual number of births for the years 1800, 1801, and 1802 and calculated the ratio of the population to annual births as 28 plus 352,845/1,000,000—a precise figure. Thus far, Laplace had not distinguished himself from men such as La Michodière. It is his next statement which set Laplace apart: "But what is the probability that the population thus determined will not deviate from the true population beyond a given limit?"[70] By raising this question Laplace admitted that the results of his precise calculation were not completely accurate, and he sought to define the limits of inaccuracy. Thus Laplace's ideas of precision were predicated on specific ideas about error.[71]

Conclusion

This study shows that quantitative ideas of accuracy and precision emerged from the social and institutional nexus of science and politics during the eighteenth century. The needs of government, coupled with the issues raised by the writings of depopulationists, promoted methods for calculating the population of France. Individuals who sought to evaluate the truthfulness of population figures frequently employed the terms precise and accurate. There was near universal agreement in the eighteenth century that a census would provide the most precise and exact account of population; however, the limitations royal officials faced in the old regime—their inability to carry out a census—forced them to adopt the method of the universal multiplier to ascertain the population. Nonetheless, government officials such as the intendant La Michodière and the Controller-General Terray continued to concentrate their efforts on record-keeping to secure accuracy. In a radical break with this tradition, the intendant Montyon argued that a complete and accurate census was fundamentally impossible, and that government officials needed only approximate figures for the population. But his use of the term approximate applied only to the results of a census; in his calculations, Montyon continued to use very precise figures. It was the natural philosopher and academician Laplace who developed eighteenth-century methods of determining population to their most refined level by employing the calculus of probabilities. Yet even in his work, the concepts of precision and accuracy remain virtually indistinguishable.

The fact that the French government undertook a partial census specified by Laplace speaks of the new possibilities open to government in the wake of the French revolution. Vauban's appeals in the late seventeenth century had not been heeded, in large part due to the administrative structure of the old regime. Whether they created or enhanced centralization, the changes in government enacted during the revolutionary period allowed for more efficient administration and for the possibility of considering the precision and accuracy of population accounts.[72]

Notes to Chapter One

1. Tore Frängsmyr, J. L. Heilbron, and Robin E. Rider, eds., *The Quantifying Spirit in the 18th Century* (Berkeley: University of California Press, 1990).

2. J. L. Heilbron briefly addresses the quantification of population in the eighteenth century in his "Introductory Essay," *The Quantifying Spirit in the 18th Century*, 1–23. For general histories of censuses, demography, and the French population, see: Jacques Dupâquier, *Histoire de la Population Française* (Paris: Presses Universitaires de France, 1988); Jacques Dupâquier and Eric Vilquin, "Le Pouvoir Royal et la Statistique Démographique," *Pour une Histoire de la Statistique* (Paris: Institut National de la Statistique et des Études Économiques, 1976), vol. 1, 83–104; Jacques and Michel Dupâquier, *Histoire de la Démographie* (Paris: Perrin, 1985); and Jacqueline Hecht, "L'idée de Dénombrement jusqu'à la Révolution," *Pour une Histoire de la Statistique* (Paris:

Institut National de la Statistique et des Études Économiques, 1976), vol. 1, 21–81.

3. In Rochas d'Aiglun, ed., *Vauban: Sa famille, ses écrits* (Paris, 1910), vol. 1, 488–89. Cited in Nannerl O. Keohane, *Philosophy and the State in France—The Renaissance to the Enlightenment* (Princeton, N.J.: Princeton University Press, 1980), 328.

4. Jean-Jacques Rousseau, *The Social Contract*, trans. Maurice Cranston (Penguin Books, 1982 [1762]), 130.

5. This tendency was not limited to demographic inquiries. For a discussion of Lavoisier's use of long decimals, see Jan Golinski, "'The Nicety of Experiment': Precision of Measurement and Precision of Reasoning in Late Eighteenth-Century Chemistry," in this volume.

6. Guy Cabourdin and Jacques Dupâquier, "Les Sources et les Institutions," in *Histoire de la Population Française*, Jacques Dupâquier, ed., vol. 2 (Paris: Presses Universitaires de France, 1988), 11–13. The following account of vital registration in France is largely drawn from this helpful article.

7. Cabourdin and Dupâquier, 15–16.

8. The number of *généralités* changed during the eighteenth century; in 1789 there were 33.

9. For the classic discussion of centralization during the old regime and the role of the intendants, see Alexis de Tocqueville, *The Old Régime and the French Revolution*, trans. Stuart Gilbert (New York: Doubleday Anchor, 1955 [1856]), esp. pt. 2, chap. 2.

10. Cabourdin and Dupâquier, 16–17.

11. For an excellent discussion of these sources, see Bertrand Gille, *Les Sources Statistiques de l'histoire de France—Des enquêtes du XVIIe siècle à 1870* (Paris, 1964).

12. For a list of these see Cabourdin and Dupâquier, 29.

13. For his biography, see P. Lazard, *Vauban 1633–1707* (Paris, 1934). On the *Capitation*, see Vauban, "Projet de Capitation," in Vauban, *Projet d'une Dixme Royale suivi de deux écrits financiers*, E. Coornaert, ed. (Paris, 1933), 254–73.

14. Lazard, *Vauban*, 544–49.

15. Vauban, *Note sur le recensement des peuples*, cited in Cabourdin and Dupâquier, 32. The historical demographer E. Vilquin first published this manuscript in 1972. Unfortunately, the original manuscript is not dated. See E. Vilquin, *Vauban et les méthodes de statistique au siècle de Louis XIV* (Paris: IDUP, 1972), cited in Cabourdin and Dupâquier, 32; see also E. Vilquin, "Vauban, inventeur des recensements," *Annales de démographique historique* 1975, 207–57.

16. The historian Edmond Esmonin rediscovered this pamphlet in the early 1950s. For his account of the discovery, see "Quelques données inédites sur Vauban et les premiers recensements de population," in E. Esmonin, *Etudes sur la France des XVIIe et XVIIIe siècles* (Paris: Presses Universitaires de France, 1964), 260–66. (This essay was initially published in *Population*, 1954.)

17. See Vauban, "Description géographique de l'élection de Vézelay contenant ses revenus; sa qualité; les moeurs de ses habitants; leur pauvreté et richesse;

la fertilité du païs; et ce que l'on pourroit y faire pour en corriger la sterilité et procurer l'augmentation des peuples et l'accroissement des bestiaux," (1696) in Vauban, *Projet d'une Dixme Royale suivi de deux écrits financiers*, ed. E. Coornaert (Paris, 1933), 274–95.

18. Vauban, "Description de l'Election de Vézelay," 284.

19. Cabourdin and Dupâquier, 35.

20. For example, while Philibert Orry, comte de Vignory (1689–1747), was Controller-General of France from 1730 to 1746, he ordered a general survey of the state of commerce and manufacturing. In a letter to the intendants dated December 1744, Orry requested information on the manufacturers and factories in each city, the number of persons employed in these industries, the number of inhabitants of each town, the number of young men available for the military draft and capable of bearing arms, as well as an account of the resources in that area. A reply to his request from a Parisian intendant can be found in the British Library. "Situation Actuelle des Provinces du Royaume de France en l'année 1746," containing "Mémoires concernant la Situation actuelle des provinces des Royaume par rapport au Commerce et à l'industrie des Villes aux manufactures, au dénombrement des Peuples et autres Mémoires," British Library, Add. MS 8757. A portion of this document is reprinted in *Mémoires des Intendants sur l'Etat des Généralités*, ed. A. M. de Boislisle, vol. 1—"Mémoire de la Généralité de Paris" (Paris, 1881), 444.

21. Cited in Fernand Faure, "The Development and Progress of Statistics in France," in *The History of Statistics—Memoirs to Commemorate the 75th Anniversary of the American Statistical Association* (New York: The MacMillan Co., 1918), 263.

22. The column headings on this form were: parishes, sénéchaussées, hospitals/communities, births (boys, girls, total), marriages, deaths (men, women, total), religious professions (men, women, total), religious deaths (men, women, total). At the very end of this form there was a table entitled "Récapitulation Particulière de l'élection de _______." Here totals of the above categories were to be filled in. Archives Nationales, Paris F^{20} 441. See, for example, "État de la Population de la Généralité de Moulins—Année 1773." Similar booklets exist for Château du Loire, Château Gantier, and de Mans.

23. Archives Nationales F^{20} 104. "Mémoire sur la nécessité et les moyens d'apprécier la population."

24. See J. F. Bosher, "French Administration and Public Finance in their European Setting," *The New Cambridge Modern History*, vol. 8, "The American and French Revolutions, 1763–1793," ed. A. Goodwin (Cambridge: Cambridge University Press, 1965), 565–91, esp. 566.

25. Montesquieu, Charles de Secondat, baron de, *Lettres persanes*, trans. John Davidson (London: George Routledge & Sons [1721]), Letter CXII.

26. See Montesquieu, *Esprit des Lois* (1748), Book XXIII.

27. J. C. Perrot, "Les économistes, les philosophes, et la population," in *Histoire de la Population Française*, ed. Jacques Dupâquier (Paris: Presses Universitaires de France, 1988), vol. 2, p. 517.

28. One historian has noted that Montesquieu's "conclusions themselves, for the most part, were contradicted by facts. . . . In spite of this affirma-

tion it appears that in treating the question of population, he shared in many respects the prejudices of his contemporaries." Antonin Puvilland, *Les Doctrines de la Population en France au XVIIIe siècle—De 1695 à 1776* (Lyon, 1912), 76.

29. Victor de Riqueti, marquis de Mirabeau, *L'Ami des Hommes ou Traité de la Population*, 2 vols. (Darmstadt, Germany: Scientia Verlag Aalen, 1970 [Avignon, 1758–60]), 11.

30. Mirabeau, 12.

31. Ibid., 12, 16.

32. [P. Jaubert], *Des causes de la dépopulation et des moyens d'y remédier* (Londres & Paris, 1767), 5.

33. Cited in John McManners, *Death in the Enlightenment* (Oxford: Clarendon Press, 1981), 108.

34. François Quesnay, *Tableau Économique*, edited with new material, translations, and notes by Marguerite Kuczynski and Ronald L. Meek (London: Macmillan, 1972 [1758/9]), 19.

35. Quesnay, in Victor de Riqueti, marquis de Mirabeau, *L'Ami des Hommes, ou Traité de la Population*, vol. 2, pt. 4, "Précis de l'organisation, ou mémoire sur les états provinciaux, avec des Questions intéressantes sur la population, l'agriculture et le commerce, par François Quesnay." (Darmstadt, Germany: Scientia Verlag Aalen, 1970 [Avignon 1758–60]), 27.

36. For a brief history of the development of this method, see Cabourdin and Dupâquier, 40. Also see Jacques and Michel Dupâquier, *Histoire de la Démographie* (Paris: Perrin, 1985), 91–93; Jacqueline Hecht, "L'idée de Dénombrement jusqu'à la Révolution," 58–61.

37. Jacques and Michel Dupâquier, *Histoire de la Démographie*, 175.

38. In 1709, Saugrain had issued *Dénombrement du Royaume par Généralitez, Elections, Paroisses et Feux*, a two-volume work which provided the number of hearths (*feux*) for all communities, classified by elections and *généralités*. In 1720, a revised edition appeared that incorporated numbers collected in various *généralités* in the interim. The last page of the book provided a "Récapitulation alphabétique," which listed the *généralités*, the number of parishes, and the number of hearths. *Nouveau Dénombrement du Royaume par Généralitez, Elections, Paroisses et Feux* (Paris: Saugrain, 1720).

39. Claude Saugrain, *Dictionnaire Universel de la France Ancienne et Moderne*, vol. 1 (Paris, 1726), 1040.

40. Historical demographers have concluded that he multiplied the number of hearths by 4.5 to arrive at the figures he presented. Cabourdin and Dupâquier, 38–39.

41. Jean J. d'Expilly, *Dictionnaire Géographique, Historique et Politique des Gaules et de la France*, vol. 1 (Liechtenstein: Kraus Reprint, 1978 [1762]), "Avertissement," [6 pp.], n.p.

42. See C. C. Gillispie, "Probability and Politics: Laplace, Condorcet, and Turgot," *Proceedings of the American Philosophical Society*, vol. 116, no. 1, 1972, pp. 1–20, p. 11; C. C. Gillispie, *Science and Polity at the End of the Old Regime* (Princeton: Princeton University Press, 1980), 47, n. 141; and Jacques and Michel Dupâquier, *Histoire de la Démographie*, 177–78.

43. Louis Messance, *Nouvelles Recherches Sur la Population de la France* (1788). Cited in Jacques and Michel Dupâquier, 178.

44. Louis Messance [La Michodière], *Recherches sur la population des généralités d'Auvergne, de Lyon, de Rouen, et de quelques provinces et villes du Royaume* (Paris, 1766), 1.

45. Messance [La Michodière], 2–3.

46. Ibid., 2.

47. For evaluations of Montyon's contributions to demography, see Jacques and Michel Dupâquier, *Histoire de la Démographie*, 171; J. C. Perrot, "Les économistes, les philosophes, et la population," 530; and William Coleman, "Inventing Demography: Montyon on Hygiene and the State," in Everett Mendelsohn, ed., *Transformation and Tradition in the Sciences—Essays in Honor of I. Bernard Cohen* (Cambridge: Cambridge University Press, 1984), 215–35.

48. E. Esmonin, "Montyon, véritable auteur des *Recherches et Considérations sur la Population*, de Moheau," *Population* 13 (1958): 269–82; J. Lecuir, "Deux siècles après: Montyon, véritable auteur des *Recherches et Considérations sur la population de la France*, de Moheau," *Annales de démographie historique* 1979, 195–249. See also Jacques and Michel Dupâquier, *Histoire de la Démographie*, 180–81.

49. Moheau [Montyon], *Recherches et Considérations sur la Population de la France*, intro. René Gonnard (Paris: Librairie Paul Geuhner, 1912 [1778]), 127.

50. See chap. 10, "De la Fécondité," Moheau [Montyon], 127–51.

51. See chap. 11, "De la Mortalité," Moheau [Montyon], 152–243.

52. Moheau [Montyon], 17.

53. Ibid., 24.

54. Ibid., 25.

55. Ibid., 15.

56. Ibid.

57. Ibid.

58. Ibid., 26.

59. Pierre Simon de Laplace, "Sur les Naissances, les Mariages et les Morts, à Paris, depuis 1771 jusqu'en 1784; & dans toute l'étendue de la France, pendant les années 1781 & 1782," *Mémoires de l'Académie Royale des Sciences de Paris* 1783 (Paris, 1786), 693–702; *Oeuvres complètes* (14 vols., ed. l'Académie des Sciences 1878–1912), vol. 11, 35–46. For a discussion of Laplace and his work in demography, see C. C. Gillispie, *Science and Polity in France at the End of the Old Regime*, 40–50. Regarding Laplace's use of probability theory in demography, see C. C. Gillispie, "Probability and Politics: Laplace, Condorcet, and Turgot," *Proceedings of the American Philosophical Society* 116 (1972): 1–20; Lorraine Daston, *The Classical Theory of Probability* (Princeton, N.J.: Princeton University Press, 1988), 275–78; Jacques and Michel Dupâquier, *Histoire de la démographie*, 186–88.

60. In these writings, Laplace developed what statisticians today call inverse probability—that is, reasoning from effect to cause. See Stephen M. Stigler, *The History of Statistics—The Measurement of Uncertainty before 1900* (Cambridge, Mass.: The Belknap Press of Harvard University Press, 1986), esp. chap. 3. For a

detailed discussion of Laplace's conception of probability, see Lorraine Daston, *Classical Probability in the Enlightenment*, esp. 267–78.

61. For a brief summary of Laplace's method, see Stephen Stigler, *The History of Statistics*, 163–64, who cites the following works for technical discussions of Laplace's method: Karl Pearson, *The History of Statistics in the 17th and 18th Centuries*, ed. E. S. Pearson, (London, 1978), 462–65; O. B. Sheynin, "P.S. Laplace's work on probability," *Archive for the History of Exact Sciences* 16 (1976): 137–87; W. G. Cochran, "Laplace's ratio estimator," in *Contributions to Survey Sampling and Applied Statistics*, ed. H. A. David (New York: Wiley, 1978), 3–10. See also Laplace, *Théorie Analytique des Probabilités*, 1st ed. (1812), 391–401.

62. Laplace, "Sur les naissances," 37.

63. Ibid., 38.

64. Ibid.

65. Pierre Simon de Laplace, *Essai philosophique sur les probabilités* (1814). In *Oeuvres*, vol. 7, i–cliii. Trans. Frederick Wilson Truscott and Frederick Lincoln Emory, *A Philosophical Essay on Probabilities* (New York: Dover, 1951), 66.

66. Laplace, *Essay*, 66.

67. Similar reports were included each year in the *Mémoires de l'Académie Royale des Sciences* until 1788. See: De Séjour, Condorcet, and Laplace, "Pour connoître la Population du Royaume, & le nombre des habitans de la Campagne, en adaptant sur chacune des Cartes de M. Cassini, l'année commune des Naissances, tant des Villes que des Bourgs & des Villages dont il est fait mention sur chaque carte," *Mémoires de l'Académie Royale des Sciences* 1783 (Paris, 1786), 703–18; ibid., 1784 (Paris, 1787), 577–93; ibid., 1785 (Paris, 1788), 661–89; ibid., 1786 (Paris, 1788), 703–17; ibid., 1787 (Paris, 1789), 601–10; ibid., 1788 (Paris, 1791), 755–67. Another example of the Academy's interest in population is found in Jean Morand, "Récapitulation des baptêmes, mariages, mortuaires et enfans trouvés de la ville et faubourgs de Paris, depuis l'année 1709, jusques et compris l'année 1770," *Mémoires de l'Académie Royale des Sciences* 1771 (Paris, 1774), 830–48. In this article, Morand indicated that beginning in 1709, the city of Paris drew up abstracts of the numbers of baptisms, marriages, and deaths by month and by parish.

68. A steady stream of correspondence from La Michodière to the Académie is recorded in the Procès Verbaux. Procès Verbaux, Académie Royale des Sciences, T. 104 (8 Jan. 1785–23 Dec. 1785), entry for 23 July 1785. See also entry for 31 Aug. 1785, and T. 105 (10 Jan. 1787–23 Dec. 1787), entry for 5 May 1787.

69. La Michodière, "Tables alphabétique des villes, bourgs, villages dans différentes régions de France, Moyenne des naissances, mariages, morts sur les années 1781, 1782, 1783," 2ème cartonnier 1785, Académie des Sciences.

70. Laplace, *Essay*, 67.

71. Kathryn M. Olesko develops this theme of the relationship between probability calculus and the meaning of precision in the early nineteenth century. See Olesko, "The Meaning of Precision: The Exact Sensibility in Early Nineteenth-Century Germany," this volume.

72. For discussions of statistics during the Revolutionary and Napoleonic periods, see: Marie-Noëlle Bourguet, *Déchiffrer la France: La statistique départementale à l'époque napoléonienne* (Paris: Editions des Archives Contemporaines, 1988); id., "Décrire, Compter, Calculer: The Debate over Statistics during the Napoleonic Period," in *The Probabilistic Revolution*, ed. Lorenz Krüger, Lorraine J. Daston, and Michael Heidelberger, vol. 1 (Cambridge, Mass.: MIT Press, 1987), 305–16; and Stuart J. Woolf, "Towards the History of the Origins of Statistics: France 1789–1815," in *State and Statistics in France 1789–1815* (New York: Harwood Academic Publishers, 1989), 79–194.

TWO

A REVOLUTION TO MEASURE:

THE POLITICAL ECONOMY OF THE METRIC SYSTEM IN FRANCE

Ken Alder

Why is a Meter a Meter?

ACCORDING to the most recent textbooks, a meter equals the distance traveled by light in a vacuum in 1/299,792,458 of a second. One might well ask, why so bizarre a number? A possible answer goes as follows: the meter acquired that present value so as to equal a platinum bar built in the 1870s and housed in the Parisian *Conservatoire des arts et métiers*. This bar was built to match an earlier platinum bar manufactured at the time of the promulgation of the metric system; that is, during the French Revolution. And that first bar was based on a geodetic survey, conducted during the 1790s, which measured a portion of the length of a quarter of the earth's meridian (the distance from pole to equator) and divided that distance by ten million. The ostensible purpose of this geodetic survey—which took six years and cost half-a-million *livres*—was to create a standard measure based on invariant nature that just so happened to be on the human scale and roughly equal to the *aune* of the Old Regime.

The very elaborateness of this charade, however, suggests a different sort of answer. To uncover it, I will need to show that at the core of "universal standards" commonly taken to be the products of objective science lies the historically contingent, and further, that these seemingly "natural" standards express the specific, if paradoxical, agendas of specific social and economic interests.[1] In particular, I will show how a technocratic elite during the period of the French Revolution used a new system of measures to mediate a fundamental tension between state authority and the construction of a market economy. This then is the story of how a rational language—the metric system—was deliberately crafted to break the hold of the Old Regime's political economy and serve as the universal idiom of the modern mechanism of exchange.

The Paradoxes of Standardization

The hobgoblin of uniformity haunted the philosophical mind of the Enlightenment. Already in his *Esprit des lois*, Montesquieu had warned against the seductive aesthetic of a leveling rationalism.

> There are certain ideas of uniformity which sometimes seize great minds (as they did Charlemagne's), but which invariably strike the petty. They find in them a kind of perfection which they recognize because it is impossible not to discover it; the same weights and measures in commerce, the same laws in the state, the same religion in all parts. But is uniformity always appropriate without exception . . . ? So long as people obey the law, what does it matter if they obey the same one?[2]

Montesquieu became known in the late eighteenth century for his sly willingness to accommodate the existing order, a pragmatic and moderating spirit which feared the consequences of its own rational criticism. Not all eighteenth-century thinkers were so deferential; for them, rational criticism sounded an unequivocal call to action. The above passage in Montesquieu greatly offended Condorcet. "Here we have one of the most curious chapters of the book. It has earned Montesquieu the indulgence of the prejudiced, of those who hate enlightenment, of those who defend abuses, etc. . . . Ideas of uniformity, of regularity, please all minds, and especially just minds. . . . A good law ought to be good for all men, as a true proposition [in geometry] is true for all men."[3]

That nature's laws were everywhere the same meant, for Condorcet, that the hodgepodge of human laws must be realigned with universal principles. By reducing the legal code to its essential form, he expected to render it comprehensible to all literate men and women, and to diminish the unfair advantage that those in authority held over the powerless. This program of social justice depended, in Condorcet's view, on a scientific study of society to be conducted by means of new rational languages for the facts of the social realm, including symbolic languages that would be universally valid and "bring to all objects embraced by human intelligence a rigor and precision that would render knowledge of the truth easy and error almost impossible."[4] One such language was to be the metric system.

Keith Baker has analyzed the virulent debate between Necker's monarchical pragmatists and Turgot's free-market Physiocrats.[5] The two camps also differed on the related question of metrical reform. Whereas Necker confessed that the reform of weights and measures lay beyond the crown's power, one of the unfinished projects of Turgot's ill-fated ministry was to ascertain the precise length of a one-second pendulum with an eye toward constructing a uniform system of measures based on invariant nature.[6] As Baker has pointed out, Turgot thereby sought to enlist the authority of science to resolve a contentious political debate.[7] And in the event, the system of rational measures elaborated by his disciple, Condorcet, proved highly prophetic of the metric system adopted twenty years later by the French Republic. This was, of course, no accident; as we will see, Condorcet assisted Talleyrand with his 1790 proposal for revolutionary measures, and guided the Academy of Sciences toward its final formulation of the metric system. His goal in both instances was twofold and paradoxical: on one hand, to facilitate free exchanges between French citizens by introducing a uniform language for the objects of daily economic life; and on the other, to enable the government to collect accurate information for the purposes of rational policy-making. Condorcet expected decimal division to enable ordinary citizens to calculate their own best interests "without which they cannot be really equal in rights . . . nor really free."[8]

This sort of liberatory rhetoric has dominated discussion of the metric system, both then and now.[9] And certainly many of the advocates of metrical reform upheld a democratic vision of equal access to knowledge. In the year II of the Revolution, the Jacobin mathematician Gaspard Monge would tout the decimal system as being "within the reach of everyone; all children will know it, and it will be a means of reducing the inequality among men."[10] This essay is intended to point out the paradoxical context in which such liberatory "rationalization" operated. As the Physiocrats themselves recognized, replacing the Old Regime's society of particularist interests with a scientific and natural social order meant in some sense erecting reason as "sole despot of the universe." In practical terms this also meant making the citizen knowable to the state. Nor was this the only irony. So long as Condorcet believed he spoke in the name of the common good (since what is true is necessarily good), he felt justified in castigating his opponents as purveyors of falsity who sought only to protect their private interests. "The uniformity of weights and measures cannot displease anyone but those lawyers who fear a diminution in the number of trials, and those merchants who fear anything that renders the operations of commerce easy and simple."[11]

To ensure the triumph of free trade, Condorcet and his followers came to adopt an authoritarian rhetoric which excluded defenders of the "Gothic" measures from the realm of virtuous and legitimate discourse. This dual irony of intent and word was to reverberate through the wider polity when the Revolutionary state later deployed the metric system as part of its effort to impose a free-market economy on France.

Recent historiography of the French Revolution has focused on the supremacy of the word.[12] This essay is, among other things, an attempt to broaden this fashionable analysis of political discourse to include the political work done by *non-verbal* languages explicitly concerned with material objects and property relations. I intend to examine both the rhetoric *about* the metric system and the metric system *as* a politicized language about the material world. My goal is to help reconnect an increasing rarefied academic analysis of political signs and symbols with the daily economic lives of French men and women in the 1790s. The revolutionaries themselves saw the metric system in just this light. Bureaux de Pusy, military engineer and representative to the National Assembly, assured the legislature that in a nation cut off from its past, the metric system would provide a unifying language that tied citizens to the economic life of the country. "What means are more capable of bringing together the *esprits* and diverse interests into that precious unity which is the strength of government than a common idiom, common symbols, and identical rules for all the objects necessary or useful for the daily needs of individuals; and how much the uniformity of measures will fulfill this goal!"[13]

For Bureaux de Pusy the advantages of such a uniform and rational language were also material: the system would increase exchanges between distant parts of the nation, and hence specialization in more valuable types of production; it would end the wasteful and fraudulent practices of merchants who purchased a commodity at a large measure, and then sold it at a smaller one; and finally, it would enhance the productivity of agriculture—the basis of national prosperity—by enabling the comparison of yields on different plots of land, and hence specialization in suitable crops. For all these good Physiocratic reasons, he assumed such a system was sure to be embraced by a grateful nation—according to the President of the Assembly, in no more than six months.[14] In fact, each of these advantages involved a fundamental misconception of the meaning of measurement in the Old Regime and resulted in a woeful misreading of the source and strength of the resistance to the new measures. And again, historians have reproduced that misreading to the point where most major accounts of the Revolution have assumed that the metric system was one of the few unrescinded successes of the period.[15] But the Revolutionary governments failed to impose the metric system on France. In 1812, Napoleon, the "great systematizer," effectively returned France to the old standards, and only under Louis-Philippe was the metric system reinstated in the 1840s.

Twentieth-century Americans are inevitably sensitive to the problems associated with converting to a new set of units. And unlike the French historians noted above, John Heilbron, in his recent study of the metric system, sees through the self-serving "grantsmanship" of the Revolutionary savants, and with mordant wit exposes their naive faith in the common people's willingness to join this orgy of rationalization.[16] But Heilbron too implies that the slow spread of the new measures can serve as a yardstick (meter stick?) for the spread of reason among an innumerate public—a people whose resistance he attributes to four causes: 1) consumers' fears that shopkeepers would raise the price of goods when they rounded up to the new units; 2) the fact that the decimal system does not lend itself easily to the familiar division of commodities by twos and threes; 3) the unexpected difficulties, particularly among an innumerate people, of mastering calculations even in the decimal system; and 4) most people's attachment to particular numbers as qualitative judgments about, say, what constitutes a "tall" person. Clearly, these are the sort of translation problems that afflict citizens asked to adjust from one set of "imperial" units to the metric system—and they certainly operated in Revolutionary France. But French citizens of the Revolutionary period were also being asked to confront the additional and very different problems associated with converting to the metric system from a world of premodern measures that was deeply embedded in the practices of the Old Regime, a world with its own autonomous moral economy and its own consistent "reasonableness." What *was* the meaning of measurement in Old Regime France? What was the savants' "grantsmanship" meant to cover up? And what can explain this strange equivocation in the history of French modernization?

The "Infinite Perplexity" of the Old Regime

The diversity of measures in pre-Revolutionary France astounded visitors. In his travels of 1789, Arthur Young was continually infuriated by a country "where the infinite perplexity of the measures exceeds all comprehension. They differ not only in every province, but in every district, and almost in every town. . . ."[17] In the eighteenth century there existed as many as 700–800 different metrical names which expressed a mind-boggling total of some 250,000 local variants![18] The diversity of measures within the province of Forez (now the department of the Loire) bears out this numbing complexity.[19] Why should this have been so?

Measurement is indispensable to exchange. As performed within a modern system, it is an operation where objects are described in abstracted, commensurable units that relate to a conventional standard. As performed in early modern France, measurement was an operation almost inseparable from the object measured and a specific measuring device.

In the Old Regime, different commodities were measured with different units, whose values differed from parish to parish. In the Saint-Etienne district of the Forez there existed separate units for grain, wine, oil, salt, hay, wood, coal, and other products. Likewise, different trades employed different units, a reflection of the particularism of Old Regime corporate practices.[20] But the fundamental reason for this variation was that most measures had meaning only in reference to a specific receptacle or ruler. The famous iron *toise* of Paris was set in the wall of the Grand Châtelet. And the *bichet* (bushel) of Saint-Etienne referred not to an abstract capacity of grain, but to a specific cylindrical receptacle housed in the municipal offices.[21] The dimensions of this vessel influenced its capacity because, depending on local custom, the grain was measured either "heaped" or after being "struck and combed" in a prescribed manner. Thus measurement in pre-Revolutionary France did not just represent an abstract quantity, it embodied a whole mesh of physical objects, ritualized custom, and the practices of artisanal producers.

But far from being irrational, the hodgepodge measures of the Old Regime made real sense to the artisans, peasants, and shopkeepers who used them. Almost all premodern weights and measures had at their origin an anthropomorphic meaning; that is, they referred to the human scale and human needs. By this I do not mean those units which had simply taken their name from human anatomy (*pied*, *pouce*, etc.); these units had long since become fixed by reference to specific rulers—at least in the major market towns. Instead, I mean the many other units associated with human labor and production. For instance, the *aune,* a measure of cloth, was defined as the width of the local looms.[22] And coal in the Saint-Etienne region was measured in *charges* equal to one-twelfth of a miner's daily output, while the unit that measured the length of his progress varied with the difficulty of the face being worked.[23] The measurement of land area in particular was closely tied to human labor. In many regions of France, the Forez among them, arable land was still measured in *bicherées*, a value based on the number of *bichets* (bushels) of grain required to sow that field; while in the vineyards, land was measured in the *homme* or *journalier,* the area of grape-growing land that a peasant could pick in a day. The Saint-Etienne district also possessed a separate measure for pasture land, itself divided into five *dégrées* based on the quality of the land.[24]

As Witold Kula, the great Polish economic historian, has pointed out, these anthropomorphic measures expressed both a quantity of human labor and a qualitative evaluation of the land (rich land was more densely sown than poor land; plentiful vines took longer to pick), features of primary concern to those who worked the soil.[25] (And for those who think there's no difference between this and modern measures, there's five hectares in Florida I'd like to sell you. . . .) These "natural" measures are analogous to the "natural" time that E. P. Thompson says workers reluctantly surrendered in emerging European manufactures of the early modern period. There, enterprising owners replaced the task-oriented time of their proto-industrial work force with the new mechanical clock-time, recasting labor in terms of an abstract quantity (time) so as to enforce greater shop floor discipline and appropriate any increase in the productivity of their operatives.[26] As we will see, the same process seems to have taken place when the natural measures of land-labor were converted into abstract units of surface area.[27] This task-oriented sense of the word "natural" has the exact opposite meaning of that advanced by Enlightenment scientists for their "natural" units, which were the antithesis of anthropomorphism.[28]

But despite their anthropomorphic origin, most units of measure had stabilized during the early modern period, as lawsuits and the protracted dealings of artisans, peasants, commercial traders, and seigniors established local convention.[29] Particularly important were the struggles between peasants who paid rent in kind and their seigniors, who sought to increase the dimensions of their measures as a way of extracting greater rent.[30] Consumers also had an interest in stable measures as a way of controlling price. This applied especially to the paramount issue of bread. As the cost of foodstuffs soared on the eve of the Revolution, angry townspeople in Saint-Etienne complained—justifiably as it turned out—that bakers had decreased their weights so as to sell smaller loaves.[31] So the crowd of irate citizens forced their aldermen to tour the town bakeries armed with the local standards and required the bakers to conform to the local standards.

This incident reveals one of the most important functions of metrical diversity. In a "fair price" economy, traders used differences in standards of weight to compensate themselves for their services. In 1754, the government's agent in Tours admitted that variations in local measures allowed dealers to make a profit by buying grain at one measure and selling it (for the same price) at a lesser measure.[32] But rather than condemn this practice, he noted that it encouraged commerce in the region, since attempts to raise prices risked the wrath of the local populace. Obviously, these differences opened the door to considerable fraud and chicanery, but as the Assembly of Sens noted in 1788, "the establishment of a uniform measure would ruin this genre of commerce, destroying at the same time an infinity of little markets which subsist only on these differences and, though of no great importance, supply the needs of the nearby consumers. . . ."[33] The ability to manipulate measures in this way depended on an intimate familiarity with the local metrical "dialect" and enabled local traders to protect their market niche.

The flexibility of measures in the Old Regime also enabled workers to skim off some of the materials that passed through their hands and thereby extract a "customary appropriation" to supplement their meager wages. As Peter Linebaugh has shown in the case of the eighteenth-century transatlantic tobacco trade, English merchants used hogshead containers of standard volume to try to end this sort of "depredation."[34] In France, too, ports and other transfer points were notorious for incurring such carrying charges; and dock workers often bitterly resisted the standardization of measures. While the metric system did not put an end to such practices—and may even have simply invited government officials to take a "cut" instead—regular standards made it easier for traders to criminalize these popularly sanctioned customs.[35]

Standards of measurement are inevitably arbitrary; but enforcement can be pursued through various mechanisms. In modern societies, standards are set by scientists and enforced by government bureaucracy. In premodern societies, local communities have the primary role in safeguarding measures, and by this means they attempt to protect their livelihood and insulate local economies from outside disruption and competition.

"One Law, One King, One Weight, and One Measure"

French sovereigns and their administrators had periodically resolved to assert their authority in metrological questions—precisely to extend their ability to govern and extract revenue at the local level.[36] In practice, however, control over weights and measures lay predominantly in the hands of seigniors who kept copies of the local standards and extracted a fee for all market weighings.[37]

Nevertheless, eighteenth-century royal bureaucrats, besieged by innumerable disputes over measurement of cloth, iron, and wine, persisted in their attempts to coax France into some semblance of metrical order.[38] In 1766 Trudaine de Montigny, the Intendant of Finances, sent eighty models of the Parisian units to the major towns of the realm, and enjoined provincial administrators to draw up tables comparing local and national measures.[39] Without such a key, royal bureaucrats could not be certain of the local grain prices listed in the *mercuriales*. Control of the all-important food supply was often hindered by the diversity of weights and measures; a dangerous weakness in the administration of enlightened absolutism.[40] During the crisis of the year II, even the "all-powerful" Committee for Public Safety had great difficulty translating the *maximum* for bread prices into the thousands of local measures, and similar problems attended the sale of the *biens nationaux*.[41] The technical branches of the bureaucracy—particularly within the military—were also eager to exploit the benefits of uniform measures. Vauban, the Sun King's fortification expert, expressed an interest in standardized measures and in the decimal system of division.[42] And early in the century the artillery service introduced a single set of measures throughout its arsenals of production. Certainly the system of interchangeable parts manufacturing, conceived by artillery general Jean-Baptiste de Gribeauval in the last decades of the Old Regime, was unimaginable without some standardization of measures.[43]

This insight into the value of uniform measures was not new of course. As is well known, a number of illustrious savants had proposed various systems of measures in the seventeenth century, some based on a regularly beating pendulum and others on the size of the earth.[44] The research program of these natural philosophers had been directed toward a search for constancy in nature, and they wished to communicate their results with a hitherto unheard-of precision. Yet even these were not the first proposals for rationalized measures. Decimal division had been advocated by the Flemish engineer, Simon Stevin, a century before.[45] And Renaissance functionaries, mindful of their classics, drew up equivalencies for Roman and Greek measures, which they offered as the basis for national standards.[46]

But the primary impetus for the spread of uniform measures under the Old Regime was the transforming economy. During the early modern period Parisian measures gained appreciable ground in major urban areas.[47] By the end of the eighteenth century, townsfolk in Saint-Etienne used a single *livre*, that of Lyon, the region's major trading center—and the source of most of the town's balances.[48] Changing patterns of land tenure also spurred new "exact" assessments by surveyor-geometers.[49] In the Lorraine, for instance, peasants' methods of estimating land area gave way to the methods of "geometers" at the same time that open-field cultivation was ending and tracts were increasingly defined as the absolute property of a single owner.[50] Even in relative backwaters like the districts around Saint-Etienne, surveyors bragged that "we have put all these defective [measures] in good order, so that in each district their content is regulated in either *perches*, *pas* or *pieds*. . . ."[51] (These same surveyors cautioned, however, that for the actual partitioning of fields "it is best to stick to the report of those who sow the land. . . .") Although the rationalization of farming advanced more slowly in France than in England, redefining holdings in abstract units of surface area, rather than units of labor, improved the chances for gains in productivity—gains that presumably accrued to the landowner.[52]

In the decade before the Revolution the mood among the propertied and literate portion of the population swung decisively against the diversity of measures. Contemporaries complained of their immense variety, and pleaded for a Legislator who would give France a uniform code of measures. Alexis Paucton, the foremost metrical compiler of the Enlightenment, wrote: "They are the rule of justice which must not vary, and the guarantee of property which must be sacred."[53]

This resentment seethed in the famous *Cahiers de doléances*: 128 of the third estate's regional *Cahiers* demanded uniformity, 32 of the nobility's, and 18 of the clergy's.[54] In the Forez all three orders demanded reform of weights and measures, and the plea for "one law, one king, one weight, and one measure" figured in the complaints of 18 of its surviving parish *Cahiers*.[55] In these complaints, metrical reform was typically coupled with reducing the impediments to free trade. For instance, in La Cotte en Couzon (Forez) metrical "multiplicity" was explicitly cited as a "hindrance to commerce."[56]

But here we must confront a paradox. If the desire for uniform measures was so widespread among the elite, why did the metric system fail? One answer lies in the *type* of metrical reform the French citizens were asked to adopt.

From "Rational" Measures to "National" Measures

The metric system was the active creation of savants such as Condorcet, Prieur de la Côte-d'Or, Monge, Borda, Lagrange, Haüy, Delambre, Méchain, Lavoisier, and a score of others. Between 1790 and 1799, they sat on the planning committees, performed the scientific research, and administered the agencies that monitored the reform.[57] Their right to define and execute policy in this area stemmed to a considerable degree from the Royal Academy's traditional authority in metrical questions.[58] And like the royal academicians they had so recently been, these republican academicians were eager to provide the intellectual instruments that would assist the central bureaucracy in its attempt to fashion France into a modern state—one that would be revenue-rich, militarily potent, and easily administered. To these men, the Revolution seemed to clear the way for a full-fledged renewal of Turgot's enlightened reforms. Academicians like Jean-Baptiste Le Roy quickly noted that the fall of feudalism and seigniorial privilege in August, 1789 removed the daunting juridical barrier that had soured Necker on metrical renewal.[59] So when in 1790 Talleyrand, in consultation with Condorcet, secured passage of legislation initiating a review of French measures, the Assembly authorized the Academy to review the various proposals. But though they developed the system on behalf of their patrons in the state, as technical experts, these academicians possessed considerable latitude to give shape to the vague demands expressed in the *Cahiers*.[60]

Between 1790 and the year III they added in succession four elements to the simple demand for a uniform system of measures. Each was proposed independently, and each called for choices among alternatives. (Even the demand for uniform measures came in a variety of forms; for instance, many *Cahiers* had called for only province-by-province uniformity.) A schematic calendar showing their order of introduction is presented in table 1. First, they insisted that the single invariable standard be *taken from nature* (though the earliest legislative proposal urged the adoption of the Parisian or Roman measures, and scientists long debated whether to derive their natural standard from a pendulum or the size of the earth). Second, the various measures of length, area, capacity, and weight were to be linked into an interconnected *system of measures* (though here too scientists disagreed over how to define these relationships, especially for units such as weight). Third, the *decimal scale* would divide these units into multiple and fractional units (though a duodecimal, base 12 system was also seriously considered). And fourth, a *systematic nomenclature* would express the relation of the fractional units to the principal unit (though proposals for simple monosyllabic names were also advanced).

By deriving the system from "nature" the scientists hoped to elevate their creation above the politics of self-interest and make it seem nonarbitrary and independent of its creators.[61] Grandiloquently accepting the definitive meter on behalf of the Council of Ancients in 1799, President Baudin admitted that while any standard is ultimately chosen by convention, one taken from "inviolable" nature was surrounded by "all the authority derived from such a source."[62] In much the same way, Revolutionary politicians such as Danton and Grégoire used the familiar idea of "natural frontiers" to give France's territorial ambitions in Belgium a patina of reason and respectability.[63]

This claim of "naturalness" was principally invoked to justify the (expensive) choice of basing the standard of length upon a fraction of the earth's meridian.[64] However, this same rhetoric was used to justify all the features of the new system. Though well aware that the preference given, say, the decimal system of division was in some sense arbitrary—a convenience for arithmetic calculations—metric propagandists such as the crystallographer René-Just Haüy insisted that division by base 10 approached a "natural" base because of finger counting.[65] More to the point, Laplace argued that the universality of the decimal system made it the vastly superior choice.[66] But even Laplace acknowledged that a system founded on the duodecimal division offered real advantages for daily commercial transactions because the number 12 possessed many divisors, and hence enabled merchants and customers to partition commodities easily.[67] For the same reason, some reformers preferred base 8 because it would enable commodities to be divided in half again and again and again.[68]

These claims of "naturalness" and "universality" served practical purposes. In the early years of the Revolution the claim that the new measures had universal properties that transcended the hurly-burly of daily politicking helped build consensus around the plan as myriad proposals for new measurement systems poured into the legislature from provincial

TABLE 1

Calendar of Legislative Proposals Regarding the Metric System, Revolutionary Period, 1790–1799

1. UNIFORMITY

1788—The *Cahiers* call for the adoption of uniform weights and measures. Some urge national uniformity, others regional or local uniformity.

August 1789—Two weeks after the abolition of feudalism on August 4, Le Roy suggests to the Academy of Sciences that there are no legal impediments to national measures.

February 1790—Tillet and Abeille of the Royal Agricultural Society are the first to propose national standards of measures to the National Assembly.

2. A MEASURE "TAKEN FROM NATURE"

February 1790—Tillet and Abeille ask the Assembly to adopt the "original Roman measures," or if these cannot be determined, those of Paris.

March 1790—Talleyrand (on advice from Condorcet) suggests the national standard be based on the length of a one-second pendulum at 45° latitude north (a site conveniently near Bordeaux).

March 1791—The Academy committee (Borda, Lagrange, Laplace, Monge, and Condorcet) win approval for a standard based on a fraction of the meridian that passes through Dunkerque and Barcelona.

April 1792—Interior Minister Roland insists that the Convention adopt a temporary measure of length while the nation awaits the completion of the meridian survey.

June 1799—Calculations of the International Commission on Weights and Measures based on the six-year geodetic survey of Delambre and Méchain, result in a "definitive" platinum meter bar, which is presented to the Council of Ancients.

3. SYSTEM OF MEASURES

March 1790—Talleyrand suggests all measures (area, volume, weight, etc.) be derived from length.

1793—Lavoisier defines a gram as the weight of a given volume of distilled water at the melting temperature of ice.

1799—Lefevre-Gineau defines a gram as the weight of a given volume of distilled water at the temperature of maximum density.

4. DECIMAL DIVISION

April 1790—Le Blond proposes to the Academy of Sciences that France adopt a duodecimal system, and invents two new characters to represent "10" and "11."

TABLE 1 *(cont.)*

March 1791—Borda makes a formal proposal for a decimal system of division, which had been first reported in committee in October 1790.

5. SYSTEMATIC NOMENCLATURE

1790–93—The committees of the Academy consider various proposals for a litany of simple monosyllabic names.

August 1793—Arbogast publicly introduces the first version of the systematic nomenclature.

1794—During the Terror, Prieur advocates using low Breton prefixes.

April 1795—Prieur substitutes Greek prefixes in the near-definitive metric law of 18 germinal, year III (April 7, 1795). Minor modifications follow in the next months.

societies, professors of philosophy, and citizen-surveyors.[69] Significantly, the system offered a national standard without resorting to the adoption of the Parisian measures at a time when provincial federalists suspected the capital of harboring the same centralizing ambitions as its royal antecedents. Whereas the first men to address the Assembly on the subject begged the legislature to simply adopt the Parisian measure so as to "scrupulously distinguish our ordinary measures from our scientific measures [and] not take us beyond our desires and hopes," Talleyrand spurned the Parisian measures as not answering "the importance of the issue, nor the aspirations of enlightened and exacting [*difficile*] men."[70]

This posture of "disinterestedness" also explains the pronounced internationalism of the early proposals of Talleyrand and others. What Britain and France did together could hardly be accused of being in the interest of any single nation or group.[71] The United States and Spain were invited to participate as well. But when, in 1791, at the urging of the Academy, the National Assembly rejected a pendulum standard in favor of a survey of a meridian that traversed France alone, it signaled to even such sympathetic observers as Thomas Jefferson that the academicians' show of internationalism was of a particularly Gallo-centric stamp.[72]

In fact, the meridian project proved a real plum for French science and its main source of financial and institutional support during the Terror.[73] On August 1, 1793, in an effort to forestall the dissolution of the Academy of Sciences, Lavoisier hurriedly pushed the metric system through the legislature, even though the scientists could only offer a temporary estimate of the meter while the meridian survey was still in progress.[74] The Jacobin Convention closed the Academy anyway one week later, but permitted the unfinished metrical research to continue. Thanks in large part to the extravagance of the meridian project, research dragged on for six more years, with a budget roughly three times the annual operating costs of the *entire* pre-Revolutionary Academy of Sciences.[75] Funds soon flowed into the coffers of the community of scientific instrument-makers who had been hit hard by the disruption of the luxury trade.[76] And when the legislature seemed to lose interest in funding the project, the department of military cartography, under General E. N. de Calon, stepped in with money.[77]

Personal and professional ambition also set the agenda for the metrical research. The astronomer Borda threw his weight behind the meridian project in the hope that it would enhance the reputation of his "repeating circles."[78] True, the oblate shape of the earth was a question about which the scientists could always use more data (though the fundamental issues had been resolved thirty years before),[79] but even at the time, observers saw the monumental geodetic survey as an unabashed exercise in grantsmanship. Louis-Sébastien Mercier, chronicler of Paris, accused the savants of having "preserved their pensions and salaries . . . under the pretext of measuring the arc of the meridian."[80] His common sense told him, "it should not have been necessary to go so far to find that which lay so near."

Ultimately, the standard "taken from nature" came to be vested in a platinum bar presented to the Council of Ancients in 1799 and housed in the legislative chamber as a symbol of the new metrical order and as a surety against all damage and decay—just as the Athenians had kept their measures in the Acropolis and the Israelites kept theirs in the Temple.[81] But this attempt to imbue the standard with the sort of sacral "solemnity" that Talleyrand had thought necessary,[82] only undermined the scientists' claims that the measure was "taken from nature." And furthermore, the International Commission that performed the final calculations was inevitably forced to gloss over numerous uncertainties. These included errors in the determination of the base lengths, the angles of the meridian, and the kilogram weight,[83] plus disputes over the value chosen to represent the oblate shape of the earth,[84] difficulties calibrating the temperature-dependence of the metal rods used to measure the triangle bases,[85] unanswered questions regarding the astronomical observations,[86] and significant last-minute changes in the latitude calculations.[87] In any case, the platinum bar itself could not be manufactured with absolute precision.[88] Recounting how he discovered an error in the angle corrections of a colleague, Delambre, the scientist chiefly responsible for the survey, said he offered up the story to "disabuse his contemporaries of the existence of some sort of chimerical perfection that mankind has never obtained and probably never will."[89] Ten years later, Delambre had revised his best guess as to the "true" length of the meter, in the process shaving off two significant figures.[90] It hardly mattered. Faced with the elaborate web of procedures already codified in the geodetic survey, neither the French nor foreign delegates to the international committee had had much scope to suggest alternatives.[91] The measure "taken from nature" had become a human construct. None of which is meant to imply that the international commission practiced "bad" science—only that they misrepresented their undertaking in the name of political expediency.

Keith Baker argues that the Revolutionary governments, basking in their newfound claims to sovereignty, no longer needed science to legitimate their administrative reforms.[92] On the contrary, Revolutionary governments through the Directory period continued to enlist the authority of science to make palatable unpopular legislation—particularly, as we will see, those acts that ushered in a market economy. This claim for objective knowledge has since become the familiar chant of experts wandering the corridors of representative government, their way of covering political interests with the veneer of logic and number.[93] Consider, for instance, the manner in which the French scientific elite defended the metric law as an ideal system beyond the reach of political debate altogether.

In the year II, comparing the various elements of the metric reform to a chain, Haüy argued that "once the principle had been established that the unit of weight and measure was to be taken from nature, the whole plan of the system was traced in advance in prescribed order by the sequence of ideas. . . ."[94] Intended to emphasize the conceptual simplicity of the reform, such a declaration had the property of stifling dissent by condemning critics as the enemies of reason. And when the generally sympathetic editors of an agricultural journal dared to suggest that the new Greek prefixes sounded foreign to most French citizens and would be difficult to understand,[95] the scientists at the Agency of Weights and Measures effectively denied their opponents the right of criticism altogether.

> You cannot attack a part of the system without endangering the whole. Otherwise many different objections will follow: some will want a new nomenclature; others will want the meter to be based on the full circumference of the earth; still others will prefer the pendulum, etc. . . . Now that the law is promulgated (after long deliberation), it is best not to attack it, but to give it the respect it is due. . . . There must not be any doubt about the goodness of the law.[96]

Where Condorcet had expected liberated citizens to assent to the self-evident truth of the new measures, the metric system was now presented as a doctrine that demanded uncritical obedience. I doubt Montesquieu was smiling. For his part, Prieur de la Côte-d'Or—former member of the Committee for Public Safety and primary author of the metric nomenclature—vehemently denounced those vacillators who raised "petty" objections against his new names: "One cannot change the plan of a building once it is underway."[97] As we have seen, however, the metric system was not an indivisible edifice; it had been constructed over several years out of disparate and individually controversial components. The republican scientists always claimed their work was undertaken to safeguard equality and liberty, but when confronted with dissent they spoke the language of technocratic absolutism.

And the scientists' tone only grew more shrill as the population's disregard of the metric system became more evident. In the end, advocates of the metric system were driven to discover a "general will" to justify state intervention as the only way to transcend each individual's reluctance to surrender his or her own familiar measures.[98] Conscious of the contradiction between their own authoritarian rhetoric and the ascendant post-Thermidorian liberalism, the scientists at the Agency of Weights and Measures tried to reassure their co-citizens that the metric system could never be "an instrument in the hands of a [tyrant]. . . ." "[The metric system] is simply a police measure to ensure the social order. . . . Neither *our good pleasure* or *full power* are part of the lexicon of a *reasonable* people who must be enlightened and convinced."[99]

The question remained, how *were* the people to be convinced? How *was* the gap between a "self-evident" rational policy and popular discontent to be closed? Though they might stifle dissent, the scientists could not compel the public to use the new measures. The problem, as Baudin despairingly observed, was that, in the words of Jean-Jacques: "Men will always prefer a bad way of knowing to a better way of learning."[100]

The Metric of the Market

The law of 1 August 1793 had declared the metric system obligatory in eleven months. But in the face of popular indifference and hostility, the government began to lower its sights; Paris would be the proving ground.[101] Even so, police reports from Paris in the mid-1790s indicated the new measures had hardly penetrated the marketplace there.[102] And after the imposition of the metric system for the entire department of the Seine, complete metrical confusion reigned.[103] Storekeepers, trapped between the administration's insistence on the metric law and their customers' preference for the old measures, now illegally stocked two sets of weights. This invited the very abuses that the system had been intended to end.

It was a bitter irony. The ostensible purpose of the metric reform had been to facilitate free and transparent exchanges (fig. 1). Interior Minister Roland had hastened the promulgation of the earliest metric system in 1792 because he believed the diversity of measures represented the main obstacle to the national circulation of grain.[104] (And because he expected that uniform measures would help him write a new tax code.) Numerous

FIG. 1. In this caricature of 1800 the customer in a milliner's shop insinuates that he "prefers the meter to the *aune*." The *aune* was a linear measure of cloth in the Old Regime. *Revue Encyclopédique Larousse* 9 (1899), 847.

petitions had insisted on uniform measures as a prerequisite for open commercial transactions.[105] And the scientists who administered the metric system in the Directory had expected their system would transform France into "a vast market, each part exchanging its surplus."[106] Indeed, the metric proponents went on to argue, their system was particularly designed to facilitate long-distance exchanges, which were the most beneficial kind of trade because they connected regions with complementary resources. Their hope was that this would encourage specialization in production and an increase in yields. Henceforth, those merchants who survived on the differences in local measures, would pursue "a more useful course" and "speculate on differences in *productivity*, which is the natural basis of commerce."[107] In other words, by abstracting measurement from objects and labor, scientists sought to break the protection that particularistic measures afforded local economies, and end those practices that enabled the "fair price" market to function. In its place, they would erect price as the paramount variable.

How could this be accomplished? Initially, the central government expected to "reconquer the unity of executive power" in the realm of measurement much as the monarchy had ended feudal control of the money supply; that is, by circulating new standards that would gradually be preferred by the citizenry.[108] This liberal approach, first proposed by Condorcet in the 1770s,[109] operated in concert with an intransigent rhetoric which denounced opponents of the new measures for their mercenary self-interest. During the Revolution, the scientists administering the reform bitterly condemned those "fraudulent" speculators who bought and sold in different measures. Distinguishing these "parasitic" cheats from honest businessmen making a legitimate profit on trade justified the use of state power to transform the economy. "Those who say that differences in measures aid commerce are just talking about their personal profits. It is not industry that profits. Just because speculators profit from diversity, do we need to keep it?"[110]

But so long as artisanal and peasant producers continued to measure their output in terms of its value (usually, the value of their labor), differences in productivity would not come to the fore. The problem was to remake the mentality of citizens. For this purpose, a uniform set of measures—such as those of Paris—did not go far enough. As the scientists in charge of administering compliance with the metric system put it: "A well-made language is essential for proper reasoning; the former [metrical] nomenclature had none of these properties."[111] Having provided a language for the rational analysis of ordinary objects, "right thinking" about exchanges would soon follow. From "right thinking" about exchanges, proper economic action would follow. That is what Condorcet meant when he said he expected the decimal division to enable ordinary citizens to calculate their own best interests "without which they cannot be really equal in rights . . . nor really free."[112] The republican scientists saw in the metric system a rational language that would remake French citizens into rational economic actors. As mathematics was the language of science, so would the metric system be the language of commerce and industry. No wonder the Minister of the Interior saw no paradox in his simultaneous assurance that "uniformity of measures has always been desired by the people," and his boast that "[it] is a splendid instrument for the molding of public opinion [*la raison publique*]."[113]

That is why the state took the lead in educating its citizens about the new system (fig. 2). Between April and December of 1795 alone, the Agency of Weights and Measures printed seventy thousand publications providing tables and graphical keys of translation.[114] Large numbers of privately printed pamphlets also explained the principles of the metric system and offered conversion tables.[115] But this did not guarantee that anyone was learning the new system. A visiting Danish astronomer observed in 1800 that "most of the people living in France do not even know the new terms, much less understand them."[116]

Nor was it enough to master this new idiom. Citizens also needed to be able to translate the old vernacular of local measurements into the new national metric vocabulary, and vice versa. Where citizens had once needed a "dictionary" to go from one town to the next,[117] they now needed one to travel into the future. The "definitive" edition of the *Tables*

FIG. 2. This 1800 illustration of the "uses of the new measures" shows workers performing traditional tasks as a way of introducing the new names for measures of capacity, weight, length, area, currency, and volume. Hennin Collection, Bibliothéque Nationale.

des rapports in 1810 confidently declared that "[a]t last the French will no longer be strangers in France."[118] The danger was that citizens would not feel at home in their own parish. In Saint-Etienne, local officials confessed that the tables did not even begin to cope with the mind-boggling diversity of local seigniorial units.[119]

Ultimately, the savants hoped the actual metric weights and rulers would themselves act as the patient teachers of this "right thinking" and serve as the physical embodiment of the new social order. In keeping with the sensationalist epistemology of the period, the scientists who administered the Agency of Weights and Measures believed that daily contact with rational measures would engender a rational citizenry.

> If we want the people to put some order in their acts and subsequently in their ideas, it is necessary that the custom of that order be traced for them by all that surrounds them. . . . We can therefore look upon the metric system as an excellent means of education to be introduced into those social institutions which conjure up the most disorder and confusion. Even the least practiced *esprits* will acquire a taste for this order once they know it. It will be retraced by the objects which all citizens have constantly before their eyes and in their hands.[120]

Unfortunately, putting meter sticks in the hands of ordinary citizens was easier said than done. In 1794, the Commission of Weights and Measures offered financial incentives for artist-citizens who undertook the production of new measures, and described a machine that would help them do so "with precision and promptitude."[121] But when private artisans failed to manufacture these standards in sufficient quantity, the Committee of Public Safety turned to the government's own *Atelier de perfectionnement,* which Prieur de la Côte-d'Or had established at the height of the Terror to mass produce interchangeable gunlocks. The *Atelier*, originally part of the mammoth Manufacture of Paris, was now assigned the task of manufacturing one thousand meter sticks.[122] These never materialized, in part because of shortages of raw materials, inflation, fraud, and mismanagement. But in a larger sense, the Revolutionary government's attempts to mass produce metrical standards foundered on obstacles stemming from the corporate work practices of artisans, obstacles in many ways analogous to the particularistic practices that made ordinary citizens unwilling to adopt the new measures.[123]

The final irony came when the state set out to justify its re-regulation of trade in the name of a language designed to foster the development of a free market. In 1799, just as the meter was being formally presented to the Counsel of Ancients, the government established the *Bureaux des poids et mesures*, offices, run as local monopolies, which charged a fee to verify commercial weighings.[124] To many, they signaled a return to the hated feudal dues of the Old Regime and a restriction on the absolute right to trade whenever, wherever, and however one wished.[125] The Parisian monopoly granted Brillat and Co. drew particularly incensed objections.[126] In the end, the "despotic" company had to send in hundreds of troops to supplant the market weighers who thrived on the old measures, using enforcement of the metric system as their excuse for re-regulating the marketplace.[127] In this way, the new weights and measures became the wedge by which the government revived the distinction, familiar to the Old Regime, between the regulated public *marketplace* (limited in time and place, so that all might have equal access) and the chaotic *free market*.[128]

But even the sanctions of the state were not enough to force compliance. As the inspector for the Department of the Loire discovered when he went undercover, the bright new metric measures that shined in his honor on the countertops during his official visit, could vanish in a day.[129]

Standards and the State

The French elite always assumed that the resistance to the new measures was born of ignorance. This overlooked the disruption caused to long-established community norms negotiated between peasants and seigniors, customers and shopkeepers, artisans and merchants—plus the actual or feared pain of opening up local markets to outside competition. In many cases the old units could not even be adequately translated into new terms, since that would involve abstracting commodities from the labor and materials that had gone into their making. This was something that many peasant and artisanal *producers* were understandably reluctant to do—whatever gain they might expect as *consumers*. Indeed, the whole thrust of the metric reform was to replace an economic system based on value, with one in which everything—human labor, as well as its artifacts—was translated into the single, paramount variable of price.

But it would be a mistake to conclude that opposition to the metric system came exclusively from the lower orders. Ironically, the system was spurned even by the government bureaucrats, those men whose duties it was supposed to ease. In 1796, provincial notaries had not switched to the new system; in 1798, the Treasury was still not using the decimal system of money; and in 1799, Interior Minister François de Neufchâteau, complained that even the Parisian administrators continued to employ the old measures in their official correspondence.[130] No less than the common people, functionaries steeped in the traditional measures must have found the new units incommensurate with tasks formerly expressed in good round numbers. It is in this light that we must read the absurd invoice sent in 1808 by the Parisian office of weights and measures to the Saint-Etienne office, indicating that the requested metric standards were packed inside, and that the total weight of the shipment came to 60 *livres* (*poids de marc*).[131]

This equivocation even applied to the artillery service, an agency committed since the middle of the eighteenth century to the ideals of uniformity, precision, and modern manufacturing. In the year IV, the Directory secured the assent of the Parisian arsenal to convert the artillery to the new measures.[132] And the next month, General Aboville, head of the artillery's central committee, requested that the government print a metric edition of Gribeauval's celebrated *Tables de construction* because the "precision indispensable to the operation of the artillery" required that the dimensions of cannon now be expressed in the new measures.[133] The central War Office rejected the artillery's request on the grounds of expense.[134] But by 1801, the shoe was on the other foot, and the Minister of War was pressing the artillery to adopt the metric system; a request renewed with greater insistence in 1805.[135] In the meantime, however, the cadre of officers around Gribeauval, men who had supported interchangeable parts manufacturing and the rationalization of the workplace, had been supplanted by a new generation of artillerists.[136] Under pressure from powerful military contractors, and unwilling during wartime to risk further rebellion among the armorers, these officers retreated from Gribeauval's vision of mass production. In analogous fashion, they resisted bringing the artillery into the metric age. The new units, they complained disingenuously, would ruin the mathematically precise relation between the cannonball's weight and its caliber, and would undo the uniformity and interchangeability "which we took so many pains to establish."[137] Even in 1822, the artillery, by its own admission, had hardly broached the subject of metric conversion.[138]

In the face of this widespread hostility—even among its own bureaucracy—the state was gradually forced to temporize. Intended to minimize fraud and eliminate barriers to the free exchange of goods, the metric system had in fact thrown marketplaces into confusion. Some unscrupulous bureaucrats now profited from the differences in old and new measures to extract a "fee," much as merchants had formerly bought and sold commodities in units of different size.[139] The imperial administration was under pressure to reestablish order. The rationalizers had long ago retreated from their attempts to decimalize the hours of the day, and the Revolutionary calendar was abolished 1 January 1806.[140] On 4 November 1800, the Consulate had substituted simple names for the metric units, reintroducing the *pinte*, *livre*, and *boisseau*—though these still corresponded to decimalized equivalents.[141] Laplace and the other elite scientists of the Institute were able temporarily to forestall the revocation of their vaunted system in 1804 and again in 1806 by arguing that the metric system was essential to the administration of a multinational empire and enabled the enlightened classes to address the ordinances of the state to the popular classes.[142] But on 12 February 1812, before embarking on his disastrous adventure in Russia, the Emperor signaled a final retrenchment.[143] Over the noisy protests of the scientists and departmental prefects, the Empire adopted a system of "usual" measures, discontinuing both the decimal division and the systematic nomenclature entirely. The new standards continued to be ultimately derived from the platinum meter-bar; the metric system remained the sole *legal* system for administrative work and wholesale transactions; it continued to be taught in the schools; and the metric units were still to be marked alongside the "usual" units on all rulers. In essence, however, the new law retained only the centralizing features of the metric system, instituting the simple demand of the *Cahiers* for uniform weights and measures. The Interior Minister now confessed that the decimal division had primarily "helped the bookkeeper but not the ordinary man, unaccustomed as he was to endless calculations."[144] Years later, from his exile in St. Helena, the former Emperor rebuked his fellow members of the Institute—those "geometers and algebraicists"—for having bungled a simple administrative reform with their abstractions. In the exile's bitter quip, "it was not enough for them to make 40 million people happy, they wanted to sign up the whole universe."[145]

During the first restoration, Louis XVIII reaffirmed the Napoleonic compromise—in metrical matters as in so much else. In 1816, the royal administration expressly forbade the metric system for ordinary transactions. Only after the revolution of 1830, did the proponents of metric reform again make themselves heard. Finally, in 1840 the metric system was definitively readopted.[146] Even so, resistance in the countryside lasted well into the late nineteenth century, and not until the twentieth century did the last traces of the premetric measures disappear.[147]

The struggle to modernize France was a protracted affair. The gradual ascendance of *le français national* over patois and "foreign" tongues is instructive in this regard. When in his great anti-Vandalism speech of 31 August 1794, Grégoire turned to the positive deeds of the Revolution, he emphasized its techno-scientific achievements, notably the measurement of the meridian and "the project to render language uniform."[148] Thereafter, the post-Thermidorian elite continued to validate their war against local idioms by reference to a "scientific, natural, and universal" system—the *grammaire scholaire* of the Ideologues.[149] In the end, however, the triumph of *le français national* depended principally on the increased circulation of citizens within national borders and on the central control of education. In the case of the *metrical* language, the rate of change likewise depended on the transformation of the economy and the government's persistence in teaching the new system. During the nineteenth century, the national market proclaimed by the revolutionary bourgeoisie slowly become a reality. Local autarkies began breaking down. The metric system was surely a factor in this transformation and a product of it. No doubt *some* national system of French weights and measures, probably based on the Parisian standards, would have evolved in any case. Here, the British and American examples are instructive. But the specific *form* of the metric system is almost unimaginable without the circumstances of the French Revolution, a time when the repudiation of tradition gave the authority of nature a privileged position and allowed the scientific community unprecedented latitude to design a hyper-rational language of measurement.

For the generation that preceded and made the French Revolution, rationality was a license to make the world anew, a lever to pry the present from the past. Touted by its architects as one of the Revolution's "great services to mankind," the metric system was designed to "efface every trace of the [old] territorial and feudal divisions. . . ."[150] This erasure of history seemed to create startling new possibilities. Power flows from standardization. Where scales are perfectly balanced, a single thumb can move the world. Under the thumb of the new order, Benjamin Constant vividly saw this connection between symmetry and the exercise of authority.[151] Constant had a prescient understanding of the "atomizing" effect of the modern state: how it pulverizes local custom and tradition to erect itself as the central abstraction and unique embodiment of legitimacy. Among the symbols of the new tyranny of uniformity was the standardization of weights and measures.

> The conquerors of our days, peoples or princes, want their empire to possess a unified surface over which the superb eye of power can wander without encountering any inequality which hurts or limits its view. The same code of law, the same measures, the same rules, and if we could gradually get there, the same language; that is what is proclaimed as the perfection of the social organization. . . . [T]he great slogan of the day is uniformity.[152]

And that is why a meter is a meter.

Notes to Chapter Two

Abbreviations:

AP *Archives parlementaires de 1787 à 1860* (1st series; Paris: Imprimerie Nationale, 1879–).

CIP *Procès-verbaux du Comité d'Instruction Publique de la Convention Nationale*, ed. James Guillaume (Paris: Imprimerie Nationale, 1891–1907).

AN Archives Nationales.
AHG Archives Historiques de Guerre.

1. For a recent demonstration that social context shapes even the values of physical constants, see Philip Mirowski, "Looking for Those Natural Numbers: Dimensionless Constants and the Idea of Natural Measurement," *Science in Context* 5 (1992), 165–88.

2. Charles de Secondat de Montesquieu, *Esprit des lois*, in *Oeuvres complètes* (Paris: Garnier, 1875), 5:412–13.

3. Marie-Jean-Antoine-Nicolas de Caritat de Condorcet, *Observations . . . sur le 29ième livre de l'Esprit des lois*, in *Oeuvres* (Paris: Didot, 1847), 1:376–81.

4. Condorcet, *Sketch for a Historical Picture of the Progress of the Human Mind,* trans. J. Barraclough (London: Weidenfield and Nicolson, 1955), 199. On the Enlightenment view of rational language, see Michel Foucault, *The Order of Things: An Archeology of the Human Sciences* (New York: Random House, 1970), 78–124.

5. Keith Michael Baker, *Condorcet: From Natural Philosophy to Social Mathematics* (Chicago: University of Chicago Press, 1975), 62–67.

6. Jacques Necker, *Compte rendu au roi* (Paris: Imprimerie Royale, 1781), 121. Turgot to Messier, 3 October 1775, in Etienne-François Turgot, *Oeuvres* (Glashütten im Taunus: Auvermann, 1972), 5:31–33.

7. Keith Michael Baker, "Science and Politics at the End of the Old Regime," in *Inventing the French Revolution: Essays on French Political Culture in the Eighteenth Century* (Cambridge: Cambridge University Press, 1990), 153–66.

8. Condorcet, *Mémoires sur les monnoies* (Paris, 1790), 3–4, in Ruth Inez Champagne, "The Role of Five Eighteenth-Century French Mathematicians in the Development of the Metric System" (unpublished Ph.D. dissertation, Columbia University, 1979), 60.

9. Louis Marquet, "Condorcet et la création du système métrique décimal," in *Condorcet, mathématicien, économiste, philosophe, homme politique*, ed. Pierre Crépel and Christain Gilian (Paris: Minerve, 1989), 52–62.

10. Gaspard Monge, "Adresse de la Commission des poids et mesures à la Convention Nationale," in *CIP* (6 January 1794), 3:249.

11. Condorcet, *Observations*, 376–81.

12. This revisionist historiography is often said to have begun with François Furet, *Penser la Révolution française* (Paris: Gallimard, 1978).

13. Bureaux de Pusy, *AP* (8 May 1790), 15:441.

14. *AP* (8 May 1790), 15:443. The theory of political economy under which these savants operate seems to be largely within Physiocratic doctrine, hence their emphasis on agricultural yields as the source of prosperity. Georges Weulersse, *Le mouvement physiocratique en France (de 1756 à 1770)* (Paris, 1910).

15. Georges Lefebvre acknowledges that the metric reform was not complete at the end of the Revolutionary period; see *The French Revolution from 1793–99*, trans. John Hall Stewart and James Friguglietti (New York: Columbia University Press, 1964), 296. However, he never mentions, and nor do the other major French historians, that the metric system was revoked in the early nineteenth century.

16. John L. Heilbron, "The Measure of Enlightenment," in *The Quantifying Spirit in the Eighteenth Century*, ed. Tore Frängsmyr et al. (Berkeley: University of California Press, 1990), 207–42. John L. Heilbron, "The Politics of the Meter Stick," *American Journal of Physics* 57 (1989), 988–92.

17. Arthur Young, *Travels during the Years 1787, 1788, and 1789* (2d ed.; London, 1794), 1:315–16.

18. Adrien-Marie Legendre et al., *L'Agence temporaire des poids et mesures aux citoyens rédacteurs de la Feuille du Cultivateur (Paris: Imprimerie de la République, year III [1795]), 11. Roland Zupko, French Weights and Measures Before the Revolution: A Dictionary of Provincial and Local Units* (Bloomington: Indiana University Press, 1978), 113.

19. J.-B. Galley, *Le régime féodal dans le pays de Saint-Etienne* (Paris: Imprimerie de la Loire Républicaine, 1927), appendix. Jean Merley, Charles Vincent, and P. Charbonnier, "Les anciennes mesures de la Loire," in *Les anciennes mesures locales du Massif Central d'après les tables de conversion*, ed. P. Charbonnier (Clermont-Ferrard: Institut d'Etudes du Massif Central, 1990), 143–77.

20. William Sewell, *Work and Revolution in France: Language of Labor from the Old Regime to 1848* (Cambridge: Cambridge University Press, 1980), 28–29.

21. Galley, *Régime féodal*, 287.

22. Zupko, *French Weights*, 11–14.

23. A. de Saint-Léger, ed., *Les mines d'Anzin et d'Aniche pendant la Révolution* (Paris: Leroux, 1935), 2:364–65.

24. Galley, *Régime féodal*, 315–16, 326.

25. Witold Kula, *Measures and Men*, trans. R. Szreter (Princeton: Princeton University Press, 1986).

26. E. P. Thompson, "Time, Work-Discipline, and Industrial Capitalism," *Past and Present* 38 (1967), 56–97.

27. Jean Peltre, *Recherches métrologiques sur les finages lorrains* (Lille: Atelier Reproduction des Thèses, 1977).

28. In his article, " 'Nature' and Measurement in Eighteenth-Century France," *Studies on Voltaire and the Eighteenth Century* 87 (1972), 277–309, Maurice Crosland overlooks this anthropomorphic definition of "natural."

29. Kula, *Measures*, 161–264.

30. Etienne Fournial and J.-P. Gutton, *Cahiers de doléance de la province de Forez* (Saint-Etienne: Centre d'Etudes Foréziennes, 1974), 353.

31. Archives Municipales de Saint-Etienne (Loire) HH 9 Municipal minutes, 8–18 April 1789. J.-B. Galley, *Saint-Etienne et son district pendant la Révolution* (Saint-Etienne: Imprimerie de la Loire Républicaine, 1903–9), 1:51–52.

32. Robert Vivier, "Contribution à l'étude des anciennes mesures du département d'Indre-et-Loire," *Revue d'histoire économique et sociale* 14 (1926), 196.

33. Charles Porée, ed., *Département de l'Yonne, Cahiers de doléances du Bailliage de Sens* (Auxerre: Imprimerie coopérative ouvrière "l'Universelle," 1906), 177–78. Constancy in price was also a response to a chronic shortage of small coin for change.

34. Peter Linebaugh, *The London Hanged: Crime and Civil Society in the Eighteenth Century* (Cambridge: Cambridge University Press, 1992), 153–83. On customary practices among French artisans in the same period, see Michael

Sonenscher, *Work and Wages: Natural Law, Politics, and the Eighteenth-Century French Trades* (Cambridge: Cambridge University Press, 1989), 208–9, 256–66.

35. AN F12 1289 Paque to Fauchat (Secretary General of Commerce) 3 February 1815. Heaped measures also invited such "fraud"; see Saint-Léger, *Mines*, 364–65.

36. E. Clémenceau, *Le service des poids et mesures en France à travers des siècles* (Saint-Marcellin-Isère: Ateliers Graphiques de Sud-est, 1909), 89–92. Georges Picot, ed., *Histoire des Etats Généraux* (Paris: Hachette, 1872), 2:256–57; 3:30, 204; 4:130.

37. Pierre Jacquart, *Traité des justices de seigneur et des droits en dépendants* (Lyon: Reguilliat, 1764), 250–51.

38. AN F12 1287 Guilloton (Inspector of Manufactures, Rennes), 1768; Montaran to Intendant of Grenoble, 19 February 1786; "Potier d'Etain" to Calonne, 22 May 1785.

39. *Histoire de l'Académie pour 1772, Mémoires*, pt. II, 501.

40. Vivier, "Contribution," 196. Kula, *Measures*, 173. Steven Kaplan, *Provisioning Paris: Merchants and Millers in the Grain and Flour Trade during the Eighteenth Century* (Ithaca: Cornell University Press, 1984).

41. D.M.G. Sutherland, *France, 1789–1815: Revolution and Counter-Revolution* (New York: Oxford University Press, 1986), 203–3. For the *maximum*, see Ch. Lorain, *Département de la Haute-Marne, les subsistances en céréales dans le district de Chaumont* (Chaumont: Cavaniol, 1911), 356–58. For the *biens nationaux*, see René Caisso, *La vente des biens nationaux de première origine dans le district de Tours* (Paris: Bibliothèque Nationale, 1967), 71.

42. Sébastien Le Prestre de Vauban, "Description géographique de l'élection de Vezalay," *Projet d'une dixme royale* (Paris: Alcan, 1933).

43. For a full treatment of this subject, see Ken Alder, "Forging the New Order: The Origins of French Mass Production and the Language of the Machine Age, 1763–1815" (unpublished Ph.D. dissertation, Harvard University, 1991), 403–7.

44. René Taton, "Jean Picard et la mesure de l'arc de méridien Paris-Amiens," *La découverte de la France au XVIIe siècle, neuvième colloque de Marseilles* (Paris: CNRS, 1980), 349–61.

45. René Taton, "La tentative de Stevin pour la décimalisation de la métrologie," in *Acta Metrologiae Historicae*, ed. Gustav Otruba (Linz: IIIe Congrès International de la Métrologie Historique, 1983), 39–56.

46. François Garrault, *Recueil des nombres, poids, mesures et monnoyes anciennes et modernes* (Paris: Mettayer et l'Huillier, 1595).

47. Armand Macheby, "Aspects de la métrologie au XVIIe siècle," *Les Conférences du Palais de la Découverte,* Series D, 33 (1955), 8. Armand Macheby, *La métrologie dans les musées de province* (Troyes: CNRS, 1962), 19.

48. Galley, *Régime féodal*, 274, 278–30.

49. Edouard Gruter and Yannick Marec, "Des anciens systèmes de mesures au système métrique," in *Actes de l'université de l'été sur l'histoire des mathématiques*, Le Mans, France, July, 1984 (Université de Maine, 1986), 111–13, 116.

50. Peltre, *Recherches*, 90–91, 200–206.

51. J.-B. Galley, *L'élection de Saint-Etienne à la fin de l'ancien régime* (Saint-Etienne: Ménard, 1903), 188. Galley, *Régime féodal*, 282, 307. On the changing class-basis of land ownership at the end of the Old Regime in Forez, see Josette Barnier, *Bourgeoisie et propriété immobilière en Forez aux XVIIe et XVIIIe siècle* (Saint-Etienne: Centre d'Etudes Foréziennes, 1982).

52. Bureaux de Pusy, *AP* (8 May 1790), 15:440.

53. Alexis Paucton, *Métrologie, ou traité des mesures, poids et monnoies des anciens peuples et des moderns* (Paris: Veuve Desaint, 1780), 7, 11.

54. Beatrice Fry Hyslop, *French Nationalism in 1789, According to the General Cahiers* (New York: Columbia University Press, 1934), 56.

55. Fournial, *Cahiers*, 57, 106, 122, 127, 141, 149, 151, 160, 170, 179, 182, 217, 263, 311, 314, 319, 334, 353. The clergy, however, only wanted uniformity of measures "in each province"; the nobility advised a "general uniformity"; and the third estate demanded universal uniformity.

56. Fournial, *Cahiers*, 149.

57. Champagne, "Role."

58. Roland Zupko, *Revolution in Measurement: Western European Weights and Measures since the Age of Science* (Philadelphia: American Philosophical Society, 1990), 114–35.

59. Archives de l'Académie des Sciences, Register of the Academy, vol. 108, J.-B. Le Roy (27 June 1789), 171; (14 August 1789), 207.

60. For a nuanced discussion of the latitude possessed by experts in their dealings with the state, see John Carson, "Army Alpha, Army Brass, and the Search for Army Intelligence," *Isis* 84 (1993), 278–309.

61. Crosland, "'Nature' and Measurement," 286–89.

62. Baudin, 4 messidor, year VII [22 June 1799], in Jean-Baptiste-Joseph Delambre and Pierre-François-André Méchain, *Base du système métrique* (Paris: Baudouin, 1806–10), 3:651.

63. Danton, *AP* (31 January 1793), 58:102–3. Grégoire invoked natural frontiers to make manageable the missionary claims of "la République universelle." *AP* (27 November 1792), 53:610–15.

64. Heilbron, "Measure," 216–24. See also Alder, "Forging," 451–55. For an opposing view, see C. C. Gillispie, "Laplace," *Dictionary of Scientific Biography* (New York: Scribners, 1978), 15:334–35.

65. [René-Just Haüy], *Instructions sur les mesures déduites de la grandeur de la terre* (1st ed.; Paris: Imprimerie Nationale, year II [1794]), xxvii–xviii.

66. Pierre-Simon Laplace, "Mathematiques," in *Séances des Ecoles Normales [de l'an III], Débats* (Paris: Reynier, [1795]) 1:10–23

67. A. G. Le Blond, *Sur la fixation d'une mesure et d'un poid—lu à l'Académie des Sciences, 12 Mai 1790* (Paris: Demonville, 1791); Rollin, *CIP* (12 frimaire, year II [2 December 1793]), 3:90–91.

68. Gueroult, *Observations sur la proposition faite par le cit. Prieur* (Paris: Guerin, [year III-IV]), 5.

69. AN F12 1288 M. Fontaine, 1792; Société populaire de Livry (Bayeux), year II; Simon, arpenteur, 24 vendémiaire, year III [15 October 1794].

70. Tillet and Abeille, *AP* (6 February 1790), 11:466. Talleyrand, *AP* (9 March 1790), 12:106.

71. Condorcet and Talleyrand seized on a speech in Parliament in 1790 advocating international standards of measurement. However, no legislation was ever directly submitted to the House of Commons. See John Riggs Miller, *Speeches in the House of Commons upon the Equalization of the Weights and Measures of Great Britain* (London: Debrett, 1790).

72. Jefferson to Short, 28 July 1791, in C. Doris Hellman, "Jefferson's Efforts toward the Decimalization of United States Weights and Measures," *Isis* 16 (1931), 286.

73. Archives de l'Académie des Sciences, Dossier: Lavoisier, Lavoisier to Arbogast, 16 April 1793. Roger Hahn, *Anatomy of a Scientific Institution* (Berkeley: University of California Press, 1971), 252–85.

74. *AP* (1 August 1793), 70:70–74. The "temporary" meter had been announced the previous year at the insistence of Interior Minister Roland, who cited the urgent need of French business for "some standard, whatever it might be." *AP* (3 April 1792), 41:100.

75. The decree of 20 August 1790 set the Academy's budget at 93,458 *livres* (*CIP,* 1: 260n); whereas the initial grant for the meridian survey alone came to 300,000 *livres*. For a memorandum proposing that sum, see AN F12 1289 [Académie des Sciences], 19 March 1791.

76. AN F12 1289 Borda (President of the Commission des poids et mesures) to Pavé (Interior Minister) 12 brumaire, year II [2 November 1793]; also various contracts for the year IV with Le Noire, Mercklein, etc.

77. Delambre, *Base*, 1:57; Heilbron, *Measure*, 230–31.

78. Jean-Baptiste-Joseph Delambre, *Grandeur et figure de la terre* (Paris: Gauthier-Villars, 1912), 202–3, 213.

79. Laplace, "Mathematiques," in *Séances*, 5:203–14. For the earlier period, see Mary Terrall, "Representing the Earth's Shape: The Polemics Surrounding Maupertuis's Expedition to Lapland," *Isis* 83 (1992), 218–37.

80. Louis-Sébastien Mercier, *Le nouveau Paris* (Brunswick, 1800), 3:44.

81. Delambre, *Base,* 3:581–655.

82. Talleyrand, *AP* (9 March 1790), 12:106.

83. Thomas Bugge, *Science in France in the Revolutionary Era*, ed. Maurice Crosland (Cambridge: MIT Press, 1969), 205–11.

84. Estimates varied from Laplace's value of 1/148 to Legendre's 1/320. Delambre, *Base,* 3:92, 554–55.

85. The international commission reset Borda's calibration, but he was vindicated in 1870. Guillaume Bigourdan, *Le système métrique des poids et mesures* (Paris: Gauthier-Villars, 1901), 86–87, 147.

86. Méchain discarded a whole set of "accursed" (*maudite*) results that, in fact, provide a value much closer to the one presently accepted. Delambre, *Grandeur*, 222, 238–34. Bigourdan, *Système métrique*, 152–54.

87. William Hallock, *Outline of the Evolution of Weights and Measures and the Metric System* (New York: Macmillan, 1906), 59.

88. Delambre, *Base,* 3:447–62.

89. Delambre, *Grandeur,* 224–36.

90. Delambre, *Base,* 3:101–3, 545–46, 557.

91. In his article, "The Congress on Definitive Metric Standards, 1798–1799: The First International Scientific Conference?" *Isis* 60 (1969), 230, Maurice Crosland denies that foreign delegates were under pressure to acquiesce to a French *fait accompli.* But see Laplace's letter to Delambre of 10 pluviôse, year VI [29 January 1798], in which he assures his colleague that the meeting is a "mere formality." Yves Laissus, "Deux lettres inédites de Laplace," *Revue historique des sciences* 14 (1961), 287–88.

92. Baker, "Science and Politics."

93. Theodore Porter, "Objectivity as Standardization: The Rhetoric of Impersonality in Measurement, Statistics, and Cost-Benefit Analysis," *Annals of Scholarship* 9 (1992), 19–59.

94. Haüy, *Instructions*, (1st ed.), xii.

95. *Feuille du Cultivateur* 38 (9 messidor, year III [27 June 1795]), 227–28. Though the journal had originally supported Tillet and Abeille's proposal for standards based on the Parisian measures, it welcomed the law of 18 germinal, year III. *Feuille* (15 January 1791), 118; (12 floréal, year III [1 May 1795]), 160.

96. Legendre, *L'agence temporaire*, 5. The chastened journal even reprinted this rebuke. *Feuille* 44 (7 thermidor, year III [25 July 1795]), 258.

97. *CIP* (24 thermidor, year III [11 August 1795]), 6:532–37. Prieur had initially suggested using Breton names for the measures.

98. Baudin, in Delambre, *Base,* 3:651. Baudin was engaged at the time in a campaign against political factions (motivated by a feared renewal of Babeuf's revolt), in which he justified elite governance by reference to the transcendent neutrality of science. See M. Staum, "Public Relations of the Second Class of the Institute in the Revolutionary Era," *Proceedings of the Annual Meeting of the Western Society of French Historians* 16 (1989), 213–14.

99. Legendre, *L'agence temporaire*, 1, 18. Emphasis in original.

100. Baudin, in Delambre, *Base,* 3:651.

101. C.-A. Prieur, *Rapport . . . sur la nécessité et les moyens d'introduire dans toute la République les nouveaux poids et mesures*, in *CIP* (10 ventôse, year III [28 February 1795]), 5:551–63. See also *Projet de décret*, ibid. (1 vendémiaire, year IV [23 September 1795]), 6:671.

102. François-Alphonse Aulard, ed., *Paris pendant la réaction thermidorienne et sous le Directoire* (Paris: Cerf, 1898–1902), (25 February 1798), 4:556–57; (30 December 1798), 5:287; (June–July 1799), 5:632.

103. François-Alphonse Aulard, ed., *Paris sous le Consulat* (Paris: Cerf, 1903), (November–December 1799), 1:65; (12 September 1801), 2:521.

104. AN F12 1288 Roland (Interior Minister) to President of the National Assembly, 19 May 1792.

105. AN F12 1288 Amis de la République (Carcassonne) to General Assembly, 11 December 1792.

106. Agence temporaire, *Notions élémentaires sur les nouvelles mesures* (1st ed.; Paris: Imprimerie de la République, year IV [1795]), 1, 3–4.

107. Agence temporaire, *Notions élémentaires sur le nouvelles système des*

mesures (2d ed.; Paris: Imprimerie de la République, year VI [1797]), 4. Emphasis added.

108. *AP* (9 March 1790), 12:106.

109. Condorcet, *Observations,* 377.

110. Agence temporaire, *Notions élémentaires sur les nouvelles mesures* (1st ed.; Paris: Imprimerie de la République, year IV [1795]), 1, 3–4.

111. Ibid., 10.

112. Condorcet, *Mémoires sur les monnoies*, 3–4, in Champagne, "Role," 60.

113. Interior Minister to Departmental Administrations, 23 fructidor, year V [9 September 1796], in Kula, *Measures*, 241, 254–55.

114. Champagne, "Role," 208–27.

115. Scores of these appeared in the 1790s and early 1800s. For instance: Bonnin, *Vocabulaire étymologique des poids et mesures de la République française* (Paris: Fournier, year VII).

116. Bugge, *Science*, 204–5.

117. Legendre, *L'agence temporaire*, 2.

118. François Gattey, *Tables des rapports des anciennes mesures agraires avec les nouvelles* (2d ed.; Paris: Michaud, 1810), 6. Kula, *Measures*, 247–48.

119. Marquet, "Anciens mesures, anciens poids," *Amis du vieux Saint-Etienne* 35 (1957), 4.

120. Legendre, *L'agence temporaire*, 9.

121. Agence temporaire des poids et mesures, *Avis instructif sur la fabrication des mesures de longueur à l'usage des ouvriers* (Paris: Imprimerie de la République, year III [1795]). *CIP* (2 messidor, year III [20 June 1795]), 6:314–15. Monge, "Adresse," 249.

122. AN F12 1310 *Extrait des Registres du Comité du Salut Public*, 27 floréal, year III [16 May 1795]. AN F12 1311 Council on Weights and Measures to [*Atelier*], 8 floréal, year IV [27 April 1796].

123. Alder, "Forging," 555–608. Prieur blamed the failure of the new system principally on the shortage of meter sticks, etc. C.-A. Prieur, *Rapport sur l'exécution des lois rélatifs aux poids et mesures* (Paris, Imprimerie Nationale, year VI), 7, 8, 11–13.

124. Law of 27 brumaire, year VII [17 November 1798]. Bigourdan, *Système métrique*, 186–87. The law was amplified on 7 brumaire, year IX [29 October 1800], and again on 16 June, 1808. Désiré Dalloz, ed., *Jurisprudence générale* (Paris: Bureau de la Jurisprudence Générale, 1845–70), 35:983–85.

125. A.B.J. Guffroy, *Avis civique contre un projet liberticide* (Paris: Everat, year VII).

126. Pérès, *Rapport . . . relative aux peseurs publics* (21 vendémiaire, year VIII [13 October 1799]). Pérès professed himself a supporter of metrical uniformity.

127. Brillat et al., *Mémoire . . . sur le rapport fait par le représentatif Pérès* (Paris: Bailleul, [year VII]), 6, 17. Bigourdan, *Système métrique*, 187–89.

128. On the distinction, see Kaplan, *Provisioning Paris*, 47–48, 68–69.

129. Marquet, "Anciens mesures," 35:6–8; 36:8–11. On the Bureaux of inspection, see D. Roncin, "Mis en application du système métrique (7 avril 1795–4

juillet 1837)," *Cahiers de métrologie* 2 (1984), 33. And Robert Vivier, "L'application du système métrique dans le département d'Indre-et-Loire," *Revue d'histoire économique et sociale* 16 (1928), 211–14.

130. Reveillière-Lépeaux et al., 7 pluviôse, year IV [27 January 1796], in Antonin Debidour, ed., *Recueil des actes du Directoire-Exécutif* (Paris: Imprimerie Nationale, 1910–17), 1:492. Aulard, *Réaction thermidorienne* (August–September, 1798), 5:99. American bureaucrats in the 1990s have been having similar problems complying with legislation that insists that they convert to the metric system. House Committee on Science, Space and Technology, *Metric Conversion Activities of Federal Agencies in Compliance with P. L. 100-418, Section 5164, Metric Usage: 1992 Update* (Washington, D.C.: U.S. Government Printing Office, 1992).

131. Marquet, "Anciens mesures," 36:9.

132. Carnot et al., in Debidour, *Recueil* (11 frimaire, year IV [2 December 1795]), 1:162.

133. AHG 4c3/2 F. M. Aboville, *Mémoire*, 1 nivôse, year IV [22 December 1795]; 14 pluviôse, year IV [3 February 1796].

134. AHG 4c3/2 General Drouân, *Mémoire*, 3 vendémiaire, year V [24 September 1796].

135. AHG 4c3/2 War Minister to Central Committee of Artillery, 19 vendémiaire, year X [11 October 1801]. Chief Inspector of Revenue to Artillery General Songis, 25 fructidor, year XIII [12 September 1805].

136. Alder, "Forging," 704–27.

137. AHG 4c3/2 Central Committee of Artillery, *Observations*, 29 March 1806.

138. AHG 4c3/2 Anon., *Mémoire*, 1822. This pattern is corroborated by Heilbron's documentation on the hesitant reception that architects and technical writers accorded the new system in the early nineteenth century. Heilbron, "Measure of Enlightenment," 238–42.

139. AN F12 1289 Prefect of Department of Rhône to Interior Minister, 28 floréal, year VIII [18 May 1800]; Paque to Fauchat, 3 February 1815.

140. James Friguglietti, "The Social and Religious Consequences of the French Revolutionary Calendar" (unpublished Ph.D. dissertation, Harvard University, 1966).

141. Dalloz, *Répertoire* 35:983–84.

142. AN F12 1296 Anon., 2 March 1811. See also Bigourdan, *Système métrique*, 192.

143. Lucotte and Noiret, *Nouveau système des poids et mesures* (Dijon: Noellat, 1813), 89–100.

144. Interior Minister Montalivet, in Kula, *Measures*, 260.

145. Napoleon Bonaparte, *Dictionnaire, ou Recueil alphabétique des opinions et jugements de Napoléon I* (Paris: Au club de l'honnête homme, 1964), 179.

146. Bigourdan, *Système métrique*, 205–6, 222–25.

147. Eugen Weber, *Peasants into Frenchmen: The Modernization of Rural France, 1870–1914* (Stanford: Stanford University Press, 1976), 30–33. Arthur E. Kennelly, *Vestiges of Pre-Metric Weights and Measures Persisting in Europe, 1926–27* (New York: Macmillan, 1928).

148. Grégoire, *AP* (14 fructidor, year II [31 August 1794]), 96:153–54; also (16 prairial, year II [4 June 1794]), 91:318–27.

149. Patrice Higonnet, "The Politics of Linguistic Terrorism and Grammatical Hegemony during the French Revolution," *Social History* 5 (1980), 41–69.

150. Fourcroy, *AP* (11 September 1793), 73:669–70.

151. Benjamin Constant, *Cours de politique constitutionnelle* (Paris: Guillaumin, 1872), 2:170–75.

152. Benjamin Constant, *De l'esprit de conquête* (Paris: Librarie de Médicis, 1813), 53–54.

THREE

"THE NICETY OF EXPERIMENT": PRECISION OF MEASUREMENT AND PRECISION OF REASONING IN LATE EIGHTEENTH-CENTURY CHEMISTRY

Jan Golinski

> Such pretensions to *nicety* in experiments of this nature, are truly laughable! They will be telling us some day of the WEIGHT of the MOON, even to *drams*, *scruples* and *grains*—nay, *to the very fraction of a grain*!—I wish there were infallible experiments to ascertain the *quantum* of *brains* each man possesses, and every man's *integrity* and *candour*:—This is a *desideratum* in science which is most of all wanted.
> *(Robert Harrington, 1804)*[1]

MUCH of the historiography of eighteenth-century science has been concerned with quantification. In a range of disciplines, from the physical to the social sciences, progress has been charted toward increased precision of measurement and the achievement of a mathematical form of expression. Precision measurement has been seen as the favored route by which other sciences sought to emulate the certainty and predictive power represented by Newton's *Principia*. A recent collection of studies has provided a broader context for this by surveying the expression, in many fields, of "the quantifying spirit."[2] Statistical methods, precision measurement, mathematical formalization, and purportedly geometrical styles of reasoning are found to have been widely distributed across the Enlightenment landscape. The contributors to this volume have explored the connections between these practices in activities as diverse as the compilation of meteorological data and cartographical surveying. They have also considered how models of algebraic analysis were extended into such realms as artificial languages and plant taxonomy. And they have pointed to the influences of refinements in instrumentation and the burgeoning state bureaucracies in providing the contextual conditions for much of this activity.

In light of this, it seems worthwhile to reopen the question of the introduction of quantified methods into chemistry in the "Chemical Revolution" of the 1770s and 1780s. Antoine Lavoisier used such measuring instruments as the balance (to determine the weights of substances involved in reactions) and the calorimeter (to gauge the accompanying heat-exchanges). He was thereby able to establish and to secure widespread acceptance for his claim that oxygen, not phlogiston, was the entity exchanged in combustion, calcination, and respiration. Precise measurements of the quantities of reactants and products, which both assumed and exhibited the conservation of weight in chemical processes, gave Lavoisier the proofs he sought. One way of looking at this has been to see him as a vehicle for the influence on chemistry of an already mathematized experimental physics. Chemistry is taken to have achieved maturity by assimilating the methods endorsed by the more exact sciences. Such accounts are, however, weakened by rather generalized characterizations of the nature of experimental physics at this time and the questionable assumption that it had already achieved a mathematical form.[3]

In other recent research, Lavoisier's use of weight-measurement was shown to have been deeply rooted in his own mode of investigative practice. In his penetrating study of Lavoisier's work on the chemistry of life, Frederic L. Holmes has shown him securing measurements of quantities at every point, driven by the unvarying conviction that the total weights of the substances involved would be conserved in any chemical reaction. In many instances, Holmes displays the adjustments Lavoisier made in recorded data in light of this conviction. The balance sheet, used repeatedly by Lavoisier, both embodied his own tenet of the conservation of weight and served to display its confirmations to readers of his work.[4] Extending this insight, Norton Wise and Bernadette Bensaude-Vincent have explored Lavoisier's application of the method of the balance sheet in other fields, such as political economy. Wise argues that the balance comprised a "mediating technology" with wide application in Enlightenment rationalism. Lavoisier exemplifies this in his use of particular kinds of instrumentation (including the calorimeter, which can be viewed as a type of balance), and in his use of equations to link reactants with products through the relation of conserved weight.[5] By yielding numbers whose relations could be expressed in equations, measurements of quantities fed into the representation of chemical change in the form of algebra, which was thought to provide a particularly effective means of rapid and unambiguous communication.

My purpose here is to develop what I take to be one implication of these studies: that Lavoisier's precision measurement achieved its significance in a particular context of use. He sought to obtain measurements of weights and other quantities in order to deploy them in balance sheets that displayed the transformations of chemical species within a general order of nature in which the total amount of substance was conserved. Quantification was thus intimately connected with argument, and it is this connection that I particularly want to examine. I shall investigate how these measurements worked (or were meant to work) to persuade others that Lavoisier was right, and how they sometimes failed to do so. By suggesting that precision measurement formed part of Lavoisier's mode of argument, I do not imply that it was of only superficial significance, or that the rhetorical uses of precision were unconnected with its uses in experimental practice. On the contrary, I argue that Lavoisier's rhetoric of precision derived much of its power from its intimate connections with a particular mode of experimental practice and the resources (instrumental and social) that it mobilized.

Notwithstanding this, a consensus regarding the meaning of Lavoisier's measurements was not readily achieved at this time. His work provoked a prolonged controversy among chemists in which issues of quantification were explicitly raised. Insofar as quantification was a persuasive resource it was, in this instance, a controversial one. Throughout the 1780s, Lavoisier's claims continued to be rejected by a number of British chemists. This was so despite purportedly precise measurements of weights of reactants and volumes of gases, yielded most notably by a public demonstration of the analysis and synthesis of water in Paris in 1785. Joseph Priestley, Richard Kirwan, James Keir, William Nicholson, and others remained unpersuaded by these measurements. They continued to cast doubt on what Lavoisier had stipulated as experimental "facts." In this instance, quantification was strenuously resisted when it was yoked to factual claims that were themselves matters of dispute.

To the British chemists, Lavoisier's use of the language of precision appeared particularly clearly *as a rhetoric*. They discerned it as such because they operated within an alternative set of discursive conventions, not because they were themselves free from the constraints of any rhetorical framework. For Priestley and Keir particularly, the expectation was that chemists would communicate in purportedly factual narratives, uncontaminated by interpretive theory. Because Lavoisier did not seem to adhere to this convention, they were obliged to take apart the novel construction of instrumental and discursive elements he had assembled to make his case. The calorimeter and the balance were connected to a network of material and persuasive resources that together constituted a radical break with what Holmes has called the "longue durée" of the eighteenth-century chemical laboratory.[6] Precision measurement was one of these elements, and consequently problematic for Lavoisier's opponents.

In its broadest terms, what was in question was how chemical practice was to be carried on.

I describe the resistance of the British phlogistic chemists not to endorse their point of view but to discern how problematic the connection between measurement and argument could be. They contested the claimed link between precision measurement and demonstrative (sometimes called "geometrical") reasoning. While Lavoisier asserted that accuracy in his experiments enabled him to deduce the consequences directly, his critics questioned the purported connection between measurement and demonstration. A focus on the controversy therefore allows us to discern how this connection was constructed by Lavoisier and deconstructed by his opponents.

Thus Lavoisier's ally, the mathematician Jean Baptiste Biot, wrote of the antiphlogistic chemists: "One felt the necessity of linking accuracy in experiments to rigor of reasoning."[7] While the anti-Lavoisian Keir noted of the same group that "the precision of method in their experiments may have led to a belief that an equal precision and justness prevailed in their inferences and reasonings."[8] Such a belief, he implied, was entirely unwarranted. The link between precision of measurement and precision of reasoning was, precisely, the point at issue. With some justification, critics took Lavoisier to be claiming that exact measurements had yielded conclusive proof of the antiphlogistic theory, and they bridled at such a claim. Precision of measurement and precision of reasoning, two elements of the "quantifying spirit" that were intimately bound together in Lavoisier's practice, fell apart on the other side of the Channel.

From the beginning of his scientific career in the 1760s, Lavoisier had been using thermometric and barometric measurements in geological surveys and applying hydrometric methods to mineral-water analysis. In the 1780s he began to exploit methods of precision measurement in his campaign against traditional chemical theory. His collaboration with Pierre Simon de Laplace, leading up to the composition of the "Mémoire sur la Chaleur" in 1783, was critical for this move. Lissa Roberts has recently described how the two men fabricated and described a novel, initially unnamed, machine. They presented it as an unproblematic measuring device for heat exchanges in reactions, professing that it had no implications as to the theory of heat.[9] The naming of the apparatus as a "calorimeter" occurred subsequently in Lavoisier's *Traité élémentaire de chimie* (1789). Meanwhile, other experimenters such as Josiah Wedgwood and Adair Crawford experienced difficulties in replicating Lavoisier and Laplace's experiments and disputed the working of their machine. An anonymous English writer appears to have reflected the general appraisal, when he wrote in 1797 that "little reliance . . . can be placed on the accuracy of this much-boasted process of the French chemists," notwithstanding that their results had been presented "with all the precision of the new school."[10]

The incident of the calorimeter shows how Lavoisier began in the early 1780s to use highly crafted instruments to try to secure conviction among his audience by precision measurement. At the beginning of the decade he had no allies among leading chemists. Most remained sure that phlogiston existed as a material entity released from burning bodies. The Irish chemist Richard Kirwan articulated an alternative to Lavoisier's theory of combustion that won considerable support. In Kirwan's view, the phlogiston released by a burning body combined with dephlogisticated air in the atmosphere to form fixed air, which then united chemically with the residue of the solid to form a calx or acid. Kirwan's account had the appeal of accommodating the weight gain that had been agreed occurs in instances of combustion and calcination while maintaining the existence of phlogiston.[11]

In the face of this alternative to his theory of combustion, Lavoisier's fortunes turned on a new issue introduced into the debate—the composition of water. In 1781, Henry Cavendish produced water from a mixture of inflammable and dephlogisticated airs ignited by an electric spark. He canvassed a couple of explanations, suggesting that the more likely one was that water was part of the composition of both airs and was released on their combination by a kind of condensation reaction.[12] Lavoisier repeated the experiment in June 1783, before the long-delayed publication of Cavendish's paper. He announced a new interpretation of the reaction, stating that water was the sole product of combination of the two gases and hence not an element, as Cavendish and all other chemists had maintained, but a compound.[13] This interpretation explained two classes of phenomena that had previously constituted troublesome anomalies for his theory. The inflammable air generated by metals when they dissolved in acids could now be explained as a product of the decomposition of water, while the reduction of lead calx and other calces by inflammable air could be understood in terms of the combination of the gas with oxygen from the calx (or oxide) to synthesize water.

Reporting his result, Lavoisier admitted that, as yet, he lacked accurate measurements of the quantities of the gases used in the reaction. Nonetheless, he wrote, one might assume that the amounts consumed would be found to equal the weight of water produced, since "it is no less true in physics than geometry that the whole is equal to its parts."[14] This stipulation indicated the most important way in which Lavoisier would attempt to make precision measurement demonstrative of his claim of the compound nature of water. The equality between weights of products and of reactants, established as accurately as possible, would carry the burden of this proof. The equations of algebra, rather than the proportions of geometry, provided the model for the kind of demonstration in which determinations of weights would be embedded.

Lavoisier's construal of the reaction was not, however, accepted by others. Cavendish, when he finally published the account of his own experiment in 1784, referred to Lavoisier's antiphlogistic explanation but professed himself unconvinced. James Watt, the Birmingham steam-engine manufacturer and friend of Priestley, similarly insisted that alternative ways of "solving the phenomena" remained open.[15] Cavendish and Watt had no difficulty severing Lavoisier's description of the experimental phenomena from what they took to be its theoretical interpretation; and they dismissed the latter. In its place they offered conjectural explanations that indicated a potential difficulty with the algebraic proof from an equation of weights. In Watt's version, dephlogisticated air was taken to be water deprived of phlogiston with bound latent heat; inflammable air was phlogiston with a little water and latent heat. When the two airs united, water was released along with heat. The implication was that heat and phlogiston were components of the reactant substances that would have to be included in any equation representing the transformation. It was thus possible that relations among the weights of ponderable substances contained in the reaction vessels did not encompass all of the components of the reaction. Other entities, perhaps capable of passing through the walls of the vessels, might affect the weights of substances with which they were combined.

Lavoisier was thus given cause to realize that a more persuasive proof than the June demonstration was needed, if his contention of the compound nature of water were to be accepted. In the months of autumn and winter of 1783–84 he labored to provide such a proof. His approach was to attempt more accurate measurement of the quantities of reactants and products. In this respect, he followed the example of Gaspard Monge, instructor in experimental physics at the military engineering academy, the École Royale du Génie, at Mézières. In June and July 1783, independently of Lavoisier's work, Monge had conducted his own experiments on the synthesis of water, measuring the volumes and specific weights of the reactant gases and establishing their (almost exact) equality to the weight of water produced.[16]

To repeat this experiment, Lavoisier enlisted the help of Jean-Baptiste Meusnier, a former pupil of Monge at Mézières, who set about designing vessels capable of measuring the volumes of the gases they would contain. Meanwhile the two worked on an experiment to separate the component parts of water by analysis. They passed steam through a red-hot, iron gun-barrel. The steam was taken to be decomposed, its oxygen uniting with the iron to form an oxide and its inflammable air emerging from the pipe to be collected along with undecomposed water. Lavoisier reported the success to the Académie des Sciences in April 1784 and incorporated the results in the (still unpublished) account of the synthesis experiment of the previous year. Without exhibiting its derivation, he produced a figure for the proportions of the two gases in the composition of water: 12 parts of vital air (oxygen) to 22.924345 parts of inflammable air (hydrogen). From this he calculated the weights of the component parts of a pound of water to eight places of decimals: 0.86866273 pounds of vital air and 0.13133727 pounds of inflammable air, making a total of 1.00000000 pounds of water.[17] Although he indicated the need for further experiments to determine the proportions of composition more reliably, Lavoisier was deploying a further weapon in his campaign of persuasion. The algebraic form of proof, which displayed an equation between weights of reactants and products, was now supplemented by figures for weight measurements given to up to eight places of decimals. Such lengthy figures were of questionable persuasive value (as we shall see), but it seems likely that they were presented to suggest a command of highly precise techniques of measurement and to enhance the plausibility of Lavoisier's claims.

Again, however, dissension continued. Kirwan and Priestley denied to Lavoisier's experiments the implication their author sought to give them. Both insisted that the proposed analysis of water was no such thing. What had happened was that phlogiston (inflammable air) had been displaced from iron by combination of water with the metal. In February 1785, Priestley described to the Royal Society his own replication of Lavoisier and Meusnier's experiment, in which he argued that the source of the inflammable air was the iron, not the water. As had Cavendish and Watt, Priestley charged Lavoisier with transgressing the convention that experimental philosophers should describe what they observed without imposing hypothetical rationalizations on the phenomena.[18]

Facing this persistent opposition to his claim that water was a compound of two gases, Lavoisier continued his experimental work to produce a more stringent and compelling proof. He sought to push back the boundary that his critics had erected between the "facts" of experiment and what they insisted could only be an interpretation or "hypothesis." He worked to make the compound nature of water into a fact—a direct, unmediated inference from experiment, permitting no possibility of doubt. This was to be done by employing new, more refined apparatus to yield quantitative weight measurements of an unprecedented accuracy. His efforts culminated in the large-scale set-piece demonstration of the analysis and synthesis of water in the Paris Arsenal on 27 and 28 February 1785. On that occasion, all the elements of Lavoisier's form of experimental practice were assembled to convey a demonstrative proof of his claims—accurate apparatus and methods were deployed in a setting designed to maximize their persuasive efficacy.[19]

The analysis part of the new experiment was relatively little changed from what Lavoisier and Meusnier had accomplished the previous year. But the operation to synthesize water was performed with unprecedented care and very sophisticated new apparatus. Working to Meusnier's design, the instrument-maker Pierre Mégnié produced the new pneumatic vessels toward the end of 1783. Around this time he also built two balances for Lavoisier, using new techniques to suspend the beams and damp their oscillations. The larger of these was estimated to be capable of weighing one pound with an accuracy of about 1 in 100,000 (i.e., to five places of decimals).[20]

The new gas holders were derived from the traditional pneumatic trough, in which gases were stored in a tank suspended over water, with the addition of a scale to enable the volume of the gas to be measured. An adjustable counterweight could be set to maintain a constant flow of gas out of the vessel. After calibration, readings on the scale could be converted to the volume of the gas in the vessel, and hence the weight of gas could be calculated from previously prepared tables of the densities of oxygen and hydrogen under various conditions of temperature and pressure.[21] Accuracy was striven for in all the measurements taken. The position of the pointer against the volume scale was read to two places of decimals of a degree of arc, by use of a vernier. Verniers seem also to have been fitted to the thermometer, which was read to one decimal place of a degree Réaumur, and to the barometer, which was read to one decimal place of a line of mercury (one line being 1/12 of an inch). Lavoisier had been interested in the accuracy of thermometers and barometers since his early mineralogical work, and he now put to use the most advanced precision versions of these instruments.[22]

Having completed the calibrations, with the assistance of Laplace, Meusnier, and Monge, Lavoisier invited about thirty savants, including a dozen witnesses nominated by the Académie, to attend the demonstration. In the course of two days, two analysis experiments and one synthesis were performed. Hydrogen produced by the analysis was used in the synthesis, along with oxygen obtained by heating red mercury calx. The weights of gases used and that of the water produced by their combination were carefully calculated.

Many of those who took part in the demonstration were convinced. The academicians inspected the apparatus, took measurements, and signed their names to the records of the results. The procedures of witnessing, recording, and certifying both guaranteed the authenticity of the results and gave the participants a stake in their validity. Perhaps because they were more practiced in the use of precision measurement, Monge and many of the mathematicians and physicists in the Académie confirmed their support for Lavoisier's doctrine. Some chemists who were present, on the other hand, such as Balthazar-Georges Sage and Antoine

Baumé, remained opposed to Lavoisier's theory. Crucially, the chemist Claude-Louis Bethollet was converted to Lavoisier's point of view, announcing a little later that he had been convinced by "the beautiful experiment" that water was indeed a compound.[23]

Skepticism also survived among the wider chemical community. A brief account of the experiment, probably based on drafts by Lavoisier, appeared under Meusnier's name in the *Journal polytype des sciences et des arts* in February 1786.[24] Meusnier was apparently also charged with producing a comprehensive report, but this never emerged. The paper, although it included two plates showing the apparatus, gave only a very incomplete impression of the sophisticated methods used. No description was given of the calibration procedures for the gasometers or the meticulous care taken to ensure the accuracy of measurements. Figures were given for the results of one analysis experiment, including weights of water consumed, of the gun barrel before and after the experiment, and of the inflammable air collected. All weights were reported in pounds, ounces, drams, and grains, to an accuracy of whole numbers of grains. A deficit of 2 drams 3 grains was ascribed to loss of water vapor and inflammable air through cracks in the apparatus. Results were given to the same degree of precision for the synthesis experiment. In this case an excess of water over the combined weight of reactants of 30 1/2 grains was read as further testimony to the accuracy of the procedures—it was taken to have resulted from a slight warming of the vessels in which the gases were weighed.

Elements of Lavoisier's distinctive rhetoric of precision were present in this account, presumably reflecting his influence on its composition. Weights were specified to a single grain or even a fraction of a grain, and the relatively small quantities of deficit and excess mobilized the authority of accurate measurement in support of the claim that water was a compound. Meusnier announced the figure of 85:15 for the ratio of vital to inflammable airs in the composition of water—a figure that was readily memorized and came to be widely reproduced among defenders of the Lavoisian position.[25] Finally, the figures were supported with visionary assertions as to how the future of chemistry depended on the adoption of methods that had enabled the other physical sciences to achieve significant progress. Analyzing composition in terms of weights and measures offered a more objective and certain method than comparing phenomenal properties or recording affinities. Admittedly, increased labor would be required to conduct chemical experiments with enhanced standards of precision, but this would be amply compensated for by the assurance that the science would advance steadily forward along a path that would never need to be retraced.

In subsequent work, Lavoisier reiterated that his procedures had established the doctrine of the composition of water with certainty. The connection between exact measurement and exact reasoning was stated in his report on Hassenfratz and Adet's new system of chemical symbols, read to the Académie on 27 June 1787: "One of the points of the modern doctrine which appears the most solidly established is the formation, the decomposition, and the recomposition of water; and how would it be possible to doubt it, when one sees that in burning together 15 grains of inflammable gas and 85 of vital air, one obtains exactly 100 grains of water, and that one may, by way of decomposition, retrieve those same two principles and in the same proportions?"[26]

Four constituents of Lavoisier's argument for the composition of water may be distinguished in this passage: first, that the weights of the identified parts (inflammable and vital airs) add up to the whole; second, that the composition proportions are now known (15:85); third, that the weights have been measured to a high degree of accuracy ("exactly"); and fourth, that this procedure warrants the acceptance of the doctrine of the composition of water without any doubt remaining. In subsequent statements, the fourth point was regularly asserted. Thus, in comments added to the French translation of Kirwan's *Essay on Phlogiston* the following year, Lavoisier connected the use of precision methods with the the quality and directness of the proof that they had made possible. He wrote of the 1785 demonstration:

> This double experiment, one of the most memorable which was ever made, on account of the scrupulous exactness which was attended to, may be regarded as a demonstration of the possibility of decomposing and recomposing water, and of its resolution into two principles, oxigene and hydrogene, if in any case the word Demonstration may be employed in natural philosophy and chemistry. . . . The proofs which we have given . . . being of the demonstrative order, it is by experiments of the same order, that is to say by demonstrative experiments which [*sic*] they ought to be attacked.[27]

Lavoisier's critics did not see things this way. By connecting the doctrine of the compound nature of water with the notion of a demonstrative experiment requiring precision measurement, he had linked his factual claims with a particular model of scientific practice. This gave Lavoisier's statements much of their power but it also offered numerous points of attack. Opponents called into question the appropriateness and reliability of his apparatus, the replicability of the experiments, and the discursive forms in which they had been reported. Doubt was also cast on the validity of the accuracy claimed and its purported implications for producing a geometrical standard of proof.

Priestley expressed some dissatisfaction with the stipulated precision of the measurements made in the 1785 experiments. Reviewing them more than a decade later, he wrote that ". . . the apparatus employed does not appear to me to admit of so much accuracy as the conclusion requires; and there is too much of correction, allowance, and computation, in deducing the result." Precision as such was not really the issue for Priestley, however. Attempts at accurate measurement apparently failed to convince him largely because such methods curtailed the possibilities for replicating experiments. Replication, as I have argued elsewhere, was a crucial feature of Priestley's envisioned democratic and egalitarian scientific culture; it would be hindered by the adoption of elaborate instrumentation and complex methods.[28] He noted particularly that the synthesis experiment required "so difficult and expensive an apparatus, and so many precautions in the use of it, that the frequent repetition of the experiment cannot be expected; and in these circumstances the practised experimenter cannot help suspecting the certainty of the conclusion."[29]

Elements of Priestley's critique were echoed by Keir in his *First Part of a Dictionary of Chemistry* (1789). Keir fastened on Lavoisier's assertion that chemistry had been founded on such a secure factual basis that deductions could be made with demonstrative certainty.[30] On the contrary, he insisted, many aspects of the new chemical doctrine, including the composition of water, remained hypothetical. Hence, the "precision of method employed in geometry" was not applicable to chemistry. Instead, chemists should be satisfied with examining and classifying the phenomenal properties of bodies, an approach clearly modeled on contemporary natural history that promised unending progress: ". . . if, instead of aiming at mathematical demonstration, and certainty, we be satisfied with examining the various modes, in which chemical phenomena may be viewed and arranged; and with comparing these . . . constantly distinguishing certainty from probability, and hypothesis from demonstrated truth . . . our minds will be ever open to receive the improvement that time and repeated experiments alone can produce."[31]

Thus, the status of demonstrated fact should be accorded only to raw phenomena, and all interpretations should be judged equally hypothetical. By maintaining this distinction, chemistry could hope to progress in a stepwise fashion toward the truth. While Lavoisier had worked to render the composition of water a fact, and had deployed precision measurement as a tool for that purpose, Keir and Priestley were denying that such a doctrine could ever be a fact. Their epistemology sought to build upon consensually accepted phenomena that would provide the certain basis for a knowledge of nature. Interpretation was the application of some hypothetical scheme to impose order upon the phenomena; it was inevitably of secondary validity to the facts themselves.[32] For Lavoisier to assert that his doctrine had attained the status of demonstrative proof seemed like dogmatism to these English critics. They responded by mobilizing an egalitarian rhetoric of factuality against such an imposition of power. Keir and Priestley thus denied the relevance of precision measurement to Lavoisier's proof. They simply did not accept that what they saw as a gap between phenomenal facts and interpretation could be bridged by this means.

Other critics of Lavoisier were more sympathetic to the project of accurate measurement. Kirwan praised the French chemist as "the first that introduced an almost mathematical precision into experimental philosophy," although he continued through the 1780s to raise objections against the doctrine of composition of water and other aspects of the antiphlogistic theory. He seems to have accepted in principle that accurate measurement might be able to prove Lavoisier's claims, as he had proven "by direct and exact measurements" that metals gained weight on calcination.[33]

This was the line taken in a sophisticated critique of Lavoisier's practices of precision, by the scientific lecturer, writer, and instrument-maker, William Nicholson. In his 1788 translation of the *Elements of Natural History and Chemistry* by Antoine François de Fourcroy, Nicholson commented: "If it were allowable to place much dependance [*sic*] on the quantities of products in experiments of this nature, wherein different philosophers disagree, the results of M. Lavoisier, which answer to the proportions of the two principles determined in the composition of water, would add much force to his conclusions."

Precision measurement of quantities might, potentially, lend considerable strength to Lavoisier's case. But his methods could not be accepted as conclusive as long as the more general theoretical issues remained unresolved. Experimental procedures and conclusions stood or fell together—Nicholson was identifying what would later be called "the experimenters' regress." He went on to say that the defenders of phlogiston still had plausible alternative accounts of Lavoisier's experiments. Cavendish's 1784 paper, which rejected the composition of water, was itself "a masterpiece of precision." An unprejudiced reader would have to conclude that the question of phlogiston was not yet resolved, "because decisive experiments are wanting." And, in this situation, precision measurement could not in itself render experiments decisive.[34]

In his introduction to the second English edition of Kirwan's *Essay on Phlogiston*, in 1789, Nicholson again commented on the role of precision measurement in Lavoisier's arguments. He acutely noted their rhetorical function as a purported means of proof: ". . . a reference to weights in the experiments of Mr. Lavoisier is made to constitute a great part of the arguments adduced to prove the composition of water, and its decomposition." Some of the measurements given, however, particularly for the weights of gases, laid claim to what seemed to Nicholson a quite implausible degree of exactness. He noted that difficulties with weighing the containing vessels would prevent gases being weighed with anything like the precision Lavoisier had claimed. Such "an unwarrantable pretension to accuracy" had presumably come about through working from tables of specific gravities carried to too many figures, or from reducing fractions to extended decimals, or from recording weights to an accuracy beyond the limitations of the apparatus. Thus, Lavoisier had recorded weights to six, seven, or even eight figures, whereas Nicholson estimated that only three could reliably be asserted. He concluded that claims that a scrupulous exactitude had been maintained and a demonstrative proof achieved were highly questionable. Thus, by exposing the weakness of the third component of Lavoisier's argument, Nicholson undermined the plausibility of the fourth:

> If it be denied that these results are pretended to be true in the last figures, I must beg leave to observe, that these long rows of figures, which in some instances extend to a thousand times the nicety of experiment, serve only to exhibit a parade which true science has no need of: and, more than this, that when the real degree of accuracy in experiments is thus hidden from our contemplation, we are somewhat disposed to doubt whether the *exactitude scrupuleuse* of the experiments be indeed such as to render the proofs *de l'ordre demonstratif*.[35]

The campaign continued in 1790, in Nicholson's *First Principles of Chemistry*, where the author devoted a chapter to the accuracy of balances. A standard precision balance made by a skilled craftsman was said to be capable of a maximum accuracy of five places of figures. Two instruments recently constructed by Jesse Ramsden (one for the chemist George Fordyce and one for the Royal Society) were thought to permit a reasonable guess to be made as to the sixth figure. Nicholson concluded:

> From this account of balances, the young student may form a proper estimate of the value of those tables of specific gravities, which are carried to five, six, and even seven places of figures, and likewise of the theoretical deductions in chemistry that depend on a supposed accuracy in weighing, which practice does not authorize. In general, where weights are given to five places of figures, the last figure is an estimate or guess figure; and where they are carried farther, it may be taken for granted that the author deceives either intentionally, or from want of skill in reducing his weights to fractional expressions, or otherwise.[36]

Nicholson's point here was the need for an awareness of significant figures. The accuracy of a balance being known, tables of specific gravities should not be carried to more decimal places than was warranted. Experimenters should be willing to curtail their calculations, even to discard figures from data, if the result extended to more figures than justified by the limitations of the measuring apparatus. The fact that Lavoisier had not practiced such parsimony was seen as having undermined the validity of his "theoretical deductions."[37]

Lavoisier did nothing to deflect such criticism. Indeed, his remarks in the *Traité* (1789) might well have given occasion to Nicholson's comments on the errors of balances. Lavoisier recorded in his textbook that two instruments made for him in the late 1780s by Nicolas Fortin were more accurate than any others, except Ramsden's. He gave estimates of the precision of which each was capable (one could weigh up to 20 ounces with an accuracy of about 1/10 grain, the other up to 1 dram with an accuracy of about 1/100 grain); but he did not spell out the implications for the number of significant figures that could legitimately be cited for measurements made with them. He also ignored the problem of significant figures in urging chemists to give weights in decimal fractions of a pound, instead of ounces, drams, and grains. In an example of the conversion (omitted by Robert Kerr from the English translation), Lavoisier cheerfully carried the calculation to seven places of decimals of a pound, that is, to eight figures in all. In a table at the back of the volume (also omitted by Kerr), readers were given figures to convert numbers of grains to decimal fractions of a pound that were carried to *nine* places, well beyond the limits of accuracy of any balance of the day.[38]

It is unsurprising, therefore, that Nicholson felt the need to reiterate his arguments in his *Dictionary of Chemistry* (1795). Discussing the water synthesis experiment, and referring specifically to the replication in 1790 by Fourcroy, Vauquelin, and Séguin, which was reported at much greater length than the original, he showed himself quite willing to accept that water had been proved to be a compound. He nonetheless persisted in reminding the French chemists of the limitations of their procedures. Nicholson noted that gas weights had been calculated from measured volumes, corrected for temperature and pressure, and converted to weight by using figures for the standard weight per unit volume. Clearly, this made the accuracy of the calculation dependent on that of the determination of the standard, an operation performed on 810 cubic inches of gas in a globe that weighed 24,179 grains. In this container, hydrogen was found to weigh just 35 1/4 grains, a result that assumed (as Nicholson pointed out) "a considerable degree of accuracy" in the balance. An error of just 1/4 grain in this measurement would affect all but the first two or three figures in the weight of hydrogen in the reaction, so that "the result may therefore, at best, be considered as estimate in all figures but the three first." The results for oxygen, which was weighed in greater quantities, were probably accurate to four places of figures.

Nicholson emphasized that he was not (by this stage) casting doubt on the Lavoisians' basic claims: ". . . as the fidelity of these philosophers cannot be suspected, as the product of water so remarkably coincides with the weight of the air which was burned, . . . the experiment may be admitted to prove that vital and inflammable air in certain due proportions do unite at the temperature of moderate combustion, and form water."[39] His point was that results did not always have their plausibility enhanced by claims to great exactness of measurement. An assertion of excessive precision, beyond the limits of what could be justified by the available apparatus, actually made the result less persuasive. In place of Lavoisier's link between precision of measurement and precision of reasoning, Nicholson was suggesting that recognition of the limits of precision (the "nicety of experiment") could yield a more appropriate way of presenting experimental claims.

This was by no means a common stipulation in experimental science at this time. Late eighteenth-century practitioners of exact measurement, in fields such as trigonometrical surveying or the measurement of heights by barometric hypsometry, did not routinely display a sensitivity to the problem of significant figures.[40] There was a recognition that sensitive instruments were particularly liable to error, as the use of the term "nice" (or Nicholson's "nicety") in connection with experimental apparatus implied.[41] But to insist that figures be dropped from the results of calculations if they suggested an accuracy greater than the instrument warranted was highly unusual.

It was perhaps Nicholson's didactic concerns, as a mathematics tutor and a lecturer on chemistry and natural philosophy, that enabled him make this point. In an introductory account of the practices of barometric hypsometry, published in 1788, he had given examples of the kind of calculations made by the leading practitioners in the field. After showing how long decimals were routinely manipulated without regard for significant figures, he noted, " . . . though the decimals in this computation are mostly retained, . . . it will in general be sufficiently exact, and much less operose, if only the first decimal figures of any number be retained."[42] This was, of course, some way short of an articulation of the doctrine of significant figures. Nicholson was proposing that strings of decimals be curtailed according to a feeling for what was "sufficiently exact" rather than a rigorous understanding of the limitations of particular apparatus. Nonetheless, the stipulation did comprise a self-denying ordinance against the lengthy decimals usually exhibited by practitioners of barometric hypsometry. It may be relevant that Nicholson made the point in the course of presenting techniques of measurement and calculation to a nonexpert audience. Directly following this remark he commented on the accessibility of barometric methods of surveying to practitioners who lacked substantial resources of money or skill. It seems that implicit in this endorsement of "sufficiently exact" measurement was a vision of the community of experimenters in which resources were to be as accessible as possible.

To account for Lavoisier's techniques of measurement as a simple transfer of methods from physics to chemistry would be to fail to grasp the problematic nature of his achievement. Considering the matter from the perspective of his opponents allows us to recognize this. While Lavoisier succeeded in creating a local culture of precision measurement, by forging alliances with mathematicians and engineers, by constructing refined apparatus, and by organizing the audiences at set-piece demonstrations, he initially failed to convey the persuasiveness of these practices more widely. Instead, such critics as Priestley and Keir resented the attempt to yoke exact measurement to a geometrical standard of proof, a standard they insisted was quite inapplicable to chemistry.

Nicholson's critique shows that a novel understanding of the limits of proof by exact measurement could emerge from this controversy. By the mid 1790s, he and many others were willing to agree with Lavoisier's basic antiphlogistic claims and acknowledged the broad validity of his measurements, for example of the composition of water. The process of consensus formation did not, however, take the form that Lavoisier seems to have wanted. Nicholson commented that, in many kinds of controversies, religious and metaphysical as well as scientific, "converts are seldom produced by the direct force of right reasoning; but in an indirect method, from the repetition of their adversaries' arguments, with a view to confute them."[43] While Lavoisier seemed to have been aiming at something like "the direct force of right reasoning"—an exactness of procedure and deduction that would compel an audience to assent—Nicholson was indicating a less direct, but perhaps more effective, mode of persuasion.

It seems appropriate that such a critique emerged in England, where Priestley had pioneered opposition to Lavoisier from a standpoint that valued replicable experimental methods to aid the establishment of natural phenomena in the public realm. Nicholson did not follow Priestley in condemning precision methods as such, on grounds of their limited replicability and special liability for error. But he did insist that procedures be opened to public scrutiny and the potential for error be explicitly avowed. In this discourse, the value of precision and the recognition of its limits would each have their place. This implied a degree of give-and-take in consensus formation that Lavoisier's rhetoric of precision had not invited, and a lessening of the degree of exactness claimed for experimental measurements in the interests of reaching agreement. In England, Lavoisier's precision measurement was uncoupled from his purported precision of reasoning.

Notes to Chapter Three

1. Robert Harrington, *The Death-Warrant of the French Theory of Chemistry* (London, 1804), 217. (Harrington, an eccentric Carlisle surgeon, was still holding out against the antiphlogistic chemistry in the first decade of the nineteenth century.)

2. *The Quantifying Spirit in the Eighteenth Century*, ed. Tore Frängsmyr, J. L. Heilbron, and Robin E. Rider (Berkeley, 1990).

3. Arthur L. Donovan, "Lavoisier and the Origins of Modern Chemistry," in *The Chemical Revolution: Essays in Reinterpretation*, ed. Donovan (*Osiris*, 2d ser., 4 (1988), 214–31); Anders Lundgren, "The Changing Role of Numbers in 18th-Century Chemistry," in *Quantifying Spirit*, 245–66.

4. Frederic L. Holmes, *Lavoisier and the Chemistry of Life: An Exploration of Scientific Creativity* (Madison, Wisc., 1985), esp. pp. xviii–xix, 276–83, 388–402.

5. M. Norton Wise, "Mediations: Enlightenment Balancing Acts, or the Technologies of Rationalism," in Paul Horwich, ed., *World Changes: Thomas Kuhn and the Nature of Science* (Cambridge, Mass., 1993), 207–56. Bernadette Bensaude-Vincent, "The Balance: Between Chemistry and Politics," in *The Eighteenth Century: Theory and Interpretation* 33 (1992), 217–37.

6. Frederic L. Holmes, *Eighteenth-Century Chemistry as an Investigative Enterprise* (University of California, Berkeley, 1989), esp. chap. 5. See also Maurice Daumas, "Precision of Measurement and Physical and Chemical Research in the Eighteenth Century," in *Scientific Change: Historical Studies*, ed. A. C. Crombie (London, 1963), 418–30.

7. Jean Baptiste Biot, *Essai sur l'Histoire Générale des Sciences Pendant la Révolution Française* (Paris, 1803), 22.

8. J[ames] K[eir], *The First Part of a Dictionary of Chemistry* (Birmingham, 1789), vii.

9. Lissa Roberts, "A Word and the World: The Significance of Naming the Calorimeter," *Isis* 82 (1991), 198–222.

10. T. H. Lodwig and W. A. Smeaton, "The Ice Calorimeter of Lavoisier and Laplace and Some of Its Critics," *Annals of Science* 31 (1974), 1–18; [Anon.], *Critical Examination of the First Part of Lavoisier's Elements of Chemistry* (London, 1797), 20–21.

11. Richard Kirwan, "Remarks on Mr. Cavendish's Experiments on Air," *Philosophical Transactions* 74 (1784), 154–69; Michael Donovan, "Biographical Account of the Late Richard Kirwan, Esq.," *Proceedings of the Royal Irish Academy* 4 (1850), lxxxi–cxviii.

12. Henry Cavendish, "Experiments on Air," *Philosophical Transactions* 74 (1784), 119–53.

13. A. L. Lavoisier, "Mémoire dans lequel on a pour objet de prouver que l'eau n'est point une substance simple," in *Oeuvres de Lavoisier* (6 vols., Paris, 1864–93), II, 334–59.

14. Ibid., 339.

15. James Watt, "Thoughts on the Constituent Parts of Water," *Philosophical Transactions* 74 (1784), 329–53, esp. pp. 329, 333.

16. Carl Perrin, "Lavoisier, Monge and the Synthesis of Water," *British Journal for the History of Science* 6 (1973), 424–28.

17. A. L. Lavoisier and J.-B. Meusnier, "Mémoire où l'on prouve, par la décomposition de l'eau, que ce fluide n'est point une substance simple," in Lavoisier, *Oeuvres*, II, 360–73; Lavoisier, "Mémoire dans lequel," 340.

18. Joseph Priestley, "Experiments and Observations Relating to Air and Water," *Philosophical Transactions* 75 (1785), 279–309.

19. Maurice Daumas and Denis Duveen, "Lavoisier's Relatively Unknown Large-scale Decomposition and Synthesis of Water, February 27 and 28, 1785," *Chymia* 5 (1959), 113–29; Holmes, *Lavoisier*, 237–38.

20. Maurice Daumas, *Lavoisier: Théoricien et Expérimentateur* (Paris, 1955), esp. chap. 6; id., "Les appareils d'experimentation de Lavoisier," *Chymia* 3 (1950), 45–62.

21. I have given a more detailed description of the apparatus used in this demonstration in my paper, "Precision Instruments and the Demonstrative Order of Proof in Lavoisier's Chemistry," *Osiris*, 2d ser., 9 (1994), 30–47.

22. Estimates of accuracy of measurement are drawn from examination of notes from the experiment surviving in the Lavoisier MSS, History of Science Collection, Cornell University Library, Ithaca, N.Y. On the accuracy of thermometers and barometers, see Theodore S. Feldman, "Late Enlightenment Meteorology," in *Quantifying Spirit*, 143–77, on pp. 156–57 and 166.

23. C.-L. Berthollet, "Mémoire sur l'acide marin déphlogistiqué," *Observations sur la Physique* 26 (1785), 321–25, esp. p. 324; H. E. LeGrand, "The 'Conversion' of C.-L. Berthollet to Lavoisier's Chemistry," *Ambix* 22 (1975), 58–70, esp. pp. 67–68; Carleton Perrin, "The Triumph of the Antiphlogistians," in *The Analytic Spirit: Essays in the History of Science in Honor of Henry Guerlac*, ed. Harry Woolf (Ithaca, N.Y., 1981), 40–63, esp. pp. 49, 55–56, 62.

24. A. L. Lavoisier and J.-B. Meusnier, "Développement des dernièrs expériences sur la décomposition et la recomposition de l'eau," in Lavoisier, *Oeuvres*, V, 320–34.

25. For example: C. L. Berthollet, "Considérations sur les Expériences de M. Priestley," *Annales de Chimie* 3 (1789), 63–114, on p. 71; Thomas Garnett, *Outline of a Course of Lectures on Chemistry* (Liverpool, 1797), 74; Jeremiah Joyce, *Dialogues in Chemistry* (2 vols., London, 1807), I, 219. See also Holmes, *Lavoisier*, 237.

26. A. L. Lavoisier, "Rapport sur les nouveaux caractères chimiques," in Lavoisier, *Oeuvres*, V, 365–78, on pp. 370–71.

27. Lavoisier quoted in Richard Kirwan, *An Essay on Phlogiston and the Composition of Acids*, ed. William Nicholson (2d ed., London, 1789), 59–61.

28. Jan Golinski, *Science as Public Culture: Chemistry and Enlightenment in Britain, 1760–1820* (Cambridge, 1992), chap. 3. See also John G. McEvoy, "The Enlightenment and the Chemical Revolution," in *Metaphysics and Philosophy of Science in the Seventeenth and Eighteenth Centuries*, ed. R. S. Woolhouse

(Dordrecht, 1988), pp. 307–25; id., "Continuity and Discontinuity in the Chemical Revolution," *Osiris*, 2d ser., 4 (1988), 195–213.

29. Joseph Priestley, *Considerations on the Doctrine of Phlogiston and the Decomposition of Water* (Philadelphia, 1796; repr. ed. William Foster, Princeton, 1929), 34, 41; id., *The Doctrine of Phlogiston Established and that of the Composition of Water Refuted* (Northumberland, Penn., 1800), 77.

30. In addition to the passage quoted earlier, Keir might have been reacting to Lavoisier's assertion in the *Traité*, that chemistry "has not, like elementary geometry, the advantage of being a complete science of which all the parts are closely connected together; but nonetheless its current progress is so rapid, the facts are arranged in such a happy manner in the modern doctrine, that we may hope, even in our own time, to see it approach near to that degree of perfection which it is capable of achieving." A. L. Lavoisier, *Traité Elémentaire de Chimie* (2 vols., Paris, 1789), I, p. xii; cf. Lavoisier, *Elements of Chemistry in a New Systematic Order, Containing all the Modern Discoveries*, trans. Robert Kerr (Edinburgh, 1790), xx. See also Lavoisier's remarks on the failure of traditional chemistry to attain to the rigor of geometry, as it appeared to him in the course of his own education, in the undated manuscript, "Sur la manière d'enseigner la chimie," in Bernadette Bensaude-Vincent, "A View of the Chemical Revolution through Contemporary Textbooks: Lavoisier, Fourcroy and Chaptal," *British Journal for the History of Science* 23 (1990), 435–60, on p. 457.

31. Keir, *Dictionary*, ix–x.

32. Cf. Wilda C. Anderson, *Between the Library and the Laboratory: The Language of Chemistry in Eighteenth-Century France* (Baltimore, 1984), 83, 96–97.

33. Kirwan, *Essay on Phlogiston*, 3, 7.

34. Antoine François de Fourcroy, *Elements of Natural History and Chemistry*, trans. William Nicholson (4 vols., London, 1788), I, xvi–xvii. Reflecting continuing disagreements among chemists, Nicholson continued until 1796 to preserve an account of the phlogiston theory in his popular textbook, *The First Principles of Chemistry* (3d ed., London, 1796).

35. Kirwan, *Essay on Phlogiston*, viii, xi.

36. William Nicholson, *The First Principles of Chemistry* (London, 1790), 67–69, quotation on pp. 69–70.

37. Compare Guerlac's remark that "the disregard, or ignorance, of significant figures is characteristic of a period before the full grasp of error theory." (Henry Guerlac, "Chemistry as a Branch of Physics: Laplace's Collaboration with Lavoisier," *Historical Studies in the Physical Sciences*, 7 (1976), 193–276, on p. 253, n. 145.)

38. Lavoisier, *Traité*, II, 330–34, 561–63; cf. Lavoisier, *Elements*.

39. William Nicholson, *A Dictionary of Chemistry* (2 vols., London, 1795), II, 1021–23.

40. Sven Widmalm, "Accuracy, Rhetoric, and Technology: The Paris-Greenwich Triangulation, 1784–88," in *Quantifying Spirit*, 179–206, esp. pp. 195–96; Theodore S. Feldman, "Applied Mathematics and the Quantification of Experimental Physics: The Example of Barometric Hypsometry," *Historical Studies in the Physical Sciences*, 15 (1985), 127–97.

41. The *Oxford English Dictionary* records instances of "nice" meaning "requiring or involving great precision, accuracy, or minuteness" from the works of Robert Boyle to the early nineteenth century. Other meanings include implications of doubtfulness and uncertainty, and (of balances and other instruments) "finely poised or adjusted." Nicholson reflected this usage when he wrote that Lavoisier's experiments were "extremely nice; such that the utmost accuracy of observation was necessary to distinguish the result. In such cases mistakes are easily made; nay, it is scarce possible to avoid them." A. F. de Fourcroy, *Elements of Natural History and Chemistry*, ed. Nicholson (2d ed., 3 vols., London, 1790), I, xiii.

42. William Nicholson, *An Introduction to Natural Philosophy* (3d ed., Philadelphia, 1788), 320–30, on p. 328.

43. Ibid., I, vi.

FOUR

PRECISION: AGENT OF UNITY AND PRODUCT OF AGREEMENT

PART I—TRAVELING

M. Norton Wise

THE PAPERS in this section support the historiography that places the beginnings of something new in the realm of precision measurement in the second half of the eighteenth century. Only then did the rhetoric of precision acquire the power to carry conviction in virtually any domain in which it was applied; and only then was it applied in every imaginable domain. But the authors see a great deal of variety in the values ascribed to precision and they see contests everywhere over its claims. Here I want to explore briefly some of the ways in which their arguments may generalize and to point out some of the directions for future research that they suggest.

Science in the Bureaucratic State

Centralized administration of large countries requires some method of accounting for what goes on in different areas and for comparing them one with another. Such information forms the basis for organization and control in all bureaucratic states. The state needs to be able to count its people, to measure their land and its produce, and to survey manufactures in order to be able to organize, plan, defend, and tax with efficiency. Such accounting depends on uniform measurements across wide geographical areas and across a broad spectrum of family structures, farming and manufacturing practices, work organization, and types of materials and products, not to mention military units and weapons. And this uniformity depends on precision in the measuring instruments. So the emergence of centralized bureaucratic states in the eighteenth century, controlled more by information than by might, naturally served as one of the primary motors of precision measurement, and *vice versa*. That at least is the picture that emerges from recent studies presented here and elsewhere.[1]

But precision comes no more easily than centralized government. Andrea Rusnock reminds us that counting people may be a simple matter in a village, but it is immensely complex over a territory the size of France. The desirability of an accurate census for administrative purposes thus called forth both a bureaucracy spread more widely and a variety of new statistical techniques for *calculating* the population based on universal multipliers of either the number of hearths or the birthrate. Most important, Laplace gave to the French republic a method for estimating the reliability of the calculation. Pursuing government interests like those of the census takers, Lavoisier and Lagrange too rendered service to the state accounting apparatus. Lavoisier attempted to construct a "thermometer" to register the health and wealth of the entire economy with a "great balance sheet" for recording the production and expenditure of all goods. Lagrange attempted to calculate the quantity of foodstuffs required to maintain the population, by devising an instrument like the universal multiplier for population, a "calculus of nourishment." He based this instrument on the belief that the relative price of different foods gives a natural measure of their nutritional value, and on estimating the average consumption of a person by equating one man to a woman plus three children under the age of ten.[2] Such imaginative schemes suggest at once how important the numbers were to the political economy of the state and how difficult it was to obtain them with any consistency. They suggest too why Adam Smith so famously attempted to measure the wealth of nations in terms of a universal and natural unit of value—namely labor value, which would reduce all commodities to commensurability—and why he and all succeeding economists have so famously failed to attain precision in such measures.

Equally famous, and somewhat more successful, has been the metric system, whose promoters included some of the same reforming French administrators and savants as the census takers and economists. We tend to think of the metric system today as a rationalized system of measurement for the sciences, based on units of mass, length, and time derived from universal phenomena of nature; but of course it was from the beginning, and more prominently, a system of weights and measures for standardizing commerce, taxation, and the monetary system. Ken Alder teaches that its primary value was uniformity, whether looked at from the perspective of bureaucrats, large-scale merchants, or consumers. And again, it was precision that would guarantee uniformity, implying also trustworthiness in traders, taxpayers, and money itself.

As the ultimate guarantee of legitimacy, Alder also reminds us, the savant-administrators wanted the system of measures to speak the language of nature itself, just as Lavoisier and Lagrange claimed for their measuring instruments of wealth and consumption. Mary Douglas has argued that indeed every society, whether "primitive" or "modern," attempts to project its most cherished beliefs into nature, that this externalization is one of the primary means by which sameness is constructed and common institutions established.[3] In the present case, certainly, her thesis holds. The attempt to attain natural standards for weights and measures reflected a more general attempt to make uniform laws of society express laws of nature. If this program of measurement in the interests of the enlightened state also served to guarantee the place of academicians as its servants and spokesmen, and even to supply funding for their other mathematical and natural philosophical activities, that fact did not necessarily compromise the integrity of their beliefs. And they certainly believed it possible to design instruments that would evoke the eternal verities of an objective nature.

I have argued elsewhere that these instruments, whether material or mathematical, were typically designed as balances. In fact, they constituted a network of interrelated but diverse technologies which gave substance to the belief that the laws of nature, left to themselves, acted to produce equilibrium between opposing tendencies and to average out irregularity. Thus the best way to determine the natural state of a system, or to measure an entity associated with such a state, was to design an instrument which would reveal the equilibrium conditions: a balance. Rusnock describes one such balancing strategy in Laplace's use of the probability calculus to determine a universal ratio of births to population. He would average out the viscissitudes of life in different regions to yield a natural multiplier, complete with an estimate of its precision, or its probable deviation from the truth. This example should not be divorced from Laplace's similar application of the probability calculus to find the true location of a planet from a series of discrepant observations, nor from his use of the variational calculus, following Lagrange, to establish that the secular inequalities in the motions of the planets were periodic, thus averaging out in the long run to yield an eternally stable solar system.[4] Alder may be taken to generalize this point when he says that for the savants who constructed the metric system as a natural analytic language, "rationality was . . . a lever to pry the present from the past." He is implicitly citing the ubiquitously quoted Condillac, for whom reason was a language well arranged, meaning one structured according to the logic of balanced algebraic equations, the "lever of the mind," and for whom a metaphor for a good government administrator was a mechanic who could connect the balance spring of one of the newly precise pocket watches to its train of mechanism so as to make the whole run uniformly, with inequalities minimized.[5]

For many of Condillac's admirers, precision measurements, like law, language, and nature, should express this balancing and levering action of algebraic rationality. Lavoisier, in his simultaneous roles as government administrator and chemical revolutionary, perhaps realized Condillac's dream most completely. But the dream had limitations, then and now. It assumes an ease about numerical facts, about their relation to laws, and about who is qualified to judge. Such ease is rare in the history of measurement, not to speak of government policy. The contributors to this volume have been especially interested in unease and dissent. Having put the general attractiveness of precision measurement in terms of its capacity to unify extensive domains with a uniform standard, it is useful to consider dissent in terms of difficulties encountered in making precision travel.[6]

International Boundaries

Jan Golinski shows us that Lavoisier's famous accomplishments in establishing the composition of water in the 1780s did not travel well across the channel. Priestley and other British natural philosophers disputed both the validity of his vaunted precision measurements and the force of the argument he based on them. Traditionally this disagreement has been discussed in terms of Priestley's phlogiston theory versus Lavoisier's oxygen theory, and of qualitative versus quantitative chemistry. But while crediting this debate, Golinski wants us to see something more in it. It was not only a debate about which theory explained things more adequately or simply. It was also about the sorts of claims that were proper in a community of natural philosophers and how those claims should be established. In this, Golinski's paper compares interestingly to one presented at the same session by John Heilbron but published elsewhere. For it raises the great question of the place of mathematics in relation to natural philosophy (or experimental physics, including chemistry). Was natural philosophy to be more like natural history, as Priestley contended, or was it to take on a mathematical form, as Lavoisier intended? And specifically, what was the place of precision measurement in relation to either side?

Heilbron has analyzed a dispute at the Royal Society, also in the 1780s, which pitted those who counted themselves mathematicians against the supporters of Joseph Banks, who deplored the pretended hegemony of mixed mathematics and asserted the priority of general natural philosophy and natural history. Banks won; Priestley supported Banks. Heilbron makes the incident show how the mathematization of natural philosophy (in Britain) took place, not at the instigation of mathematicians, but when natural philosophers began to annex established parts of mixed mathematics while simultaneously bringing the formerly qualitative territories of heat, electricity, magnetism, and chemistry under quantitative control. This rapid expansion depended on instrument makers to invent a wide variety of newly precise philosophical instruments (thermometer, barometer, hygrometer, pyrometer, electrometer, analytical balance, etc.) which complemented their traditional mathematical ones (astronomical, optical, surveying, etc.). The new natural philosophy, then, acquired its mathematical character less from mathematicians extending their domain of mixed mathematics than as an expansionist enterprise of natural philosophers taking over parts of that domain while wielding new instruments produced in the most advanced and competitive market in Europe.[7]

Heilbron's story supports Golinski's. Natural philosophers at the Royal Society, even those like Cavendish who pioneered precision measurement, supported Banks and Priestley against claims for the superiority of mathematical argument, whether from exponents at the Royal Society or the Academy of Sciences. (We do not yet know what the mathematicians thought of this.) No wonder Lavoisier's linkage of precise measurement with demonstrative mathematics did not travel easily across the channel. Just that linkage was being rejected at the Royal Society in a bitter and public feud. In Paris, if we take the census and the metric campaign as examples, the linkage was supported by the dominant elites of the Academy and their well-placed friends: mathematicians, natural philosophers, state administrators, engineers, and other enthusiasts of Enlightenment. So there is a considerable distinction to be drawn between the quantifying spirits of Paris and London.

In contrast to the rationalizing Parisian savant-administrators, the Royal Society produced neither census calculation nor metric system. They remained what they had always been, a society of "gentlemen," full of paternal virtue, general knowledge, and classical education. They stuck to empirical observations, made no exhorbitant claims, and formed conclusions by consensus. Banks reasserted those priorities with a vengeance against the leveling tendencies of mathematics, which carried always the taint of the commercial, the mechanical, the tool-like. Algebraic analysis would continue to carry that taint in Britain through the mid-nineteenth century.[8] Only as classical geometry could mathematics rise above its suspect associations to provide aesthetic beauty, demonstrative proof, and moral example. Crucially, however, not only mathematicians but other suspect characters regularly found honor in the gentlemen's club: Priestley the unitarian democrat; Jesse Ramsden the commercial maker of incomparably accurate instruments.[9] They did so through unique contributions to either natural philosophy or the state and through respect for the social code. In natural philosophy, Priestley's democracy was not that of the individualist arguing from abstract principle, but of the community of practitioners reaching agreement by open debate, widely accessible experiments, and mutual trust. In this respect, he reinforced the code of the gentleman.

Golinski shows us how Lavoisier violated this code. With instruments less precise than Ramsden's by his own evaluation, Lavoisier claimed exactitude; with algebra as his model of mathematical logic, he claimed demonstrative proof; on the basis of an elaborate experiment, staged as a public display, he demanded agreement. Presented in this way, Lavoisier's credibility could not survive the channel crossing. William Nicholson, natural philosopher, instrument maker, exponent of precision, showed him how to transform them for British standards: relax the identity between measurement and proof; report accurately the limitations of accuracy; acknowledge alternatives; never try to enforce agreement in a consensus society. On this basis, Nicholson counseled, the oxygen theory would (and soon did) find ready acceptance. Not surprisingly, Lavoisier himself showed no inclination to submit to these conditions. The enlightened culture he sought to realize in France would recognize as innately superior a rationality based on precision measurements ordered in algebraic equations and an algebraic language.

Local Dissent

Not even in France, however, did this vision of Enlightenment go unchallenged, which is the main point of Alder's discussion of the metric system and shows up also in Rusnock's census story. In France as in Britain, detractors saw nationalized counting and measuring as a hegemonic attempt to link centralized authority to mathematical abstraction, leveling qualitative difference under quantitative uniformity while running roughshod over the real interests of highly differentiated communities and markets. Their position resonates with the present-day "diversity" movement, with its attendant critique of enlightenment images of universality and unity. Recognizing the force of this critique, Golinski and Alder want us to recognize too that the images were highly contested during the Enlightenment itself. The moral, it would seem, is that in precision as in politics, the forces of unity and diversity, the universal and the local, uniformity and difference, are engaged in eternal negotiation. Even when all of the parties involved would like to see a uniform system of measurement established, to be accepted it must express the values of the different constituencies that it affects. The metric system originally foundered because its perpetrators failed to convince large sectors of the public that it would serve their best interests.

Put in the form of the "traveling" problem, this distrust of centralization again looks similar to the problem of crossing the cultural channel from the Academy of Sciences to the Royal Society. Alder notes Legendre observing that a traveler had once needed a "dictionary" to translate between the metric language of one town and the next, or better, between the cultural traditions and values expressed in local measures, which embodied a bewildering array of materials and work habits. Legendre and his metrical colleagues attempted to aid this metaphorical traveler by circulating their central measures throughout the nation. They published their conversion scales and nomenclature, launched an educational campaign, and tried to organize the manufacture and distribution of standardized meter sticks and weights. But uniformity of standards is a mixed blessing. Its value depends on the existence of large numbers of travelers whose lives are enriched rather than impoverished by its rule. As Alder argues, the (limited) success of the metric system in France depended on an increased circulation of people and goods in an expanding economy and on the increased reach of the bureaucracy in the centralizing state.

Rusnock too observes that attempts early in the eighteenth century at making a complete census failed because the bureaucracy did not yet reach far enough or deeply enough. Its success (also limited) by the early nineteenth century depended partly on a much wider circulation of trained administrators and partly on establishing a precise instrument that could travel over the entire spectrum of parishes and provinces, namely a universal multiplier of births, supposedly taken from nature itself, which averaged out the considerable local variation in the ratio of births to population and which circumvented local opposition to being counted and controlled by the central government. It was this mobile multiplier that established the "population" as a definite object.

Indeed, Laplace's probabilistic analysis of the reliability of this number provides an excellent early example of objectification through precision. Laplace surely believed that his precise method yielded an accurate result, the true population. But from Rusnock's account he seems also to have been aware that for the purposes of government administration the precision of the number was more important than its absolute accuracy. Its precision would guarantee both consistency between successive measurements and consistency from year to year. Such consistency made the population a reliable object of state policy. Rusnock's account suggests that considerations of this kind may have been important more generally in the gradual emergence of the distinction between precision and accuracy. That view agrees rather well with Olesko's discussion in the next section of precision in weights and measures and with Schaffer's discussion of resistance standards.

Notes to Chapter Four

1. Especially in Tore Frängsmyr, J. L. Heilbron, and R. E. Rider, eds., *The Quantifying Spirit in the 18th Century* (Berkeley, 1990), chaps. 7–12. The most valuable and comprehensive treatment of science and the French state prior to the Revolution is Charles C. Gillispie, *Science and Polity in France at the End of the Old Regime* (Princeton, 1980), where the intimate relations of knowledge and power, developed institutionally, are taken as characteristic of modern states and modern science.

2. Antoine-Laurent Lavoisier, *Résultats extraits d'un ouvrage intitulé de la richesse territoriale du royaume de France* (1791), in *Œuvres de Lavoisier* (Paris, 1893), 6:403–63, on 415–16; Joseph Louis Lagrange, "Essai d'arithmétique politique sur les premiers besoins de l'intérieur de la république," *Œuvre de Lagrange*, ed. J.-A. Serret (Paris, 1877), 7:571–79; first published in *Collection de divers ouvrages d'arithmétique politique, par Lavoisier, de Lagrange, et autres*, ed. Roederer (Paris, 1796).

3. Mary Douglas, *How Institutions Think* (Syracuse, 1986), 45, 48, 57.

4. M. Norton Wise, "Mediations: Enlightenment Balancing Acts, or the Technologies of Rationalism," in *World Changes: Thomas Kuhn and the Nature of Science*, ed. Paul Horwich (Cambridge, Mass., 1993), 206–56. Of particular interest for Lavoisier is Bernadette Bensaude-Vincent, "The Balance: Between Chemistry and Politics," *The Eighteenth Century: Theory and Interpretation*, 33 (1992), 217–37. Otto Mayr, *Authority, Liberty, and Automatic Machinery in Early Modern Europe* (Baltimore, 1986), correlates balancing mechanisms in many sorts of machinery with democratizing political and economic organization.

5. Étienne Bonnot de Condillac, *La logique, ou les premiers développement de l'art de penser* (1780), in *Œuvres complètes de Condillac* (Paris, 1798), 22:2–3, 95, 138; and *Traité des système*, in *Œuvres*, 2:375–77.

6. On traveling and on the important related notions of "action at a distance," "centers of calculation," and "immutable mobiles," which are reflected below, see especially Bruno Latour, *Science in Action: How to Follow Scientists and Engineers through Society* (Cambridge, Mass., 1987), chap. 6.

7. J. L. Heilbron, "A Mathematicians' Mutiny, with Morals," in *World Changes: Thomas Kuhn and the nature of Science*, ed. Paul Horwich (Cambridge, Mass., 1993), 81–129.

8. Ibid., 86–88. The standard British objection to algebra was that it replaced thinking and judgment with mechanical operations on formulae, which indeed supplied a common metaphor for algebra in France. As Condorcet put it, "The method of developing truths has been reduced to a [mechanical] art, one could

almost say to a set of formulae," in "Reception Speech at the French Academy" (1782), *Condorcet: Selected Writings*, ed. Keith M. Baker (Indianapolis, 1976), 3–32, on pp. 5 and 15. Laplace echoed, algebraic analysis has "the inestimable advantage of transferring reasoning into a mechanical process," in *Exposition du système du monde*, 2d ed. (Paris, 1799), 339; quoted in Roger Hahn, "The Laplacean View of Calculation," in Frängsmyr et al., *Quantifying Spirit*, 363–80, on p. 378.

9. On Ramsden, see Richard John Sorrenson, *Scientific Instrument Makers at the Royal Society of London, 1720–1780* (dissertation, Princeton University, 1993), 201–24.

PART TWO

INDUSTRIAL CULTURES

FIVE

THE MEANING OF PRECISION:

THE EXACT SENSIBILITY IN EARLY NINETEENTH-CENTURY GERMANY

Kathryn M. Olesko

NO OTHER NATION in the world has been identified with precision measurement as much as Germany has. By the last quarter of the nineteenth century, German military needs especially had furthered precision technology in geodesy, optics, and electrical communication. Germany's late but rapid and complete industrialization depended in part on precision machine industries and the accurate determination of standards in such areas as electricity and optics. The crowning achievement of German precision measurement, in 1887, was the establishment of the Berlin-based Physikalisch-Technische Reichsanstalt, which David Cahan has recently called a symbol of "the young German Reich's newly acquired political power and authority" owing to that institute's ability to enhance the "cohesiveness of German political life through the establishment of nation-wide physical units and standards."[1]

The product of meticulous experimental protocols and later of more delicate and accurate apparatus, refined measurements joined the plethora of the several novel forms of quantification that permeated the sciences, daily life, and especially the state's bureaucracies across the nineteenth century. The techniques and instruments that produced more accurate weights and measures (which made social interactions as well as commercial transactions more exact) migrated early in the century to the sciences where they formed the nucleus of exact experimental practice. These numbers were not easy to generate. They eventually demanded tightly organized labor forces and financial resources of novel dimensions. By the end of the century the physicist Emil Warburg noted that because of the time, effort, and expenditure involved, precision measurements were value-laden choices and therefore it was "a task of great importance to find the right choice."[2]

The values of precision measurement for military strength, economic growth, or political power were not initially self-evident. Over the course of the nineteenth century the nature of measuring and measurement metamorphosed. *Precise* measurements did not at first exist, although refined ones could be found. By century's close, no other form of measurement or even quantification commanded as much authority as did precision measurement. The transformation in measuring and measurement was a complex process encompassing epistemological, material, legal, moral, technical, and cultural dimensions.

In order to understand how that process began we must suspend all judgment on the importance of precision measurement, and indeed even forget what precise measurements are. No matter how many decimal places eighteenth- and early nineteenth-century quantifiers could generate, consensus was lacking on what those seemingly "precise" measurements meant. How meaning was assigned, especially during the formative stages in refined measurements in the first half of the nineteenth century, is therefore crucial for understanding not only the power of precision at the end of the century, but also the values that came to be associated with it. How and under what conditions were refined measurements deemed trustworthy enough to be used in the construction of knowledge? What, indeed, were refined measurements in the first place? Natural philosophers in America, Great Britain, France, and the German states raised these questions in the decades after 1800.

German natural philosophers used probability calculus and weights and measures reform to create the terms and guidelines for deciding what kind of refined measurements could be trusted—and hence used—and what kind could not. German investigators were nearly unique in using probability calculus to assign refined measurements degrees of truth or epistemological certainty. But like their colleagues across the Rhine, the Channel, and the Atlantic, they found in standards reform guidelines for assessing an experimental investigator's integrity by transferring to experimental practice conceptions of honesty that had always been at stake in the execution, certification, and refinement of standards.[3] Trust, truth-value, and integrity—scarcely apparent in eighteenth-century refined measurements—were the most important components of the *exact sensibility* of the early nineteenth century.

By sensibility I mean the mental tools and categories, habits of thinking, and affective judgments and responses used in identifying, using, and assessing refined measurements.[4] Decisions concerning the "goodness" of refined measurements were not so clear in the eighteenth century when the means for judging measurements, especially outside astronomy, were inchoate, diffuse, or not a matter of concern. At that time, the epistemological virtues of refined measurements, as opposed to coarse ones, were a matter of open debate, as Jan Golinski has shown for the case of Lavoisier's experiments on the composition of water.[5] Neither kind of measurement was necessarily more privileged than the other, and investigators capable of creating refined measurements were not necessarily considered more skilled than those who were not. Matters changed, however, in the first few decades of the nineteenth century.

"Exact" refers to the perceptions of nature's image and experiment's character, especially in physics, that accompanied refined measurements. As the framework for discussing refined measurements became more elaborate in the early nineteenth century, parts of the natural world were viewed in increasingly fine-grained quantitative terms. (The choice of which ones was itself a value judgment predicated on this emerging sensibility.) Experimentation acquired exactitude in all its parts: in its design, its protocol, its written presentation, and of course its numerical results. In other ways, "exact" is less easy to define. Quantitative exactness in the sense of what we now call "precision" (the narrow clustering of values) or "accuracy" (the fit between a measurement and a mathematical model) was not necessarily a characteristic of measurements deemed exact. (Indeed, precision and accuracy did not, in the early nineteenth century, have the definitions we now know them by. "Accuracy" was the more common term, and it denoted both meanings—and others—interchangeably.) Moreover, both qualitative and quantitative considerations were a part of the determination of exactitude, and these considerations concerned the qualities of the investigator nearly as often as they did those of the measurements themselves.

Probability Theory, Trust, and Certainty

Recently historians have argued, as did Maurice Daumas over thirty years ago in an oft-cited and influential article,[6] that in the late eighteenth and early nineteenth centuries more accurate measurements were the product of improved instrumentation.[7] Technicians and practitioners improved common instruments such as thermometers and barometers; created new graduation and comparative devices; and developed new techniques, such as double weighing. Definitions and identifications of precise measures therefore hinged on the instrument's quality—or at least upon how that quality was perceived. The contest between Borda's repeating circle and Ramsden's theodolite in the Paris-Greenwich triangulation of the 1780s, as told by John Heilbron, is one example of how nationalistic sentiments, for instance, could color such perceptions. Speaking of Borda's repeating circle, Jean Dominique Cassini boasted that "we dared to flatter ourselves that we had on our side an instrument that yielded nothing to the English in the point of precision"; and Delambre called the circle's precision "nearly incredible."[8]

Faith in instrumentation alone proved incapable, however, of producing unequivocal confidence in the certainty of measurements or in the truth of conclusions based on them.[9] Early seekers of precision supplemented instrumental readings with a mathematical or geometrical ideal, sometimes expressed in the form of an equation, to compensate for what measurement itself could not provide. According to Henry Lowood, late eighteenth-century quantifiers in German forestry science who sought "a greater semblance of precision" relied on and used mathematical constructions to supplement and perfect measurements actually taken for the purpose of managing a forest's resources.[10] Sven Widmalm has argued that in the Paris-Greenwich triangulation of the 1780s, notions of "mathematical exactness" were used to create a "rhetoric of accuracy" whose purpose it was to "inspire faith" in measurements; the rhetoric of accuracy was intended to connote the truth-value of measured results.[11] The association of mathematical, and hence absolute, truth with precision measurements had its skeptics, of course;[12] but the mathematical ideal persisted as a desirable and attainable goal in measuring claiming to be precise. These and other eighteenth-century examples suggest that instrumentally produced refined measurements were considered weak without supplemental assigned meanings.

At the beginning of the nineteenth century, however, German investigators loosened ties between notions of precision and a mathematical ideal of absolute truth by applying probability calculus to measurement. Although at first practitioners in the three leading scientific areas of Europe—France, England, and the German states—showed interest in the application of probability theory to physical measurements, the effort was strongest in the German states, a condition that enhanced the local character of refined measuring practices. Ad hoc procedures for combining measurements and estimating errors existed in the eighteenth century, but not until the early 1800s were such procedures regularized in the independent work of Legendre and Gauss on the method of least squares.[13] Based on probability calculus, the method of least squares assumed that the true value of the measurement was unknown. Approximating the true value of the measurement was possible, however, if one could identify the most probable value of several measurements. According to Gauss and Legendre, the most likely value of a measurement was that for which the sum of the squares of the errors—the error generally taken to be the difference between the measurement itself and the average value of all measurements in its group, given a sufficiently large number of measurements—was a minimum.

Despite the advantages the method of least squares offered for combining measurements and determining their certainty, the method was slow at first to take hold in areas other than geodesy and astronomy in Germany. Although Gauss had urged in 1809 that making "the uncertainty of the result as small as possible" was "uncontestably one of the most important tasks in the application of mathematics to the natural sciences,"[14] German discussions of least squares were scarce until the 1830s, when the method gained popularity. Gauss's own work became the standard reference on the matter. From the start, investigators used probability to circumscribe the degree of truth or certainty that could be ascribed to analytical expressions of theory containing constants and relations derived from measurement. Such assessments hinged on judgments of the quality (*Güte*) of refined measurements, determined by computing Gauss's "*zufällige Fehler*," or accidental errors, by means of least squares.

Interestingly, the first application of probability theory to physics written in the German language came from the fringes of the community of scientific practitioners. Its purpose was to criticize how the French had used measurements in the construction of physical knowledge. In 1819 Magnus Georg Paucker, astronomer and *Physiklehrer* in Mitau, published a small treatise on "the application of the method of least squares to physical observations."[15] Taking Biot's measurements of three physical relations—between the expansion of heated fluids and temperature, between the specific gravity of water and temperature, and between the elasticity of steam and its degree of heat—Paucker demonstrated how Biot's resulting equations had to be modified if all his measurements were considered (Biot had used only those he believed most accurate, thus privileging one set of measurements over all others without otherwise justifying his choice) and if the accidental errors of those measurements were computed. Accuracy (*Genauigkeit*) of the measurements was of course necessary—for Paucker "the entire task of *practical physics* consist[ed] in attaining the highest possible accuracy in observation and apparatus"—but accuracy was not itself the guarantor of certainty. Instead, in order to achieve "*the smallest possible deviation from the truth*," Pauker relied on the principle of the method of least squares that "the sum of the errors originating in [the observations had to be] as small as possible."[16]

References to Paucker's article were scarcely to be found in the decade following its publication, but for one exception: Georg Muncke's 1825 entry on "*Beobachtung*" in the revised edition of *Gehler's Physikalisches Wörterbuch*.[17] Muncke was, however, skeptical. Although he considered least squares "one of the most important and significant extensions of the use of the calculus in astronomical and physical observations," he found the method "not entirely easy," but "copious" and "laborious."[18] Others agreed—at least until the early 1830s when German articles and books on probability theory took up the problem of measurement accuracy through the analysis of errors by least squares.[19] At that time interest in assessing accuracy was less the result of developments in the physical sciences, including the production of better instruments, than of the perception that the numbers gathered for statistical purposes and those taken in measurements seemed to possess some of the same qualities. The famous examples of the likelihood of winning a bet, of picking black or white balls from an urn, of the number of criminals convicted and imprisoned each year, and even of the number of letters deemed undeliverable by the Parisian postal service annually—all these and more provided suggestive frameworks for interpreting the meaning of refined measurements which seemed to exhibit similar regular patterns. The method of least squares became, in J. J. Littrow's words, "indispensable" when dealing with observation and measurement.[20]

Using probability theory, investigators found that they could discuss the variations between measurements in more precise terms and could distinguish between different types of variations in ways not possible before. That measurements were afflicted with errors was long known; what those errors were and how they behaved were questions probability helped address. Imperfections of the senses and instruments created regular quantitative variations in measurements; these were called *constant errors*, and they could be estimated or calculated and then analytically eliminated. One was they left with a set of measurements, not yet homogeneous. Probability calculus turned these remaining irregularities into an asset; for, as Bessel put it, probability calculus "teaches how one should determine the quality of observations from their differences."[21]

Beginning with Gauss, investigators dubbed those remaining irregularities the accidental errors of the measurements. Appearing both large and small, sometimes with a positive value and then with a negative one, accidental errors were unpredictable (although they exhibited a pattern in their values) and hence truly "accidental," a word that also suggested the investigator's ignorance, or incomplete knowledge, of them.[22] Ironically this area of uncertainty was the link between raw data and a conclusion, or even between cause and effect. With the method of least squares one could compute the value of these accidental errors. Least squares also guaranteed that the arithmetical average was the most probable result as well as the result that "in all probability" would have an error "much smaller than any individual error."[23] Least squares thus provided an "estimate of probability" that gave one an "approximation of truth"; observations that were cleansed, that were "as far as possible *error free*" enabled one "to *approximate* the desired truth."[24]

Approximating truth meant understanding truth's flip side, error. In his influential 1837 textbook on probability calculus, written for technicians in Berlin, Gotthilf Hagen explained that "the purpose of probability calculus in its application to observation is to derive a certain judgement *out of the tangle of . . . accidental deceptions*" that were present in the act of measuring. These deceptions were the errors of the experiment, which could, in part, be quantified by using probabilistic methods.[25] Probability calculus thus teased nature into giving the investigator the highest degree of certainty, but did so by conquering the unknown and the uncertain: error was the measure of truth. Hagen's telling image of the tangled web of nature's deceptions revealed that with probability, one focused attention—and a significant amount of effort—not on the enhanced clarity or refinement of measurements, but rather on what remained fuzzy or indistinct in them. The significant act in refined measuring became the reduction of error, the quantification of the unknown and the uncertain.

Least squares was thus first applied in the 1830s to areas where determinations of certainty were essential for consenting to belief or ascribing reliability. For Hagen, technology profitted from the method, and so he aimed his textbook at those who constructed bridges, buildings, machines, and water lines; who measured land; and who, under any circumstances, used a level or a tool equipped with a level.[26] Here one needed an understanding, in quantitative terms, of where the strength of a material ended and its weakness began; of the conditions under which a safe construction became unsafe; and of when a reliable structure became unreliable. Determining the threshold in each of these cases could scarcely be an exact enterprise when so many factors had to be taken into consideration. So even though theoretical formulas were an issue in each one of these—Hagen demonstrated, for instance, how different the formulas for hydraulic engineering were "before the principle of least squares was developed"—they were not intended to enhance the precision of one's final results; for in each of these cases somewhat imprecise measurements were sufficient for achieving the same goals. Instead, in these applications, least squares was used to certify the epistemological value of measurements; its practical benefits were, at this time, largely an illusion.

In the natural sciences, least squares appeared in chemistry, physics, mineralogy, and physiology in the 1830s and 1840s.[27] Here the use of least squares—in chemistry to the determination of atomic weights, in physics to constants and relations in the equations of French mathematical physics, for instance—was intended to instill confidence in theories and results that were new, controversial, or in the case of French mathematical physics, sometimes perceived as just too good to be true. What Muncke earlier had considered copious and laborious, Littrow and others now found unproblematic: the method of least squares reduced observations "with certainty and ease."[28] Encouraged by the extent of the method's use, Bessel believed himself "not to be in error if I assume that in a few years the first chapters of all textbooks for the sciences based on experience will be devoted to the application of probability calculus to the art of observation."[29] His assessment was at least partially on the mark. For a short while, at least, least squares was taught and used in the several seminars or laboratories for physics or chemistry, such as those of Gustav Magnus at Berlin, Robert Bunsen at Heidelberg, and Wilhelm Weber at Göttingen. But the production of precision measurements in Magnus's laboratory, and more still in Weber's, eventually depended more on the development of refined instruments than on least squares, which eventually played only a minor role in the assessment of measurements. The method persisted longer in Franz Neumann's physics seminar at Königsberg, Bessel's home institution, where there was an intense interest in least squares, especially at the end of the 1830s.[30] There precision was more strongly defined in terms of the reduction of error, but the approach could not be sustained as the century closed; for the overreliance on least squares and the overpowering influence of Bessel's belief that one should through analytical means "eliminate the apparatus from the result"[31] almost crippled scientific investigation by making error analysis the essence and purpose of exact experiment.[32]

The immediate effect of probability on the concept of precision in measurement, in the 1830s and early 1840s, was to begin the linguistic process of creating finer distinctions of meaning among terms used to describe the qualities of refined measurements. The word "*Präzision*" entered discussions about measurements as a technical term denoting a variable in Gauss's derivation of least squares: the probability w of an observation with an error f is $w = (g/\sqrt{\pi})\cdot e^{-(gf)^2}$, where e is the natural logarithm and g is the measurement's "*Genauigkeit*" or "*Präzision.*"[33] The term g was also viewed as the "value [*Werth*] or weight [*Gewicht*] of the corresponding observation," as well as the "*Maass der Präzision.*"[34] The rule of the arithmetical mean rested "on the assumption that all observations were of equal quality, or that the value of g was the same for all observations."[35] The value of g, however, was rarely (if ever) computed in practice. Moreover, the term "*Präzision*" was seldom used in discussions about measurement before 1850, and when it was, it did not refer to this constant in Gauss's equation.[36] Although the term "*Präzision*"—derived from the French word *précision*—entered German discussions on measurement by way of the method of least squares, only gradually over the course of the century did it become customary to use it to denote extremely refined measurements obtained by instrumental means. Instead, older terms continued to be used in describing the quality of measurements.

The first term was "*Schärfe*" (sharpness) or sometimes "*Schärfung*," a term closely tied to the protocol and mechanics of an experiment. For Muncke in 1825, it was "useful and necessary" to know the "possible sharpness of the observations which can be attained by the naked senses compared to those which instruments can yield." One could determine this sharpness, Muncke explained, if one knew the quality of the instruments used as well as their constant errors.[37] This determination was, in Muncke's case, rarely rigorously analytical in the quantitative sense, but relied instead upon accurate *verbal* descriptions or *visual* depictions of instruments. This was especially important, Muncke advised, when the instruments "did not belong to the generally used and common ones." Such descriptions were supposed to inspire "confidence" in the sharpness of the observer's measurements through the authoritative sense of mastery they conveyed, but Muncke knew that they might be insufficient for attaining that goal. So he recommended invoking other authorities, especially the manufacturer of new or unfamiliar instruments. Finally, if by these means "the observer has not sufficiently justified [his measurements and trials] before the public," then the method and protocol of the experiment, especially how the observations were taken, also had to be reported.[38] Such descriptions were not intended to instruct in experimental practice, but to inspire confidence in the measurements presented. Fourteen years later, Jahn, mercilessly plagerizing parts of Muncke's essay, urged that "an accurate description and illustration of the instruments, as well as the observational protocol, should not be lacking" in one's presentation.[39] The "*Schärfe*" of measurements was thus not automatically recognized and acknowledged; was dependent upon knowing the constant errors of the experiment; and tested the observer's ability to persuade through verbal and other means.

But even after one made a determination of sharpness, one still did not have observations "in all [their] sharpness"; for to achieve this goal, "the number of observations [had to] be infinite."[40] It was in conjunction with this impossible task that a second term, "*Genauigkeit*," attained its meaning. "*Genauigkeit*" was a troublesome term because it was used in so many different ways. It could describe a quality of laws based on observations (Paucker thought that least squares helped one ensure that laws had "the same accuracy" as the observations on which they were based[41]); of the protocol of an experiment (according to Hagen, "the observer has to try to give each operation an equal degree of accuracy"[42]); or finally of the measurements themselves. No matter how the term was used, though, the "multiplication of observation" was "the unconditional requirement for the highest possible accuracy."[43]

But how many times should observation be multiplied? When did one know to stop taking measurements? Probability theory helped solve this problem and in so doing, narrowed the meaning of "*Genauigkeit*." Early commentators acknowledged that one did not in fact have "absolutely" or "completely" accurate observations, but never explained how or why.[44] In 1837 Hagen made "*Genauigkeit*" a probabilistic concept by claiming that ascertaining accuracy was like placing a bet: one had odds, but not an exact value.[45] Littrow in 1842 tied the definition of "*Genauigkeit*" to least squares determinations and identified "*absolute Genauigkeit*" as equal to a probability of one.[46] The notion of "tolerable accuracy" ("*ziemliche Genauigkeit*") thus was taken to be the point at which enough observations had been taken to be able to justify using least squares.[47] Of no small consequence in this process of linguistic transformation was the opportunity it offered for deciding what agreement between numbers meant. Only when errors were computed could one recognize agreement: according to Muncke, "the individual results corrected for errors must be so much more accurate, the more they agree with one another."[48] In this entire matrix of linguistic changes, then, measurement attained meaning through the determination of error, and degrees of certainty became matters of error.

Probability calculus thus seemed to provide a reasonable guideline for striking a happy medium between taking a few measurements and an infinite number. Yet being content with just enough observations in order to use least squares could nonetheless be unsettling when one knew that the accuracy of one's results inevitably increased as more and more measurements were taken. For early nineteenth-century measurers, the dilemma posed by choosing between what was epistemologically desirable, greater certainty, and what was practically possible, which excluded taking an infinite number of measurements, was real, and demanded resolution. It was not, however, a problem that could be solved easily, for one was always trying to satisfy irreconcilable demands, seeking certainty in both knowing and doing. Hence, these early measurers scrutinized and tried to certify not only results of measuring, but also the measuring investigator's conduct.

For instance, Muncke's discussion of the sharpness of measurements included suggestions on how exact experiment was supposed to be conducted, especially how the investigator should report his findings. But he went further. He also thought that the investigator's character and demeanor in the course of the experiment warranted review; for it was "not merely the quality of the instrument, how it is used, [and] the method of calculation," that was significant in the conduct of experiment, but also "the sharpness of the senses, the practice and patience of the observer or experimenter, even his state of mind, his perceived ideas and impulsiveness, indeed even accidents that can scarcely be determined. It is thus easy to see that the task of determining the probable accuracy of results obtained through observations and trials belongs to the most difficult, and takes into consideration the assistance of the shrewdest psychology and the deepest calculus."[49] Muncke's description is too imprecise to tell what kind of observer he had in mind, but it is enough to identify two important features of early precision measuring. First, all aspects of the conduct of experiment had to be exposed when dealing with measurements making a claim to accuracy. Second, the authority of measurements was not simply to be found in technologies—be they instruments or the quantitative technology of least squares—but some authority was located in the personal qualities of the observer.

This bifurcated source of authority was in fact already contained in the entire matrix of linguistic changes that had accompanied the application of probability to measurement; for here the authority of measurement was not just a matter of error, but also of the observer's skill in convincingly rendering the conditions under which measurements had been taken. The early date of Muncke's writing on least squares—1825—and his classification by Kenneth Caneva as a "concretizing" scientist—as one not especially receptive to quantitative techniques—may suggest that his views on the authority of observations were outdated and old-fashioned.[50] But his discussions of least squares (interestingly, not discussed by Caneva), cautious as they were, placed him squarely in the camp of those who were trying to use the complex techniques of probability calculus to certify the refined results of exact experiment. Moveover, Muncke's views on the necessity of taking into consideration the qualities (and hence the authority) of the observer were shared by other more astute commentators following him.

Gotthilf Hagen, whose textbook on probability calculus was a landmark in the early depiction of exact experiment, emphasized how careful conduct in measuring made it unnecessary and even futile to hope that increasing the number of measurements may eventually lead to the "correct" measurement. "This hope," he wrote,

> proves itself to be entirely erroneous; for the agreement was likely produced only by an accidental coincidence of circumstances, or more likely through foreign influences, which in a later trial no longer occur. In order therefore to have complete certainty, or rather the necessary degree of probability, for the correctness of measurements, one must accurately test the entire method in all its parts, and one must make the results independent of such foreign influences which introduce in all trials almost the same error. Then one must examine whether or not the other unavoidable errors in the method could accumulate easily to a significant error in the result. It is especially important to know what certainty one has, so that the accumulation of error does not occur.[51]

Although it was probability calculus that for Hagen eliminated the need for an infinite number of observations by its quantification of the last remaining residue of error and showed one how to "use other controls to rectify the result,"[52] the course of an experiment's protocol was determined more by how one was supposed to deal with *constant* errors.

In Hagen's view it was the responsibility of "every observer and every mechanic" to diminish constant errors beforehand. For the successful execution of an experiment, "the skillful observer will never neglect, in all measurements that he performs, to undertake those tests and controls whereby he can convince himself that his apparatus has the necessary degree of correctness, or he will take the trouble to determine the value of the correction that he must introduce in order to remove the influence of the constant errors." Because these errors were not accidental but were "grounded in the instrument itself," Hagen recommended that a part of one's protocol include "special tests"—actually small, auxiliary experiments—through which one could determine the value of these errors and then modify the instrument or the environment of the experiment to compensate for them. This did not necessarily mean *perfecting* the instrument materially; for "the means which are at the disposal of the artisan in the construction of the instrument will as a rule exceed the sharpness of those which one can apply in a test of [the instrument's] soundness for one's own use." Yet even if it were the case that higher degrees of material perfection could be achieved, Hagen warned that "every instrument, even the best, still has errors even in the hands of the skilled observer." The point of this concern for constant errors was to make "the calculations involved in the reduction of observations small in comparison to the care one takes in setting up the trial." So although the analytic reduction of error was essential for judging "the certainty of the work," Hagen nonetheless seemed to have viewed the authority of experimental results as dependent, to a greater extent than did Muncke, on method, protocol, and skill. But like Muncke, Hagen believed that not everyone was reliable enough to carry out this protocol of error reduction as it should be done.[53] One could thus try to build a consensus around refined measurements, but only with the right people.

Like the infinity of measurements that an ideal accuracy seemed to demand, even this protocol of error reduction could prove endless, because there was "in fact no limit in tracing back to all individual conditions which exert an influence on the environment [of the experiment] and even on the observer." But Hagen considered it "an entirely useless waste of time and energy" to track down every last iota of error or to bother to achieve "a greater sharpness" in one part of the measurement than could be attained in any other. And, insofar as accidental errors were concerned, Hagen argued that one could stop taking measurements if the result were found to lie within the limits of the allowable error; at that point "one will not want to repeat the operation." For the case of constant errors, there was another condition that forced the conduct of the experiment into reasonable and manageable bounds: cost. Even though there were "great difficulties in trials which can scarcely be set aside," Hagen warned, "the expense of these will frighten most observers away."

An excessive concern with the elimination of constant errors (if, cost aside, such could be done) also had in Hagen's view the disadvantage that "the simplicity of the phenomena usually entirely disappears and one is no longer in the position of estimating with certainty the true value of the individual variables which one must know. The results of such trials therefore are usually entirely useless." Hagen's exact experiment, with its pragmatic and practical approach which relied on the good economic sense of the experimenter and the probabilities of least squares, was not a means of discovery, but a means of approximating nature's truth more finely with accurate measurements, but doing so within the bounds of available resources.[54]

The epistemological implications of using probability theory in judging the certainty of refined measurements were not easy to explain. It was of course a simple matter to recognize, as did J.A.W. Roeber, *Oberlehrer* at the Berlin *Gewerbeschule* in 1842, that producing only a small number of observations meant that one scarcely knew enough, especially about their errors, to know what kind of conclusions could be based on them.[55] Other implications were not so simple. Hagen and other commentators on measurement from Paucker onward viewed nature, the construction of knowledge, causality, truth, and certainty from a perspective in which were united precision and probability, a quantitatively fine-grained view of nature with approximate certainty. For Hagen that meant that although nature could be measured to finer and finer degrees, "we are not in a position to comprehend phenomena directly and purely," nor "are we able to be completely convinced of the correctness of our judgment."[56] Finding this inexactitude difficult to explain, Hagen, like others following him,[57] resorted to a story. Presented with different reconstructions of events by witnesses, it was difficult, he admitted, "to give an account of all possible deceptions and their influences, and after that to determine how large the probability is that the event actually happened just so compared to how one believed it to have happened." Ending his story with the remark that "this subject of the judgment of the witness's reliability is of great importance,"[58] he then turned to how a surveyor's measurements could attain greater accuracy, and so certainty, through probability. Like the storyteller, the surveyor was a witness whose reliability could only be judged by how well he managed deceptions and cordoned them off—that is, by how well he quantified error.

What was not completely absorbed by those who learned to trust precise measurements by using least squares was the doctrine of chance in probability. Measurements and their errors were subject to the laws of chance, to be sure: a measurement had one probability that it would be one value, another probability that it would be a different value. The same held for accidental errors. But it was not necessarily measurement per se that was the primary epistemological focus of those who used probability with precision measurement in the 1830s and 1840s. A far more pressing concern was the status of the laws of nature; for generalizations about nature, these commentators felt, had gotten out of hand with speculation and hypotheses.[59] Matters could be controlled somewhat, they believed, by using the method of least squares.[60] As Paucker had explained, and illustrated through his examination of Biot's work, least squares was invoked to guarantee that laws had "the same accuracy" as the observations on which they were based.[61] Still, laws were only "approximate"; they expressed "the most probable theory."[62] Only with the method of least squares, Hagen argued, could "one will be in a position to judge how, under somewhat changed conditions, the derived law changes, and with what certainty one may count on it happening that way."[63]

Because these commentators were more interested in degrees of certainty rather than in chance, and were convinced that nature's laws had limited, not absolute, certainty, it is not surprising that they neither discussed determinism nor held up mechanics as the conceptual model for the construction of knowledge based on precise measurements, even though almost all worked in the physical sciences. So although Roeber believed that through the method of least squares the other branches of science "more and more attained the authority of mechanics,"[64] he merely meant that probability theory helped one to attain a confidence in the conclusions and results based on accurate observation, similar to the confidence one had in mechanics. The method of least squares helped one to identify causes, which were "all the more certain with every increase in probability."[65] Assisted by least squares, precision measurements could indeed reveal the ordinary course of nature at finer levels, but what turned out to be "ordinary" could not be known with complete certainty.

Through probability theory, German investigators differentiated quantitative components of measurement; began the linguistic process of creating finer distinctions of meaning between the terms used to describe the qualities of refined measurements; helped create standards of conduct of exact experimental practice; and helped define a particular epistemological approach to the study of nature. The path to truth with refined measurements reduced by least squares was thus strewn with asperities and patches of darkness shrouding the unknown. It was also neither straight nor finite; in the end one had to be satisfied with not traveling its full course, with an answer that was only approximately true.[66] Probability calculus gave meaning to refined measurements by focusing on errors, and hence on the unknown and uncertain. Meaning having been established, discussions could take place and agreement reached. It was not exactness per se that was at stake in these discussions, but rather consensus about how accuracy and sharpness could be talked about and judged. Probability calculus thus helped in defining the exact sensibility characterizing the first community of practitioners in the German states who concerned themselves with refined measurements before 1850.

Standards Reform and Integrity

German investigators used probability theory to evaluate the trustworthiness of refined measurements. Along the way, they also began to articulate standards for the conduct of experiment, and hence for evaluating the trustworthiness of the observer. What proved most influential in shaping the ethics of conduct in precise measuring, however, was state-initiated weights and measures reforms. Authority in metrological matters had been a symbol of sovereignty since medieval times. By the end of the eighteenth century, states had set up elaborate procedures, codified in edicts, for assuring both the regularity of measures and the proper conduct in measuring. A Prussian edict from 1796 on the measurement of grain, for instance, specified both the mechanical means for measuring grain (the container had to be "perfectly circular," made of wood with iron hoops going around to prevent the deformation of the container) and the means for filling and "topping off" the vessel (that is, how the grain would be heaped or striked).[67] Any deviation from that customary behavior or infraction of the rules governing weighing was fraudulent and dishonest, and so punishable under the statutes of the law.

After 1815 a widespread reform of weights and measures in the Germanies was initiated due to the postwar retention of the metric system in several western German states and independent towns, where it had been imposed by Napoleon, coupled with the need to manage and control finances more carefully, especially in the context of trade relations. In Prussia following the Napoleonic Wars, a new financial system, administered by reformed bureaucracies, was created. The formation of a state budget and the collection and management of revenue—over half the state's revenue was from taxes and trade tariffs—became the bureaucracy's most important tasks, as well as its instruments for controlling social processes.[68]

Societies differ in how they police standards, but their objective is always the same: honesty. For the American standards reformer of the 1830s, Ferdinand Hassler, standards were "among the means of distributing justice in a country."[69] For the American Secretary of the Treasury, Louis McLane, discrepancies in standards were "a serious evil."[70] For the British commentator Patrick Kelly, weights and measures "have been very properly defined the foundation of justice. . . . Even the laws of honour particularly abhor any fraud in this respect, as a base violation of confidence."[71] Weights and measures regulations such as Prussia's in 1816 and Baden's in 1829 set up elaborate bureaucratic organizational lines for the honest testing and preservation of standards.[72]

Such invectives and protections against fraud and dishonesty in weights and measures were all the more important to German scientific practitioners in the early nineteenth century when it became necessary to certify the quality of the standards being used. German investigators began their own examinations into the inaccuracies in standard weights and measures, thus challenging prevailing customs in standards from positions outside the state's bureaucracies. Wilhelm Weber, writing in 1830 to the Prussian ministry of arts and trade from Halle, where he was professor of physics, bemoaned the inaccuracies in, and lack of agreement between, various standards. This was a pressing issue, he explained, because he was involved in an investigation that he claimed "demanded great accuracy in the determination of weight, and where not only comparisons of weight, but also the absolute determinations of weight are of importance."[73] Over a decade later Roeber still found it necessary to remind his audience that natural phenomena were represented in space and time, and so "the accurate conception of measure and number is absolutely necessary."[74]

In 1835 the Berlin physicist Heinrich Wilhelm Dove, in his classic essay *Ueber Maaß und Messen*, condemned the deplorable condition of standards in the Germanies. He blamed both standards determinations and the inadequacies of exact experimentation for the problems. "Small changes which conventional measures have undergone in revisions" as well as "the unequal sharpness of the means of observation," he argued, have produced "uncertainties" which made it "difficult to relate one measure to another," complicating not only trade, but also scientific work.[75] Although Dove, like some of his predecessors, had once believed that the quality of instruments guaranteed the quality of the measurements produced, he felt he could no longer rely on faith in instrumentation alone to improve measurement. His reviews of various metrological techniques and protocols and of the technology of instrumentation were excellent portrayals of the state of the art of measurement in physics.[76] But protocols and technologies were inadequate, Dove believed, for rectifying standards. Instead, to attain higher levels of certainty and accuracy, Dove turned elsewhere. He explained that his own examination of the present condition of standards demonstrated

> that through ingenious arrangements and combinations one can succeed in carrying the accuracy of measurements a great deal further than previously imaginable. But the most accurately set up measurement does not give the

> true spatial extension, the true interval of time, or the true weight. We could have found these accidently, but could not have proven that they were [true]. Because it is the case that through the highest possible sharpening of the senses we can make only the difference between the standard of measure and what is measured smaller [but not completely eliminate it], it follows that every observation is to be viewed as inflicted with error. . . . The probability of a determination derived out of many observations must therefore be the result of a calculation which teaches one to derive out of patently incorrect observations not the correct result—for that is impossible—but one which is more probable than any single observation is. The art of observation owes its high perfection in recent times more to the development of the mathematical methods on which it rests than to the technical perfection of the means of observation.[77]

So he directed his readers to works on the method of least squares, including Gauss's and Encke's, which he had not used in his own presentation, even though he could have when he carried out the reduction of one standard measure to another to a dizzying twelve to eighteen decimal places without indicating if all the digits were useful.[78]

At the time of Dove's writing, responsibility for establishing more authoritative and accurate weights and measures belonged to the Göttingen astronomer Carl Friedrich Gauss, in Hanover, and to the Königsberg astronomer Friedrich Wilhelm Bessel, in Prussia. Through their reforms the moral qualities of standards determinations—especially the ever-present concern for honesty—migrated to exact measuring practices, thus adding another dimension to the exact sensibility. But their involvement with the state in areas of exact measure also demonstrated the persistence of traditional social values regarding the authentication and validation of forms of knowledge; that is to say, their cases demonstrate that what still mattered in knowledge production were personal qualities, sometimes to the exclusion of obvious objective skills.

Take the case of Gauss in Hanover. When in 1836 the Hanoverian ministry wanted "to diminish the uncertainty" in the Hanoverian pound and relate it more exactly to measures used in other states, it turned to Gauss, who was interested but apparently not enthusiastic about the extra work. There were, after all, the bureaucracies and administrative details to tend. Gauss read untold numbers of acts and made contact with the relevant government authorities in Göttingen and Hanover.[79] Not one to be easily diverted from his astronomy and mathematics, Gauss tried in 1836 to reassign the tedious work of double weighing to his mechanic, Meyerstein. Hanoverian officials immediately objected, revealing their sense of how a standard's authenticity and authority were not merely matters of the quality of the precise measurements that set its value (no matter who took those measurements), but were invested partly in the authority of the person either making or supervising them. Gauss should indeed do the testing, the ministry urged, but "every measurement" done by Meyerstein had to be "conducted under the supervision of Hofrath Gauss."[80]

When Gauss's academic work, by his own account, again became too pressing three years later, in 1839, and he wanted to turn the standards work over to Meyerstein again, the ministry this time allowed the mechanic "to be responsible for the correction" it sought, but insisted that the *Landesnormalgewicht*, which was for the ministry's official depository, "be justified and standardized directly by Hofrath Gauss himself."[81] Thus, from the government's point of view, precise measurements could not be fully trusted unless they were certified by someone who himself embodied a form of authority. Moreover, the mechanic's measurements would not be trusted at all in the determination of the state's standard, a symbol of its sovereignty. The German states' creation in the nineteenth century of a *Beamtenstaat*—where authority resided in the office—notwithstanding, aspects of the earlier traditional *Herrschaft*—where authority was vested in a person—lived on.[82]

Gauss, in fact, called the state's bluff on this issue of the certification of precise measurements. If the state would not accept Meyerstein's work, then how could it allow individuals totally unskilled in the techniques of precise measuring to make customs determinations? To be sure, Gauss argued, a customs agent did not have to know the delicate procedures involved in determinating how a body lost weight in air.[83] But, he continued, the agent did need "to apply great diligence and accuracy" to customs determinations. Gauss's suggestion was direct: henceforth in customs determinations "only such persons will be trusted who possess enough skill to carry [the determinations] out in a way corresponding to the current standpoint of one of the technical sciences, and from which one may expect that they proceed with required accuracy and carefulness."[84] And how, according to Gauss, would one know in the end how accurate a customs agent was? By looking at how he managed the errors of his measurements. Knowledge of these errors—an admission of the agent's shortcomings in measurement—was an outward sign of his integrity.

When Gauss reported on the determination of another Hanoverian unit, the foot, in 1841, he addressed issues central to the normalization and dissemination of standards of measure: how they could be reproduced and compared. Reproduction and comparison were important in scientific and social uses of standards, and they required for their solution the techniques of error analysis, both of constant and accidental errors, that had been developed and applied to precision measurements in the preceding two decades. Thinking of a standard's practical use, Gauss argued that a standard unit of length had to have a degree of accuracy such that a skilled artisan could copy it within the limits of time and resources (at a certain unspecified, minimum level) available for the task. He admitted, however, that mathematical exactitude was not the point; for "the accurate knowledge of the *value of the difference* remaining [between the standard and the copy] is just as good as the perfect equivalence [of the standard and the copy]."[85] The integrity of the copy, and hence of the means of reproduction, Gauss thus seemed to say, was not found in the exactitude of the copy itself, but rather in the creator's (or the user's) knowledge of the copy's errors and its limits of accuracy. Standards thus most easily spread not by their perfect reproduction but rather by an agreement on the means (the errors) by which they could be compared. The reproduction of standards of measure could thus be viewed as a ritualistic practice for disseminating the principles of exact experimentation which incorporated error analysis.

Bessel's experience with Prussian standards reform was a bit more complicated than Gauss's in Hanover. But here, too, the supervisory practices and moral implications of standards determinations became those of precise measuring as well. Although Prussian weights and measures reforms were part of the early discussions on tariff reform after 1815, the completion of that reform in 1839 came too late to be of significance in the establishment of the *Zollverein* in 1834. A more accurate determination of the standard of length did serve, however, to strengthen the tariff policy of the *Zollverein*, which had adopted Prussia's practice, instituted in the 1820s, of charging duties on the basis of the weight or size of an item, rather than its value.[86] Weights, measures, and coinage remained diverse throughout the *Zollverein*, but the various states' continued efforts to improve upon the accuracy of locally determined weights and measures made commercial transactions more exact by enabling more accurate conversions. It was the combination of weights and measures reform with precise measuring practices, not the uniform distribution of standards, that aided Prussia in achieving financial solvency and administrative consolidation in customs and trade.[87] So strategic, in fact, was weights and measures reform to the national well-being of Prussia in the 1830s that Berlin philologists studied ancient systems of weights and measures in earnest, viewing them as the key to the dreams and fantasies of a bygone age.[88]

By the time the Prussian government commissioned Bessel to redetermine the state's unit of length in 1833, the accuracy of Prussian standards had long been in dispute. In 1816 the state had admitted that there was an "uncertainty in measures and weights" that "impeded" trade and that had to be remedied "through firm determinations" of standards. Three copies of a new standard were to be produced. The first, the "single authoritative original," was to be deposited in the Ministry of Finance and Trade; the second, a "believable exemplar," in the *Oberbaudeputation*; and the third in the mathematical class of the Berlin *Akademie der Wissenschaften*, where the copy's "mathematically accurate correctness" would be maintained "for all subsequent time." The copies were supposed to be competently compared and calibrated with a protocol made known through publication. These redetermined standards were to serve trade, construction, factories, mining, triangulation, stonemasonry, carpenters, liquid measures, and wood fuel, as well as scientific investigation.[89]

Also in 1816, for the first time in its history, the Prussian government commissioned members of the *Akademie der Wissenschaften* to participate in the reform of weights and measures. Although at first glance the union of science and the state appears praiseworthy, the project in fact floundered for almost two decades, leaving little behind but confusion as to whether science could help the reform effort at all. For the standard of length, the foundation of all other Prussian measures, state officials from the Ministry of Arts and Trade asked J. G. Tralles to perform investigations of the seconds pendulum in Berlin; he died in 1822, his metrology left undone. In 1825 and 1826 Tralles's colleague in the academy, Johann Eytelwein, wrote about how one might achieve "mathematically accurate correctness" in standards; but his trials were not clear, he gave no rigorous determination of errors and only rough ideas of accuracy, and based his calculations on some of Biot's measurements for the specific gravity of water, which were exactly the ones Paucker had earlier questioned. It was of no matter: accuracy was not his goal. "It may not remain unnoticed," he concluded, that the difference between the prescribed new measures of length and weight and the old ones is of "no influence on trade."[90]

By the time Bessel was chosen for the project in 1833 he had already five years earlier completed extensive studies of the seconds pendulum, which was to be the foundation of Prussia's unit of length, the foot. His investigation became an exemplary model of exact experimental technique that far supassed earlier *physical* investigations (not only those on pendula) in its assessment of errors. Bessel gave a meticulous quantitative assessment of constant errors, most often by providing a theoretically based, analytical expression of their effects. The most important correction—the one that became emblematic of his investigation and its methods—was his hydrodynamical correction for the motion of the pendulum in air. Moreover, Bessel fervently used the method of least squares and considered it essential in all quantitative experiments. Although a master at instrument design and analysis, Bessel assigned priority to error and data analysis in experiment. In his experimental design, he focused not on determining the length of his pendulum, but on the difference in length of two pendula, thereby eliminating the need to consider certain errors and creating an apt template for future metrological determinations.[91]

Error and data analysis also shaped Bessel's views on the epistemological significance of precision measurement. Although aware that his results were precise, he was never so arrogant as to claim he had achieved *mathematical* precision or absolute truth. In his view knowledge based on observations, even precise ones, was always approximate. Greater precision (even that achieved by the calculation, reduction, or elimination of error) did not necessarily lead directly to higher degrees of epistemological certitude because there were "continual oscillations within the limits of unavoidable imperfections" as the number and precision of observations increased.[92]

Between 1835 and 1837, with the assistance of the Berlin mechanic Thomas Baumann, Bessel constructed an original standard of measure (*Urmaass*) for the Prussian foot. His determination was based on a series of exact experiments with detailed calculations of those errors "which could influence the desired fineness (*Feinheit*) of the measurements." To avoid what he believed to be errors in previous determinations Bessel chose not to measure between engraved lines, but to measure end to end—his bar was constructed of steel with sapphire endpoints—taking into account the temperature of the bar, which the 1816 regulation had not mentioned. He repeated seconds pendulum trials for Berlin, believing that the ultimate determination of the foot would be rendered "unambiguous." Neither time or accident could change the length of the Berlin seconds pendulum, he argued, because the value of gravity at a specific geographic location remained constant.[93] On 10 March 1839 his new determination of the Prussian foot became law, remaining in effect for almost thirty years.

Bessel later said that social considerations determine standards of measure.[94] He had in fact been concerned earlier with the impact of measures upon trade. Writing to Alexander von Humboldt in 1833 about making the Prussian and Danish foot comparable, he intimated that the project "appears to me, politically taken, to possess a certain importance." Commercial and other relations between the two countries, he believed, would become closer and more solid "if we are able to communicate our weights and measures."[95] As he progressed with his investigation, however, trade turned out not to be the point; it was irrelevant especially to the issues of precision and accuracy.[96] If, in the end, there was a social connection, it was one tied to the conduct of scientific practice; for Bessel was interested in the *protocols* that should be followed in the reproduction and comparison of standards.

Bessel believed that an absolutely perfect standard could never be constructed. One had to place one's faith in qualities of the standard other than its perfection, such as its reproducibility. He maintained that reproduction of the standard was an "essential part of the regulation of a system of standards" no less important "than the determination of the unit itself." But the present state of reproduction had "left something to be desired."[97] When he finished his investigation, the reproducibility of a standard had become, in his view, *the* central issue in standards reform. An original standard of length, he argued,

> had to be bound to regulations which guarantee the reproducibility of its length. In this way, and not through the existence of an inaccessible *Urmaass*, will a standard's purpose be achieved. The more accessible the *Urmaass* can be made without any danger of alteration, the better the standard will reach this goal. When the means standing at one's disposal for the reproduction of its length can be applied easily, attaining the highest accuracy which can be demanded in the present state of practical mechanics, the standard will best achieve its goal.[98]

Thus, the *Urmaass* was, in Bessel's view, not a mathematical ideal, but the material expression of the protocols by which it could be duplicated. A standard was thus not merely a measure, but a set of reproducible actions. How good was a copy then? Like Gauss, Bessel believed that one assessed the goodness of a copy on the basis of the creator's knowledge of the copy's errors and of how much the copy deviated from the *Urmaass*. One knew from a review of the protocol that had been followed in the construction of the copy, later written down and publicly reported, how well the investigator had adhered to the regulations governing reproduction.[99] It was this aspect of standards determination that, in Bessel's view, gave meaning to precise measurements and made those measurements trustworthy.

It is easy to look upon Bessel's determination of the Prussian foot as an example of how science, in the form of exact experiment, improved the precision of weights and measures standards. But it is also the case that standards contributed something to the meaning of precision in exact experiment. At the most basic level, standards gave meaning to measurements and made measurements trustworthy; as Bessel wrote, "The value of a measurement exists only so long as the standard, on which it is based, is retained; and conversely, the preservation of a standard has value only through the measurements whose weight and meaning (*Gewicht und Be-*

deutung) depend on it."[100] Investigators shared standards and so could use them to judge each other's findings; standards thus enhanced the reliability of an investigator's work, especially where precise results were concerned. Second, the particular protocol that Bessel had followed in the determination of the Prussian foot, and that he suggested be used in its reproduction, emphasized error analysis as the means to precise determinations. Here, no less than in the case of least squares, publicizing error meant that the investigator had not deceived, but was scrupulously honest. Finally, standards enhanced the qualities of the experimental investigator by underscoring the essentially *public* nature of methods used in precise determinations: they were public in the sense that they had to be reported so as to be judged, and if necessary imitated; and they were public in the sense that they were subject to the policing and surveillance of outside agencies responsible for certifying their authenticity. This, too, enhanced the trustworthiness of precise results: Prussian officials believed that Bessel's report "guarantees to the scholarly world the precision of the determination."[101]

Of the many ways in which the construction of standards by state authorities are central to the history of precision measurement, linking that history in the eighteenth and nineteenth centuries to social concerns, it is the policing and surveillance aspects of standards, the ways in which their construction set up codes of behavior (i.e., the customary practices of measurement) and the means for judging that behavior, that are important here. Just as traditional standards of measure were rooted in the realities of life and labor (in the Germanies, for instance, the principal unit of land measure, the *Morgenland*, was that which was capable of being tilled in a day), so nineteenth-century standards had a social meaning—not in the labor of the field, but in the labor of precise measurement. Through its emphasis on honesty, reliability, and openness, standards determination was a cultural source of the categories used in judging the integrity of the precision measurer.[102]

The importance of integrity in standards determinations and exact experimental practice compels us to reconsider resistance to changes in standards. Prussian objections to the introduction of the metric system in the late 1860s seem not to have been expressions of nationalist sentiments or a general unwillingness to change, but rather a sense that no other experimenter had matched Bessel's integrity in reporting his method or in certifying the foot's degree of exactitude.[103] Over the course of the century, precision measurement remained tied to standards determinations, which became increasingly complex as standards multiplied to include new units, especially in electricity.[104] The guidelines for judging the investigator's integrity would, however, change.

Conclusion: The Meaning and Values of Precision

Precision measurement acquired culturally dependent sets of values over the course of the nineteenth century. In Germany, physics professors and secondary-school teachers used precision measurement to discipline laboratory training; here the work ethic associated with precision was used to justify the pedagogical worthwhileness of practical exercises.[105] It was not, however, just any honesty or integrity or disciplined work ethic signified in precision, but a decidedly *masculine* variety of honesty, integrity, and work. Masculine metaphors abounded in German discussions of precision measurement, as one might have expected from the ties between precision measurement and the bureaucracy and military. In 1881, Emil Warburg, then director of Freiburg's physical institute, noted: "There may scarcely be found a better means for training an earnest, masculine scientific character than these exercises."[106] About the eradication of error that was a part of the production of precise measurements, Woldemar Voigt wrote in allegorical terms about a fairy tale in which a prince struggled to reach and capture a king's daughter ensconced in a magical but treacherous castle.[107] Precision measurement, like most other forms of quantification and even science in general, thus helped cement nineteenth-century conceptions of gender and the values associated with them: the masculine was rational, disciplined, machinelike in action, while the feminine was hidden, in need of protection, a prize to be sought but not necessarily achieved.

If we were to move outside the German context, we would find other values of precision in different cultures, as the articles in the volume amply testify. In mid-nineteenth-century England, precision measurement was associated with the liberal values of accurate and rational reasoning and with the habits of exact thinking.[108] By the end of the century in America, precision measurement shed its morally uplifting disciplined work ethic and instead became associated with a machinelike industrial efficiency that cared not for the toils of *human* labor.[109] More recently and closer to home, precision measurement, especially its purported and easily manipulated accuracy, became infused with political values in the nuclear arms race.[110] The values of precision are thus many.

But before refined measurements and the activities involved in producing them could acquire these and other culturally dependent values, they had to be assigned a meaning or signification that differentiated them from other forms of measurement and even from other forms of quantification. In the German states, especially Prussia, that process of differentiation occurred largely through the application of probability calculus to measurement and through the reform of weights and measures, in which precise measurers became involved. The exact sensibility that took shape as a result focused on error and so, ironically, on the imprecise components of measurement. Here, in this area of uncertainty and free play, trustworthiness and truth-value were assigned to refined measurements and integrity to the measurer who had produced them.[111] This exactsensibility placed premiums on certain forms of conduct: the proper mental habits for certifying the authority of measurements, the proper protocol and attitudes for conducting exact experiments, the proper conductin standardization. Proper conduct and inescapable errors, not exact numerical results per se, shaped the early meaning of precision measurement.

Later meanings of precision would be different. So, too, would the values of precision. The habits of thinking associated with exact experiment and its analysis of errors, especially least squares, and the affective judgments concerning investigators and how well they could quantify error, both became out of place in a scientific world that placed its faith, once again, in the ability of instruments to produce precision measurements not in need of the "cleansing" of error analysis. By the end of the century, the method of least squares looked horribly old-fashioned, and the integrity of the experimentalist in physics could scarcely be assessed on the basis of the production of precise measurements when instruments did so much without the help of human hands.[112] The gap between precision and integrity continued to widen. In today's world where fabricated precise data seems to pass effortlessly under the eyes of referees and editors, one wonders if a return to past sensibilities may not enrich those gone awry in the present.

The author gratefully acknowledges the support of the National Science Foundation (DIR-9023476) and the National Endowment of the Humanities (RH-21005-91).

Notes to Chapter Five

1. David Cahan, *An Institute for an Empire: The Physikalisch-Technische Reichsanstalt, 1871–1918* (Cambridge, 1989), 5.

2. Emil Warburg, "Verhältnis der Präzisionsmessungen zu den allgemeinen Zielen der Physik, in *Die Kultur der Gegenwart*, ed. P. Hinneburg, vol. 3, pt. 3, nr. 1: *Physik*, ed. E. Warburg (Leipzig, 1915), 653–60, on 655.

3. Witold Kula, *Measures and Men*, trans. R. Szreter (Princeton, 1986), 9.

4. On sensibilities, mental habits, and mental equipment, see Keith Thomas, *Man and the Natural World: A History of the Modern Sensibility* (New York, 1983), 15, 16; Erwin Panofsky, *Gothic Architecture and Scholasticism* (Latrobe, Pa., 1951), 20–21, 27; Lucien Febvre, *The Problem of Unbelief in the Sixteenth*

Century: The Religion of Rabelais, trans. Beatrice Gottlieb (Cambridge, Mass., 1982), 150; Roger Chartier, "Intellectual History and the History of *Mentalités*: A Dual Re-evaluation," in Roger Chartier, *Cultural History: Between Practices and Representations*, trans. L. G. Cochrane (Ithaca, N.Y., 1988), 19–52, esp. 24–25. For an interesting example of the sensibilities associated with weights and measures, see Kula, 13–28.

5. Jan Golinski, "'The Nicety of Experiment': Precision of Measurement and Precision of Reasoning in Late Eighteenth-Century Chemistry," this volume.

6. Maurice Daumas, "Precision of Measurement and Physical and Chemical Research in the Eighteenth Century," in *Scientific Change: Historical Studies in the Intellectual, Social, and Technical Conditions for Scientific Discovery and Technical Invention, from Antiquity to the Present*, ed. A. C. Crombie (New York, 1961), 418–30.

7. *The Quantifying Spirit in the Eighteenth Century*, ed. Tore Frängsmyr, J. L. Heilbron, and Robin E. Rider (Berkeley, 1990), especially the articles by Henry Lowood, John Heilbron, Theodore Feldman, and Sven Widmalm; Zeno Swijtink, "The Objectification of Observation: Measurement and Statistical Methods in the Nineteenth Century," in *The Probabilistic Revolution*, 2 vols. (Cambridge, Mass., 1987), 1:261–86.

8. Quoted in John Heilbron, "The Measure of Enlightenment," in *The Quantifying Spirit*, 207–42, on 217, 225.

9. For examples, see Heilbron, 219.

10. Henry Lowood, esp. 328, 329 (quote), 332, 335.

11. Sven Widmalm, "Accuracy, Rhetoric, and Technology: The Paris-Greenwich Triangulation, 1784–88," in *The Quantifying Spirit*, 179–206, esp. 197, 198.

12. Lundgren and Heilbron argue in *The Quantifying Spirit* that not all found absolute truth desirable. Still, some eighteenth-century quantifiers sought a mathematical-like perfection so as to be closer to the ideal of absolute truth; see, for instance, the article by Lowood in *The Quantifying Spirit*. Anders Lundgren, "The Changing Role of Numbers in 18th-Century Chemistry," *The Quantifying Spirit*, 245–66, on 262; Heilbron, 5.

13. Adrien-Marie Legendre, *Nouvelles méthodes pour la détermination des orbites des comètes* (Paris, 1805), esp. viii, 72–80; Carl Friedrich Gauss, "Determinatio orbitae observationibus quotenunque maxime satisfacientis," in *Theoria motus corporum coelestium* (Hamburg, 1809); Gauss, "Theoria combinationis observationum erroribus minimis obnoxiae," [1821] in *Carl Friedrich Gauss Werke*, vol. 4: *Wahrscheinlichkeitsrechnung und Geometrie* (Göttingen, 1880), 1–108; Gauss, "Ueber die Genauigkeit der Beobachtung," *Zeitschrift für Astronomie und verwandte Wissenschaften* 1 (1816): 185–96.

14. Gauss, "Theoria combinationis observationum erroribus minimis obnoxiae," *Gauss Werke*, 4:95–100, on 95, 98.

15. Magnus Georg Paucker, "Ueber die Anwendung der Methode der kleinsten Quadratsumme auf physikalische Beobachtungen," *Programm*, Mitau Gymnasium, 1819.

16. Paucker, 4, 5. French assessment of the truth of theoretical formulas appears not to have been as cautious as the German assessment. See, for example,

S. H. Arnold, "The Méchanique Physique of Siméon-Denis Poisson: The Evolution and Isolation in France of His Approach to Physical Theory (1800–1840), pt. 4, Disquiet with Respect to Fourier's Treatment of Heat," *Archive for History of Exact Sciences* 28 (1983): 299–320, esp. p. 309, where Arnold cites Poisson's intent to preserve "mathematical certainty" in the analytic expression of physical theory. Kenneth L. Caneva has noted that Fourier, Ampére, and Poisson exemplify "the desire . . . for absolute, incontestable, objective, immortal certainty"; see "Conceptual and Generational Change in German Physics: The Case of Electricity, 1800–1846" (Ph.D. diss., Princeton University, 1974), 255. Despite French development of the probability theory, the method was very rarely applied to problems in the experimental verification of the mathematical laws of physics. Part of the burden of German assessments of French mathematical physics was to confirm the equations empirically, in part through the assessment of data using least squares (for example, as in the work of Wilhelm Weber on electrodynamics).

17. Georg Muncke, "Beobachtung," in *Johann Samuel Traugott Gehler's Physikalisches Wörterbuch*, revised by [H. W.] Brandes et al., 11 vols. (Leipzig, 1825–45), 1.2:884–912.

18. Muncke, 901, 909.

19. Some of the most important, in chronological order: Peter Riese, "[Review of C. F. Gauss's *Theoria combinationis* . . .]," *Jahrbuch für wissenschaftliche Kritik*, 1830, pt. 1, cols. 269–84; A. Grunert, "Wahrscheinlichkeitsrechnung," in *Klügel's Mathematisches Wörterbuch*, (1831) 5.2:890–1030; J. J. Littrow, "Bemerkungen zum practischen Gebrauch der Wahrscheinlichkeitsrechnung," *Zeitschrift für Physik und Mathematik* 9 (1831): 433–49; J. J. Littrow, *Die Wahrscheinlichkeitsrechnung in ihrer Anwendung auf das wissenschaftliche und practische Leben* (Vienna, 1833); J. F. Encke, "Ueber die Methode der kleinsten Quadrate," *Berliner astronomisches Jahrbuch*, 1834:249–312; Gotthilf Heinrich Ludwig Hagen, *Grundzüge der Wahrscheinlichkeits-Rechnung* (Berlin, 1837); Gustav Adolph Jahn, *Die Wahrscheinlichkeitsrechnung und ihre Anwendung auf das wissenschaftliche und praktische Leben* (Leipzig, 1839); F. W. Bessel, "Untersuchungen über die Wahrscheinlichkeit der Beobachtungsfehler," *Astronomische Nachrichten* 15 (1838), cols. 369–404; J. J. Littrow, "Wahrscheinlichkeitsrechnung," *Gehler's Physikalisches Wörterbuch* (1842) 10.2:1181–1251.

20. Littrow, "Wahrscheinlichkeitsrechnung," *Gehler's Physikalisches Wörterbuch*, 1200.

21. Friedrich Wilhelm Bessel, "Ueber Wahrscheinlichkeitsrechnung," in F. W. Bessel, *Populäre Vorlesungen über wissenschaftliche Gegenstände*, ed. H. Schumacher (Hamburg, 1848), 387–407, on 401.

22. Littrow, "Wahrscheinlichkeitsrechnung," *Gehler's Physikalisches Wörterbuch*, 1201.

23. Ibid., 1213.

24. Ibid., 1213, 1200.

25. Hagen, iv (emphasis added).

26. Hagen, esp. 129–44.

27. For example, Strecker, "Das Atomgewicht des Silbers und Kohlenstoffs," *Liebig's Annalen der Chemie* 59 (1846): 265; Christian Ludwig Gerling to Carl Friedrich Gauss, 23 January 1847, in *Christian Ludwig Gerling an Carl Friedrich*

Gauss: Sechzig bisher unveröffentlichte Briefe, ed. Theo Gerardy (Göttingen, 1964), 88; J. Roeber, "Das Atomgewicht des Kohlenstoffs," *Journal für praktische Chemie*, 24:451; Hermann Helmholtz, "Messungen über den zeitlichen Verlauf der Zuckung animalischer Muskeln und die Fortpflanzungsgeschwindigkeit der Reizung in den Nerven," *Müller's Archiv für Anatomie, Physiologie, und wissenschaftliche Medizin* (1850): 276–364; Albert Wangerin, *Franz Neumann und sein Wirken als Forscher und Lehrer* (Braunschweig, 1907), 18 on the application of least squares in mineralogy; Kathryn M. Olesko, *Physics as a Calling: Discipline and Practice in the Königsberg Seminar for Physics* (Ithaca, N.Y. & London, 1991), on least squares in physics instruction.

28. Littrow, "Wahrscheinlichkeitsrechnung," *Gehler's Physikalisches Wörterbuch,* 1251.

29. Bessel, "Ueber Wahrscheinlichkeitsrechnung," 398.

30. Olesko, *Physics as a Calling,* 160–61; id., "Tacit Knowledge and School Formation," *Osiris* 8 (1993): 16–29.

31. Friedrich Wilhelm Bessel to Alexander von Humboldt, 10 April 1844, A. von Humboldt Nachlaß, Staatsbibliothek zu Berlin–Preußischer Kulturbesitz, Haus Zwei, Handschriftenabteilung.

32. Olesko, *Physics as a Calling.*

33. Paucker, 3; Grunert, 1003, 1014; Littrow, *Wahrscheinlichkeitsrechnung*, 79; Littrow, "Wahrscheinlichkeitsrechnung," *Gehler's Physikalisches Wörterbuch*, 1236, 1244. Paucker used *g* to signify "*Präzision*"; other commentators used *h*.

34. Grunert, 1003; J. Roeber, "Experiment, Beobachtung," *Handwörterbuch der Chemie und Physik*, vol. 1 (Berlin, 1842): 770–90, on 777.

35. Roeber, 778–82. Of course, when the condition of equal precision among measurements was not met, special calculations had to be made and these involved more sophisticated applications of least squares. German practitioners were not the only ones to so interpret this variable, but the association between precision and the inverse of probability (which is what this identification of *g* amounted to) was not universally accepted, and in some quarters met with criticism, opening up the way to an alternative definition of precision.

36. See, for example, Helmholtz, 1850.

37. Muncke, 886. See also Roeber, 770.

38. Muncke, 886–87.

39. Jahn, 96.

40. Roeber, 771. See also Muncke, 901.

41. Paucker, 3.

42. Hagen, 31–32.

43. Grunert, 990.

44. Muncke, 887; Grunert, 986; Jahn, 97.

45. Hagen, 6.

46. Littrow, "Wahrscheinlichkeitsrechnung," *Gehler's Physikalisches Wörterbuch*, 1231.

47. Roeber, 777.

48. Muncke, 890. The concept of "agreement" seems not to have been that

important; few commentators discuss or even mention it in any way. The exception in this group is Littrow ("Wahrscheinlichkeitsrechnung," *Gehler's Physikalisches Wörterbuch*, 1204), who remarks that agreement is a measure of the "inner quality" and even the probability of observations.

49. Muncke, 911.

50. Kenneth L. Caneva, "From Galvanism to Electrodynamics: The Transformation of German Physics and Its Social Context," *Historical Studies in the Physical Sciences* 9 (1978): 63–160, esp. 79–80, 88.

51. Hagen, 2.

52. Cf. Muncke, 901, who also explains that one of the functions of least squares is to halt this dangerously endless process.

53. Hagen, 30–31, 162, 161–62, 188, 184.

54. Hagen, 22, 32, 99, 127, 127. See also Grunert (986) and Littrow (*Wahrscheinlichkeitsrechnung*, 16) for similar points of view. Hagen (154) wanted to increase the certainty of results in measurement, without adding too much time to the operation itself; he noted that time was a factor in applied operations.

55. Roeber, 777.

56. Hagen, 6.

57. Woldemar Voigt, "Der Kampf um die Dezimale in der Physik," *Deutsche Revue* 34.3 (July–September 1909): 71–85, on 85.

58. Hagen, 6–7.

59. Littrow, *Wahrscheinlichkeitsrechnung*, 40; Paucker, 4. Most commentators are surprisingly stingy in their references to what might be considered speculative or hypothetical.

60. Hagen, 129; also iv, 126.

61. Paucker, 3.

62. Littrow, *Wahrscheinlichkeitsrechnung*, 6; Bessel, "Ueber Wahrscheinlichkeitsrechnung," 406.

63. Hagen, 126, 128.

64. Roeber, 770.

65. Hagen, 14.

66. Grunert, 892, 986; Littrow, "Wahrscheinlichkeitsrechnung," *Gehler's Physikalisches Wörterbuch*, 1181; Jahn, 1; Bessel, "Ueber Wahrscheinlichkeitsrechnung," 387; Hagen, iii–iv.

67. As reported in Kula, 45.

68. James J. Sheehan, *German History, 1770–1866* (Oxford, 1989), 433–34.

69. F. R. Hassler, *Comparison of Weights and Measures of Length and Capacity, Reported to the Senate of the United States by the Treasury Department in 1832* (Washington, 1832), 4.

70. Louis McLane to the President of the Senate, 20 June 1832, rpt. in Hassler, 1.

71. Patrick Kelly, *Metrology: or, an Exposition of Weights and Measures, chiefly those of Great Britain and France* (London, 1816), vii. As these examples and the ones below illustrate, guidelines for judging the integrity of the measurer spring from cultural sources.

72. *Maaß- und Gewicht-Ordnung für die Preußischen Staaten nebst Anweisung zur Verfertigung der Probemaaße und Gewichte* [16 May 1816] (Berlin, 1816); *Gesetz-Sammlung für die Königlichen Preußischen Staaten. 1816.* Nrs. 1–19. (Berlin, 1816), 142–52; "Maas-Ordnung für das Großherzogthum Baden," *Großherzoglich-Badisches Staats- und Regierungs-Blatt*, 27 January 1829, 5–24. Baden's regulations, which used the meter as the basis of the state's weights and measures, were less elaborate in this regard than Prussia's, where uncertainties plagued standards; see below.

73. Wilhelm Weber to Ministerium des Handels und der Gewerbe, 11 April 1830, Geheimes Staatsarchiv Preußischer Kulturbesitz, Abteilung Dahlem, Rep. 120A, Abt. IX, Fach 1, Nr. 2, Bd. 1 (1814–1862), fols. 67–68.

74. Roeber, 770–71.

75. Heinrich Wilhelm Dove, *Ueber Maass und Messen oder Darstellung der bei Zeit-, Raum- und Gewichts-Bestimmungen üblichen Maasse, Messinstrumente und Messmethoden, nebst Reductionstafeln*, 2d ed. (Berlin, 1835), 28.

76. Dove, 51–166.

77. Ibid., 165–66.

78. Ibid., 30, 41, 49.

79. See the extent of Gauss's official correspondence on the matter in C. F. Gauss Nachlass, Physik 26, Niedersächsische Staats- und Universitätsbibliothek, Altbau, Abteilung für Handschriften und seltene Drucke.

80. Hanoverian Ministry of the Interior to Gauss, 11 November 1836, Gauss Nachlass, Physik 26.

81. Hanoverian Ministry of the Interior to Gauss, 19 July 1839, Gauss Nachlass, Physik 26.

82. The result should not surprise. In the Germanies, two social orders coexisted well until the end of the century. Sheehan, 782.

83. Gauss, "Bericht über die Art, wie die Hannoversche Normal-Pfunde dargestellt sind," Gauss Nachlass, Physik 26.

84. Gauss, "Bericht der zur Regulierung der Maassen und Gewichte angeordenten Commission [1836]," Gauss Nachlass, Physik 26.

85. Gauss, "Bericht über die Darstellung der Hannoverschen Normalfüsse," Gauss Nachlass, Physik 29.

86. At least in principle and until such time as prices dropped, in which case a duty still based on weight or size represented an increase when the value of the object was taken into consideration.

87. Cf. Sheehan, 434.

88. See, for example, August Böckh, *Metrologische Untersuchungen über Gewichte, Münzfüsse und Masse des Altertums in ihrem Zusammenhang* (Berlin, 1838).

89. *Maaß- und Gewicht-Ordnung* [16 May 1816].

90. J. Eytelwein, "Ueber die Prüfung der Normal-Maaße und Gewichte für den königlich-preussischen Staat und ihre Vergleichung mit den französischen Maaßen und Gewichten," *Abhandlungen der Akademie der Wissenschaften*, Mathematische Klasse (1825), 1–21, on 1, 21.

91. Friedrich Wilhelm Bessel, *Untersuchungen über die Länge des einfachen*

Secundenpendels (Berlin, 1828). By contrast, the American weights and measures reformer, Ferdinand Hassler, considered least squares "superfluous." Hassler, 74.

92. Bessel to G. B. Airy, 9 November 1833 [in English], Darms. Samml., Sig. J 1844, Staatsbibliothek zu Berlin–Preußischer Kulturbesitz, Haus Zwei, Handschriftenabteilung.

93. Friedrich Wilhelm Bessel, *Darstellung der Untersuchungen und Maassregeln, welche, in den Jahren 1835 bis 1838, durch die Einheit des Preussischen Längenmaasses veranlasst worden sind* (Berlin, 1839), on 14, 5, 94.

94. Friedrich Wilhelm Bessel, "Ueber Mass und Gewicht im Allgemeinen und das Preussische Längenmass im Besonderen," in *Populäre Vorlesungen*, 269–325, on 269–70.

95. Bessel to Humboldt, 20 May 1833, A. v. Humboldt Nachlass, Staatsbibliothek zu Berlin–Preußischer Kulturbesitz, Haus Zwei, Handschriftenabteilung.

96. Bessel, "Ueber Maass und Gewicht," 274, 275.

97. Bessel, "Ueber das preussische Längenmaass und die zu seiner Verbreitung durch Copien ergriffenen Maassregeln [1840]," *Abhandlungen von Friedrich Wilhelm Bessel*, ed. Rudolf Engelmann, 3 vols. (Leipzig, 1875–76), 3:269–75, on 269.

98. Bessel, *Darstellung*, 5.

99. Ibid., 94, 95, 96ff. (section 17 esp.), 118, 120; id., "Ueber Maass und Gewicht," 274, 288, 289.

100. Bessel, "Ueber Maass und Gewicht," 276. Naturally, standards also gave exact experimenters the means to compare results, and so to link or unite disparate findings otherwise confined to local scientific cultures.

101. "Motive," Anhang I to Bessel, *Darstellung*, 143.

102. The process was not unique to the German states. Cf. Hassler, 17, 37, 38.

103. See, for example, Franz Neumann's reaction reprinted in Luise Neumann, *Franz Neumann: Erinnerungsblätter von seiner Tochter* (Leipzig, 1904), 448–52.

104. For a general view of the connection between precision measurement and weights and measures at the end of the century, see Johann Pernet, "Ueber den Einfluss physikalischer Präcisionsmessungen auf die Förderung der Technik und des Mass- und Gewichtswesens," *Schweizerische Bauzeitung* 24 (1894): 110–14.

105. See, for example, Kathryn M. Olesko, "German Models, American Ways: The 'New Movement' Among American Physics Teachers, 1905–1909," in *German Influences on Education in the United States to 1917*, ed. Henry Geitz, Jürgen Heideking, and Jurgen Herbst (Cambridge, 1994); Graeme Gooday, "Precision Measurement and the Genesis of Physics Teaching Laboratories in Victorian Britain," *British Journal for the History of Science* 23 (1990): 25–51.

106. Emil Warburg, "Das physikalische Institut," in *Die Universität Freiburg seit dem Regierungsantritt seiner königlichen Hoheit des Grossherzogs Friedrich von Baden* (Freiburg, 1881), 91–96, on 93.

107. Voigt, "Kampf um die Dezimale," 85.

108. Gooday, "Precision Measurement and the Genesis of Physics Teaching Laboratories."

109. Olesko, "German Models, American Ways."

110. Donald MacKenzie, *Inventing Accuracy: A Historical Sociology of Nuclear Missile Guidance* (Cambridge, Mass., 1990).

111. It was not unusual for precision to acquire its meaning in that area of uncertainty and free play; K. Olesko, "The Meaning of Precision," unpub. paper presented to the Semiotic Society of America, College Park, Maryland, 26 October 1991.

112. Olesko, "German Models, American Ways."

SIX

ACCURATE MEASUREMENT IS AN ENGLISH SCIENCE

Simon Schaffer

EVERYWHERE they went, Victorians were provided with detailed instructions on the meaning of precision: "Even as you whiz along in a railway carriage, accustom yourself to time the pace at which you travel, to count the number of telegraph poles there are to a mile. . . . All such practices tend to impress useful facts on the memory. . . . It is taken for granted that [you have] a fair knowledge of arithmetic, of at least the first two books of Euclid, of plane trigonometry, of algebra as far as quadratic equations, and of permanent fortification. [You] should be able at a glance to distinguish the common vegetable productions, including the various species of timber. For facilitating the measurement of distances &c., every one should know the exact length of his ordinary pace, and be able to pace yards accurately; he should know the exact length of his foot, hand, cubit, and sword, and arms from tips of fingers of left hand to right ear; he should know the height of his knee, waist and eye, and also the exact proportion that his drinking-cup bears to a pint."[1]

These fearsome recommendations appeared under the heading "What all officers should carry in their heads" in *The Soldier's Pocket Book,* a compendium of military virtues produced in 1869 by Garnet Wolseley, veteran of most of Britain's colonial wars, immortalized by W. S. Gilbert as "the very model of a modern Major General." In Victorian Britain, exact measurement was advertised as a vital accompaniment of commercial, military, and thus imperial triumph. Its role was most vivid in the context of metrology, the process through which the British state constructed and disseminated standards. Much has recently been made of the importance of precision measurement in scientific progress. The values which such measurements yield are supposed to escape the value-system which gave them such high status. In fact, they are the results of this system. Precision is too rapidly identified solely with mensuration. Wolseley, for example, turned it from a set of quantities into a complex of qualities: a good memory, regular habits, a disciplined forearm, a reliable beer mug. Precision is the result, rather than the cause, of consensus among scientific practitioners. It is a quality granted to some claim following negotiations about the status of the work which generates the claim. Undoubtedly a Victorian value, precision badly needs a cultural history which maps its historical credibility instead of assuming its methodological validity.

The link between metrology and commercial values was much in vogue among Victorian physicists and engineers. William Thomson told James Clerk Maxwell in 1871 that new electrical units would be summoned into existence by new markets: "When electrotyping, electric light, become commercial we may perhaps buy a microfarad or a megafarad of electricity. . . . If there is a name to be given it had better be given to a real purchaseable tangible object." Maxwell agreed that metrology helped secure exchange. Cable technology and the new electric resistance unit were a good example. What worked in one place could be made to work elsewhere, the key to imperial power. "The equations at which we arrive must be such that a person of any nation, by substituting the numerical values of the quantities as measured by his own national units, would obtain a true result."[2] In a manuscript of the 1870s, he began with the thought-experiment of communicating the knowledge of some magnitude to another person. "The want of a unit is first felt in buying and selling. Hence the most ancient units are those which have been adopted by men of the same trade and residing in the same place." Universal units were just those which measured "goods which are carried from a few centres to great distances." So the task of "wise governments" was to universalize their standards, punish all others, and thus extend the range of their commerce. "We have no right to assume that any physical theory may not at length come to be necessary to us in the way of business." Physics must set up its standards as "national treasures." "The man of business requires these standards for the sake of justice, the man of science requires them for the sake of truth, and it is the business of the state to see that our . . . measures are maintained uniform."[3]

Victorian laboratory managers perceived a close link between the security of physics' values, their capacity to travel, and the role of the commercial state. In 1884 the cable engineer Fleeming Jenkin argued that "Poverty is no child of Commerce. She is the unwelcome offspring of that Physical Necessity which Sir William Thomson calls the Dissipation of Energy." So "everything we have is for ever losing its value by decay." We may add that Victorian physicists reckoned that some of this decay of values could be made up by the constant work of metrological surveillance—literally so, since the search for imperishable substances out of which standards would be made was relentless. "It is not enough to possess a standard of an abstract kind . . . the great difficulty is to preserve it unaltered from age to age," wrote Britain's leading astronomer, John Herschel.[4] The immense labor required to set up such unalterable standards was always accompanied by a deliberate effort to efface this labor.

Standards had to be represented as properties of nature rather than contingent outcomes of cultural work. This is why it is hard to see the social basis of metrology. In order to reveal the sociocultural basis of this sense of standardization and precision, H. M. Collins has drawn attention to the significance of "digitization" in the assignation of values to material objects. A system in which coins' face values are supposedly just the same as the value of their metal content is therefore highly unstable, because of wear and forgery. A system which maximizes the difference between face value and the value of the material, such as paper money, is robust. This is because the system has been digitized. Notes can only take up certain specified values, and, because of this restricted range, easy valuation is secured. So robust systems may well demand imprecision; they tolerate wide differences in their material representation. The digitization of the system is a social convention, an aspect of a given form of life.[5] We see digitization at work in the rapid development of measurement devices in fields such as electromagnetism and astronomy. Victorians' interest in their own units shows the cultural formation of the value they gave precision.

This paper is devoted to one of the most celebrated enterprises of late Victorian physics, one in which the establishment of a precise standard interacted with the development of physicists' doctrines. This is the story of the determination of the ratio between the electrostatic and electromagnetic units, often conventionally symbolized as v. In autumn 1861, while working on the third part of his paper "On Physical Lines of Force," Maxwell famously established to his own satisfaction that the coincidence between this ratio, equivalent to the propagation of magnetic action in his mechanical ether, and the speed of light, helped demonstrate that light itself was due to transverse waves in the electromagnetic ether. The coincidence in question amounted to a difference of rather more than 1 percent between v, measured in the late 1850s in Göttingen, and light speed, determined in Paris a decade earlier. Maxwell then printed his result in the *Philosophical Magazine* in early 1862.[6] This derivation and implication of the identity between the units' ratio and light speed have been subject to much subsequent, sometimes suspicious, comment. The match has occasionally seemed too good. Attention has been drawn to the troublesome approximations involved in modeling the electromagnetic ether as an array of perfectly spherical cells, to the errors Maxwell made by using the rigidity coefficient rather than the shear modulus in deriving the speed of transverse waves from this ether's properties and, last, to his careful adjustments to the ether model so that these waves' speed would be just the ratio of the electrostatic and electromagnetic units.[7] Furthermore, for three decades after 1862 Maxwell's developing electromagnetic theory of light was met with sterling criticism from the nation's leading physicist, William Thomson. Thomson's hostility depended on a set of related concerns, including the status of his theory of submarine telegraph signaling and his philosophical and practical commitment to a truly elastic solid ether capable of sustaining longitudinal waves. This attack was most famously articulated in the Baltimore Lectures of 1884, where an audience including Henry Rowland, Lord Rayleigh, and Abraham Michelson was told that "the so-called Electromagnetic theory of light" was "rather a backward step from an absolutely definite mechanical motion." In the later 1880s it seemed briefly as though Thomson had conceded the veracity of Maxwellian electromagnetism.[8] But a salient and recurrent theme in Thomson's critique was his estimate of the significance of the coincidence between the ratio of units and the speed of light. In 1871 he told the British Association that it was "premature to speculate" on the numbers' agreement. In 1876, at a conference on precision measurement held at the South Kensington instruments show, Thomson repeated his view that "still we must hold opinion in reserve before we can say" that v was equal to light speed: "The result has to be *much closer* than has been shown by the experiments already made before the suggestion can be accepted." Thus for Thomson's project the value of v was always what he called it in 1883: "this marvellous quantity."[9] The intriguing problem remains—for this notorious advocate of the role of exact measures in physics' progress, just how close would the marvelous quantity and light speed have to be before such a physicist as Thomson would be satisfied that they were the same? What was required from the cultures of astronomy, telegraphy, and physics which generated these numbers?

This long debate about the criteria of the two values' identity was explicitly a metrological issue. In summer 1861, while Maxwell was working on the last sections of his paper on physical lines of force, the British Association agreed to establish a committee, including Thomson and Jenkin, to determine an exact standard of electrical resistance. Such a standard was a prerequisite of successful telegraphy and it was also a prerequisite of most reliable methods for the estimation of v. In summer 1863 Thomson reported to the Association that the committee would include the determination of v as part of its brief and in a joint appendix to this report Jenkin and Maxwell, himself a swift and enthusiastic participant in the project, cited recent German values for v and discussed five different methods for measuring it.[10] From the early 1860s, the status of experimental determinations of v hinged on the progress of this standards project, including the mobilization of unprecedented numbers of electrotechnicians and of increasingly bulky and carefully designed instrumentation. Thomson's prolonged skepticism of the security of the value of v was countered with an accumulation of practical and cultural resources. The credit to be invested in v was to be established by using these resources. Aspects of this career are addressed in the paper which follows.

In the following section, the response of the American physicist Henry Rowland to rival British and German estimates of *v* is used to highlight the resources for precision measurement needed by electromagnetic laboratories of the 1860s and 1870s. Next, Maxwell's strenuous efforts to mobilize these resources, first in London and then, after 1871, in Cambridge, are discussed. He had a very hard time assembling the right combination of trained technicians, reliable equipment, and secure experiments within the potentially hostile milieu of the liberal varsity.[11] Finally, the outcome of these trials in the 1870s and early 1880s is shown to depend on much wider attempts to make local metrological standards of resistance produced in Cambridge count elsewhere. Precision itself may be neither necessary nor sufficient to establish conviction. In 1861–62, Maxwell reckoned that the identity of *v* and light speed was decisive evidence for his theory; a decade later, in his *Treatise on Electricity and Magnetism,* Maxwell merely claimed that comparisons between the two values provided no evidence against it. Many, including Rowland, subscribed to the electromagnetic theory of light without being completely satisfied that the identity of *v* and light speed was well founded in the laboratory. Others, such as Thomson, came eventually to concede the identity without thereby agreeing to Maxwell's model.[12] Furthermore, the complex and fragile resources required to establish such an identity with sufficient exactitude suggest that the history of precision needs to be written as a history of coordination of the multiple and separated sites of scientific labor. Alongside "precision," which, as its etymology suggests, indicates abbreviation, we need equally to attend to exactitude, a word which recalls the demands exacted from practical work, and to accuracy, a term which points to the care which that practice requires. These demands and cares are established as part of an entire form of life within and between laboratories, workshops, and the field.

"The Rough Methods of Early Pioneers"

> While in Göttingen, I had the pleasure of seeing the apparatus used by Gauss and Weber and also that more recently used by Kohlrausch in the determination of the absolute value of Siemens' unit. . . . So far it seems to me that the accurate measurement of resistance either absolutely or relatively is an English science almost unknown in Germany.
> *(Henry Rowland to James Clerk Maxwell, March 1876)*[13]

Writing from Graz in March 1876, the young American physicist Henry Rowland had good news for his Cambridge patron. German accuracy could not be trusted. During the mid-1870s Maxwell and his collaborators at the Cavendish Laboratory set out to check and repeat earlier work on the construction of a unit of electrical resistance. This earlier work, performed in King's College, London, between 1862 and 1864 by Maxwell, Balfour Stewart, and the electrical engineers Fleeming Jenkin and Charles Hockin, had resulted in the production of the well-known British Association resistance standard, embodied in resistance boxes first issued by the London makers Elliott Brothers in spring 1865 and designed to be as close as possible to an absolute value of 10 million meters, one earth quadrant, per second.[14] Maxwell had adopted a technology designed by his Glasgow friend William Thomson (fig. 1). A spinning coil revolving round its vertical axis in the geomagnetic field, driven by a Huygens gear and a new governor designed by Jenkin, would produce a constant deflection of a galvanometer needle at its center. The angle of deflection would depend on the coil's diameter, the rate of spin, and the coil's resistance. The resistance could therefore be established by measuring the other three parameters. Hence the group could in principle make a coil of rather precise resistance. This setup placed major demands on accurate engineering and experimental skill. "The temperature of the coil increases by spinning owing to the induced currents," Maxwell told Thomson in the summer of 1863. Jenkin's governor, which controlled the speed of spins, relied on an ingenious brake band mechanism. Maxwell took the chance to discuss this feedback system in an important theoretical paper for the Royal Society. The superintendent of Kew Observatory, Balfour Stewart, a master of precision measurement, designed a sound-and-clock method for the timer. The magnet was suspended on a nine-foot silk fiber. High accuracy degenerated into inconvenient oversensitivity: it was claimed that the governor was so good that Thames steamers outside the College had more effect on the galvanometer needle than did variations in the coil's speed. Careful precision turned into makeshift ingenuity. To measure the very long copper wire, "the wire was gently uncoiled and laid in a groove between two planks of the . . . floor fifty feet at a time and so measured straight but not stretched." Collaborators recalled that "when it was wound into a coil the value of π was taken to seven places of decimals, in order to determine as accurately as possible the mean radius."[15]

But it was equally well known that the British Association unit had not met with unqualified success, especially in the German lands. The great German engineer Werner Siemens, chief competitor of the British telegraph industry, had his own candidate for the basic unit of resistance, defined in terms of a column of mercury one meter in length and one square millimeter in cross-section, and already in common use in his rapidly expanding telegraph networks. Siemens simply denied the accuracy of the London trials. He observed that the King's College results differed

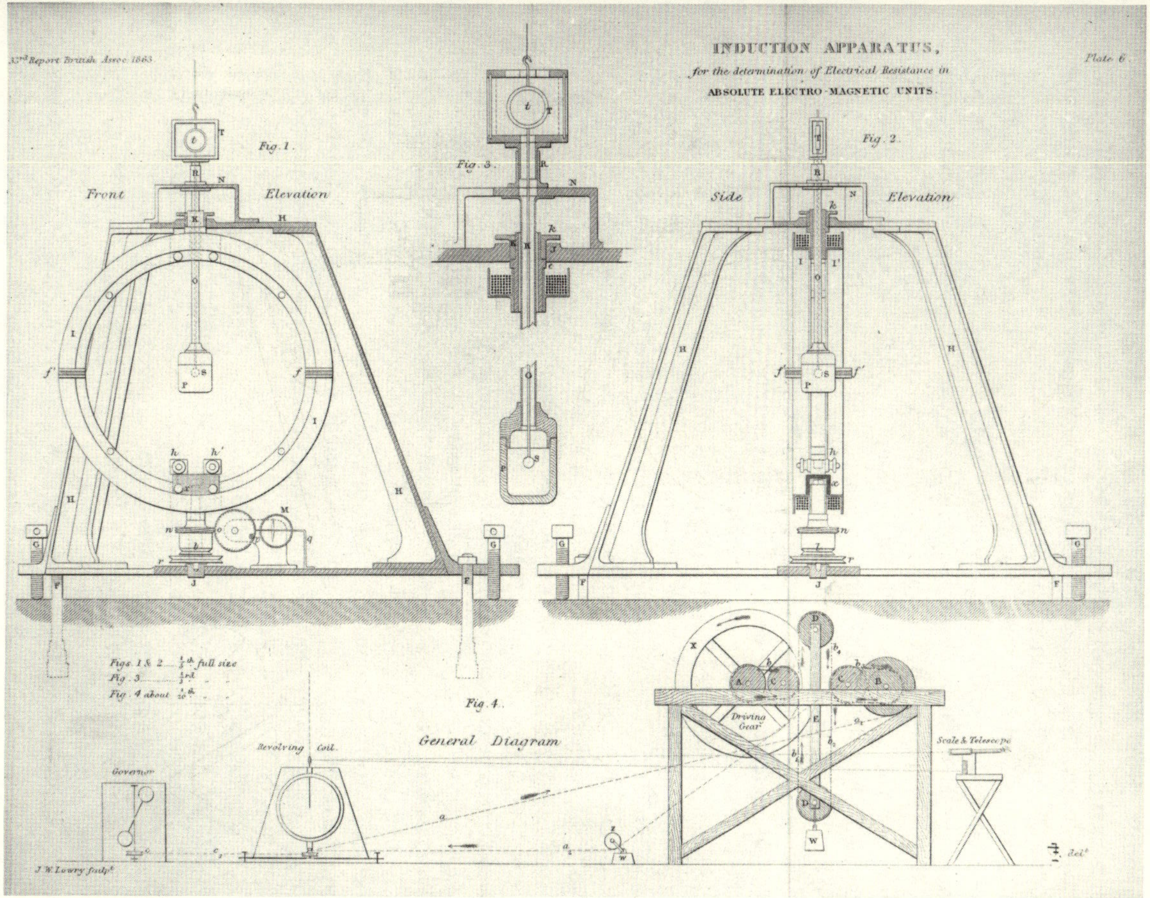

FIG. 1. The apparatus used by Maxwell and Jenkin at King's College, London, in 1862–64 to determine the resistance of a spinning wire coil. The "general diagram" gives the layout of the governor, coil, driving gear, and scale. From *British Association Reports* (1863), 176.

from each other by as much as 1.4 percent, so "by what means the [British Association] Committee holds itself justified, in the face of such differences between even the means of their single observations, in concluding upon an error of only 0.1%, I am totally unable to imagine."[16] It could be successfully alleged that the artifactual character of the Associations' boxed units, which erred noticeably among themselves and from any notional absolute, testified to the failure of laboratory standards among Maxwell's London group. They had to work out ways of making their resistance units compelling. Worse was to follow. At the historic headquarters of scientific electrotechnology, Weber's magnetic observatory at Göttingen, Friedrich Kohlrausch and W. A. Nippoldt spent 1869 working out the absolute value of resistance units specially supplied by Siemens. In autumn 1870 they reported that there was a 2 percent error in the British Association units, so that "there seems need for the greatest caution" in using the Committee's work. Kohlrausch reckoned that "we can but rejoice in the competition between the Siemens unit and that of the British Association; for thereby the best means is afforded of testing the unalterableness of both, and this is all that comes into account for scientific application." The Göttingen team backed Siemens's mercury standard as the best way of resistance testing. The British Association unit was no closer to 10 million meters/second by design than was Siemens's unit "by accident." "In which case it is self-evident that the preference would be given to mercury [and] there does not seem to be any reason whatever why a column of mercury of simple and convenient dimensions should not be chosen." Kohlrausch soon made himself an acknowledged authority on exact measurement when he released his *Leitfaden der Praktischen Physik* as a physics classroom handbook in 1870. Siemens recalled that "the great majority of calculations with electrical resistances belonged to the geometrical and not to the dynamical domain, and that the reproducible unit with a geometrical foundation proposed by me might just as well be called an absolute one as [that] which was proposed as a unit on the English side."[17]

European members of the program to make reliable resistance units had taken sides, hence the value of Rowland's intervention. Working in an ill-equipped hut at Rensselaer Polytechnic in the early 1870s, he had already performed important tests on a magnetic analogy of Ohm's Law and, inspired by the claims of his hero Michael Faraday, had designed a means of testing the magnetic effect of a moving charge by spinning a charged perforated disc. In summer 1873 he contacted Maxwell to arrange the publication of his Ohm's Law work in William Francis's *Philosophical Magazine.* Maxwell told his American correspondent that this would be "the best medium of publication for any researches in exact science. There are several other scientific periodicals but most of them circulate among a class of readers such that their Editors are apt to be suspicious of any article involving exact methods." Much preoccupied with his own team's exacting work on testing Ohm's law inside the new Cavendish laboratory, Maxwell complained that Rowland had not supplied sufficient details of his experimental methods. He welcomed the news that Rowland was planning a visit to Europe.[18]

In June 1875 Rowland was offered a post at the new Johns Hopkins University and its founder-president Daniel Gilman accompanied him on a trip to Europe. From Rowland's point of view, this would enable him to improve his German, establish contacts with the world's leading physicists, and, crucially, enable him to buy fine precision instruments for the Baltimore laboratory he hoped to set up. Armed with a total budget of over $6000, Rowland toured the physics laboratories and workshops. In March 1875 he went to Göttingen. "They have not many instruments but what they have are for work. Saw the earth inductor used by Kohlrausch in his experiments and can well understand the reason of this difference between his results and the B.A. The coil is very large or rather broad and made of large wire. If he could get the area within one per cent he would do well." During the summer, Rowland worshipped at the Faraday shrine in Albemarle Street, attended the British Association meeting, saw spinning coils made for Carey Foster's laboratory in London, and spent some time on Maxwell's country estate at Glenlair.[19] Equally significantly, in August he used the *Philosophical Magazine* to publish an attack on Kohlrausch's redetermination of the resistance of Siemens's unit. "Kohlrausch's experiments were made with such great care and by so experienced a person that it is only after due thought and careful consideration that I take it upon me to offer a few critical remarks." He concluded that the difference of 2 percent between the London and Göttingen results could be explained away by Kohlrausch's neglect of self-induction between the elements of his experimental circuit.[20]

Rowland's visit to Germany and Austria between October 1875 and April 1876 confirmed these views. In Berlin he was given the opportunity to use the resources of Helmholtz's laboratory to perform his spinning disc trials, explaining his results on the magnetic effect of moving charges with theoretical resources learned firsthand from Maxwell. He told Gilman in November that "in America we have apparatus for illustration, in England and France they have apparatus for illustration and experiment, but in Germany they have only apparatus for experimental investigation." He met Kohlrausch, who in summer 1875 had taken up the ordinary physics chair at Würzburg where Rowland found "a very poor assortment of instruments and none that I had not seen before." At Graz between 1873 and 1876 the Austrian government supplied the director of the new physics institute, August Toepler, with a huge instrument budget.

There Rowland saw "the splendid new physical laboratory and was, on the whole, quite well pleased with it." His comments to Maxwell contrasted with his reports from Helmholtz's laboratory: "I would rather have seen more instruments for use and fewer for illustration: I am surprised to find the instruments of research used here often quite poor."[21]

Rowland's distinction between illustration and research was learned from Maxwell, publicly presented in his inaugural lecture at Cambridge in October 1871. There Maxwell defined the work of illustrative training, which "tends to rescue our scientific ideas from that vague condition in which we too often leave them, buried among the other prospects of a lazy credulity." He discriminated this from the work of research, which required accurate measurement. "The history of science shews that even during that phase of her progress in which she devotes herself to improving the accuracy of numerical measurement of quantities with which she has long been familiar, she is preparing the materials for the subjugation of new regions, which would have remained unknown if she had been contented with the rough methods of her early pioneers." What Rowland seems to have admired in Britain was this expansionist combination of the two strategies of precision. Classroom techniques for rendering phenomena less vague were matched with laboratory techniques for determining their magnitude.[22] Maxwell's use of imperial language in his brief sketch of the history of laboratory measurement was especially revealing. Metrology and precision were recognized as important values of the late Victorian state. It was apt that the culmination of the success of the British Association and Cambridge electrical standards projects was the recognition of their work in protocols published by the Board of Trade in 1891, protocols which linked highly detailed recipes for the production of standard resistance boxes and batteries with moral claims about the security of the new electromagnetic system of units.[23]

"The Physical Difficulty of Keeping Up the Spirit of Accuracy"

The security of these systems of units depended on the security and confidence vested in the workforce and its conduct in their new physics laboratories. In summer 1868 Maxwell described the work he was then performing in London with the engineer Charles Hockin on the determination of the ratio between the electrostatic and electromagnetic units, a number which Maxwell's team baptized "*v*." Maxwell pointed out the intimate link between this work and Fleeming Jenkin's commercial telegraphy. He stressed the "difficulties which are always turning up" in such trials, "the hunting for sources of discrepancies, the hitches in the working of new apparatus, the mathematical difficulties, and not least the physical difficulty of keeping up the spirit of accuracy to the end of a long and disappointing day's work."[24] However disappointing, the days which Maxwell's London team spent working on their estimation of v were crucial for his program. This enterprise provides an excellent opportunity for studying what laboratory physicists meant by precise equality. The experiments in King's College, London, performed between 1862 and 1864 for the British Association Committee resulted in a temporary conciliation between the values of the laboratory and of commerce. The status of the claim that light traveled through the electromagnetic ether was made to rely on the standards program. In Saxony in 1855, before this program's launch, Weber and Friedrich Kohlrausch's father, Rudolf, had jointly determined the electrostatic and electromagnetic units' ratio. It appeared as a factor, $\sqrt{2}$ larger than Maxwell's v, in Weber's general equation for electrostatic and electromagnetic interactions. Weber's equation implied that at such a speed the net force between two moving electric masses would vanish. The German physicists measured the charge of a Leyden jar both from the product of its capacity and potential, and by discharging it through a galvanometer and recording the extreme deviation of the magnetic needle. They got a value of 439,450,000 meters/second, giving an equivalent estimate for v of 310,740,000 meters/second. Later observers, including Maxwell, judged that this method was unreliable, because the discharge was much faster than the time it took to measure the jar's potential, so the electromagnetic quantity would be underestimated, and hence the value of v would be too high. "I would use a much larger condenser than Weber," he told Jenkin in summer 1863, "and determine its capacity by more steps." In public, Maxwell was always more polite. "I have such confidence in the ability and fidelity with which their investigation was conducted, that I am obliged to attribute the difference of their result from mine to a phenomenon the nature of which is now much better understood."[25]

Gustav Kirchhoff followed up Weber's work in 1857 at Heidelberg. He showed that the velocity of propagation of an electric wave was connected with this constant velocity with which two particles of electricity must move toward each other so that they exercised no mutual force. In autumn 1861, just before joining the British Association program in which Weber's experiments would provide a key concern, Maxwell designed a mechanical ether for such electromagnetic interactions. He set its torsion modulus and its density, whose ratio was equivalent to the square of the units' ratio, so that transverse shear waves in the ether would travel with a speed equal to the value of the ratio v. He pulled off this trick by setting the density at unit value and the torsion modulus, proportional to the relative speed of waves in the medium, at a value equivalent to v^2. He came back to King's College, London, at the end of the year, and soon wrote enthusiastically to his friends Thomson, Faraday, and Cecil Monro about his discovery that this speed was close to that of light. He told Faraday that he had derived the equations "before seeing Weber's number"; he told Thomson that he "worked out" these equations "before I had any suspicion of the nearness between the two values of the velocity of propagation." So, as he insisted, "the coincidence is not merely numerical . . . whether my theory is a fact or not, the luminiferous and electromagnetic medium are one."[26]

Monro admitted that this was a "brilliant result," but added that "a few such results are wanted" before anyone would be convinced by Maxwell's mechanical models of the ether's behavior. Thomson, for one, was never convinced by the argument from numerical "coincidence." The British Association standards committee, which Maxwell now joined and which Thomson dominated, was necessarily concerned with Weber's reliability. The value of the resistance unit played a crucial role in the estimation of the ratio of the electrostatic and electromagnetic units. For several years Thomson used Weber's laboratory as a standards center for his own resistance coils and telegraphic physics.[27] In 1862 Jenkin noted that "Weber had made some similar determinations with less care some years since," and his committee reported that "in a matter of this importance the results of no one man could be accepted without a check." The following year, when it became clear that the King's College resistance values differed from the Göttingen ones by at least 8 percent, Thomson wrote that "when it is considered that the method [for resistance determination] is the simplest known, the discrepancy between the few determinations hitherto made in absolute measure will cause no surprise. The time, labour and money required could hardly be expected to be given by any one person." In June 1862 Kirchhoff also responded to Jenkin's invitation to comment on the British Association project. The Heidelberg professor gave further evidence of the variable quality of resistance measures. His determinations differed from those of Weber by "about one-seventh. . . . The reason of this want of agreement consists partly in the imperfection of the instruments which I had used," and partly in the difference in temperature between the Heidelberg and Göttingen resistance experiments. Two years later, however, Weber published results on the value of the ratio of units which matched those of Kirchhoff rather well. Thomson's appeal was persuasive. To secure a value of v on the basis of a widely accepted resistance standard would require big resources. These were resources which Maxwell still lacked, especially when he left London in 1865.[28]

The same problem affected the measures of light speed. This was a value which had historically been established by the astronomers using satellite eclipses or the transits of Venus across the Sun to establish the astronomical unit. When Maxwell began casting around for an accepted estimate of the speed of light in late autumn 1861, he first turned to a series of textbooks by the fierce antievolutionist mathematician Samuel Haughton, geology professor at Trinity College Dublin, coauthored with the Church of Ireland divine Joseph Galbraith. Maxwell used their books for his courses at Aberdeen and London: "There is no humbug in them, and many practical matters are introduced instead of mere intricacies." They quoted the best, indeed the only contemporary, terrestrial experiments on light speed, performed in Paris from 1849 by Hippolyte Fizeau using an ingenious method in which a toothed wheel was spun in front of a slit. In what Galbraith and Haughton called these "beautiful experiments" Fizeau directed light from Paris to Montmartre over 8,600 meters from this slit toward a mirror and back to an observing telescope, increased the wheel's speed until the light's image was extinguished, and thence deduced the light speed, since at this point the wheel's speed was rapid enough to carry each tooth in front of the slit in the time the light took to reach the mirror and return. Fizeau's results raised the accepted astronomers' value from 308 million meters to about 315 million meters, or 196,000 miles, per second. This result was greeted in Paris as a "miraculous agreement." But a chapter of mistakes and revisions dominated Maxwell's use of this number. He wanted it to be as close as possible to Weber's value for v, 310,740,000 meters or 193,088 miles per second. First he worked out the light speed from Galbraith and Haughton: 193,118 miles/second. This looked very good. But in Fizeau's own paper in the journal of the Académie des Sciences, the light speed was rather higher than the Irishmen reported. Maxwell made a small error in transcription from the *Comptes Rendus,* bringing Fizeau's own number down from 195,777 to 195,647 miles/second. So in his 1862 paper he published both his mistranscribed version of Fizeau's estimate and the accurately transcribed, but inaccurately reported, textbook one. The values of v and the light speed therefore differed by about 1 percent, a gap of the same order as the difference between Weber's and Thomson's resistance units, but, as we have seen, a difference ten times larger than the error claimed by the King's College experiments of 1862–64. It was on this basis that Maxwell told Thomson of the "nearness" between the two speeds.[29]

Worse was soon to follow. In summer 1862 the great Parisian physicist Léon Foucault published a new series of light speed trials which he had initiated in the previous decade, replacing Fizeau's setup with a revolving mirror and much reduced light path. Foucault got this design from experiments on the speed of electricity performed by Maxwell's King's College colleague Charles Wheatstone. Foucault's work was of importance to Maxwell since, as he explained to Stokes in May 1857, he needed data on the relative speed of light in air and in water as part of his investigation of bodies' relative displacement in the ether. Foucault reported that Fizeau had overestimated light speed, which might have been good news for Maxwell, but that he had overestimated it considerably, which was very bad news. "The speed of light is notably diminished," Foucault announced. "Following the received assumptions, this speed would be 308 million metres/second. The new experiment with the spinning mirror gives, in round figures, 298 million." Foucault's result was consistent with George Airy's reports on the eclipses of Martian satellites and other astronomers' reports on aberration measures from South Africa and Pulkovo. The Royal Astronomical Society unanimously agreed in February 1864 to abandon their received estimate of light speed. John Herschel gave the reduction his imprimatur as "a crucial fact." The gap between this fact and Weber's value for v had widened to more than 4 percent. Maxwell could not directly challenge the astronomical establishment. A world network of astronomical observatories concurred on an increase of the solar parallax on the basis of this fact. Maxwell's great 1865 paper on the dynamics of the electromagnetic field, in which Lagrangian analysis of the energy stored in the field was used to derive the wave equations for transverse vibrations, acknowledged Foucault's results but also reprinted the discredited, if more appealing, value of 308 million meters/second.[30] Maxwell told Hockin that he had now "cleared the electromagnetic theory of light from all unwarrantable assumptions, so that we may safely determine the velocity of light by measuring the attraction between bodies kept at a given difference of potential, the value of which is known in electromagnetic measure." Thomson scarcely agreed with this claim about the electromagnetic theory. He had long held that given the commercial value of the ratio of units, of crucial importance to the speed of transmission of telegraph signals, the value of v must be accurately redetermined. "I believe it may be actually estimated (roughly)," he told Stokes in 1854, and the following year reported useful information from Airy which bore on the speed of transmission of electrical signals. Maxwell needed more than a rough estimate of this troublesome number. In the later 1860s he sought to use electrotechnological interests to provide resources for the project.[31]

Maxwell's own experimental resources were compromised after he left London in 1865 because Jenkin and Hockin were recruited to the Atlantic cable team. When Hockin was released from cable duty, however, he could help. Maxwell reported that Hockin "performed everything, except the actual observation of equilibrium, which I undertook myself." And through his contacts with Elliott Brothers, Maxwell had access to the British Association condensers and resistances. The firm's engineer, Becker, designed his new electrical balances, and Maxwell could occasionally command J. P. Gassiot's London laboratory with its great batteries. Jenkin provided electrical equipment and the expert telegraph engineer Willoughby Smith lent a huge resistance of one million ohms. The celebrated London electromagnetic network might match the strength of the astronomers' values.[32] The team made the determination of *v* into an example of the way in which "purely scientific" problems could be commercially crucial. Maxwell understood from Thomson that "the importance of the determination of this ratio in all cases in which electrostatic and electromagnetic actions are combined is obvious. Such cases occur in the ordinary working of all submarine telegraph cables, in induction coils, and in many other artificial arrangements." This was conventional work for the British Association Committee, and Maxwell reported his experiments to the Association in 1869. But he also insisted that "a knowledge of this ratio is I think of still greater scientific importance when we consider that the velocity of the propagation of electromagnetic disturbance through a dielectric medium depends on this ratio and hence, according to my calculations, is expressed by the very same number."[33]

Now Maxwell set out to show that this number was "the very same." In autumn 1864 he began discussing a Weberian method for measuring *v* using a commutator and a Leyden jar. He reckoned that the security of the method hinged on the reliability of the British Association resistance boxes: "I think we might get this done this year by diligence." In the winter of 1864–65, in debates with Thomson, Maxwell worked out how to improve the measure by building "a special electrometer and dynamometer combined in which the electric attraction of two discs is balanced by the electromagnetic repulsion of two coils." Gassiot planned to supply a dynamometer designed by Weber.[34] In spring 1865 Maxwell confirmed that he would use a method which involved "equilibrating electrostatic attraction with electromagnetic repulsion derived from the same source." The advantage over Weber's method was that "the forces are applied simultaneously to a body already in equilibrium and you have no trouble about unstable equilibrium when you have equilibrium at all." Getting balance needed technicians: "Hockin is working at the galvanometer." The setup involved balancing the force developed between two oppositely charged discs against the repulsion between two current-carrying coils. The potential of the fixed disc was stabilized by surrounding it with a guard ring designed by Thomson. Maxwell changed the gap between the two plates with a micrometer until balance was reached, while Hockin varied the resistance in the circuit until a galvanometer linked to the coils was undeflected. Maxwell read the position of the plates through a microscope focused on a glass scale graduated to 1/100 of an inch, while they reckoned that "the equilibrium of the suspended disk could be made to coincide with the plane of the guard-ring to the thousandth of an inch." Sometimes the two men changed places. They reckoned that high-precision tooling would allow a good estimate of the ratio of the two units and hence the value of Maxwell's electromagnetic program: "As it seemed probable that the time occupied in the construction and improvement of these instruments would be considerable, I determined to employ a more direct method."[35]

There were reasons for haste. In autumn 1867 Thomson set his own Glasgow students W. F. King and J. D. Hamilton Dickson to measure the ratio. Thomson's prescription was to determine the potential across an electrodynamometer both from the product of its resistance and the current flowing through it, and by putting an absolute electrometer across its terminals (fig. 2). The method hinged on Thomson's high estimate of the reliability of his own electrotechnology workshop: "It was part of the laboratory training of the students to make these instruments and to adjust them." The Glaswegian team included William Leitch, subsequently dispatched to try Thomson's siphon recorder on the Eastern telegraph route to India, and the brilliant experimenter Dugald McKichan. Conditions for the initial work in winter 1867–68 were judged "very unfavourable," so new standard resistance coils were requisitioned. King reported to the British Association electrical standards committee in November 1869 that because "various causes have prevented the obtainment of as satisfactory results as the method . . . allows us to expect," the Glasgow group had now built "a new form of absolute electrometer . . . with good promise as to accuracy and convenience" (fig. 3). In spring 1871 the team moved to the new Glasgow physics laboratory and retried the measurements. McKichan admitted that the resistance of the team's standard coils needed great care: "It would have been desirable to have had more frequent measurements of the resistance of the coils, but as only two observers were generally available and the comparison of the electrodynamometer and electrometer required the attention of two observers, this was not generally practicable." So a continuous surveillance was adopted. They did not finish their work until late 1872, in time to have their results announced to Stokes at the Royal Society in spring 1873. On the basis of eleven sets of trials, they claimed that v was somewhere between 27.5 and 29.2×10^7 meters/second.[36]

Maxwell decided that these results could not be trusted. He noted that the Glaswegian procedure involved the simultaneous estimation of two different forces and he reckoned the London method (fig. 4) was better because there the two forces were opposed, cancelled in the resultant, and thus "the only measurement which must be referred to a material standard is that of [Willoughby Smith's] great resistance." Jenkin had already suggested a method using the resistance of a very bad conductor in both

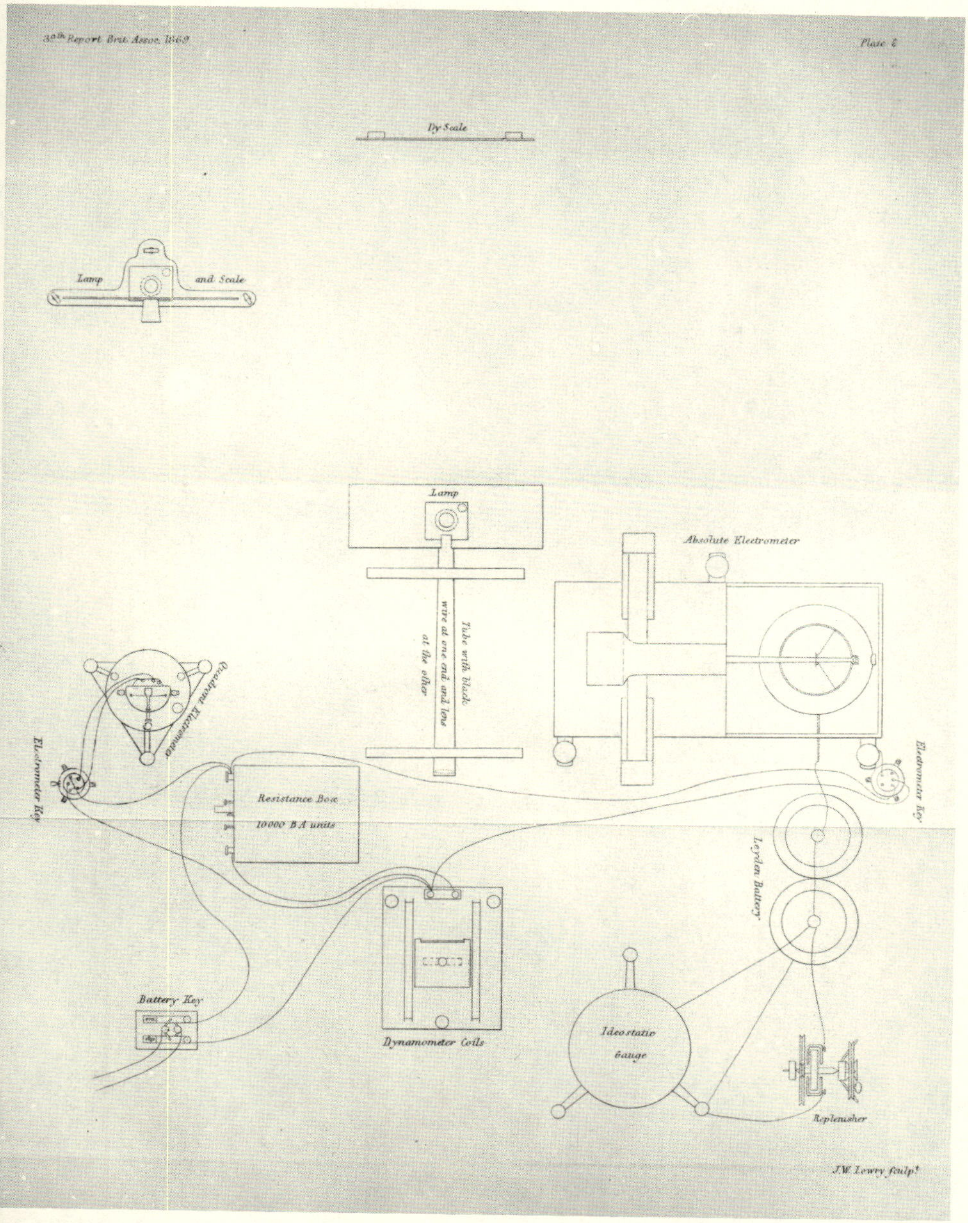

FIG. 2. Glasgow experiments to measure v designed by King and Dickson in 1867–68. Thomson's new "accurate and convenient" absolute electrometer is at the top right. *British Association Reports* (1869), 434.

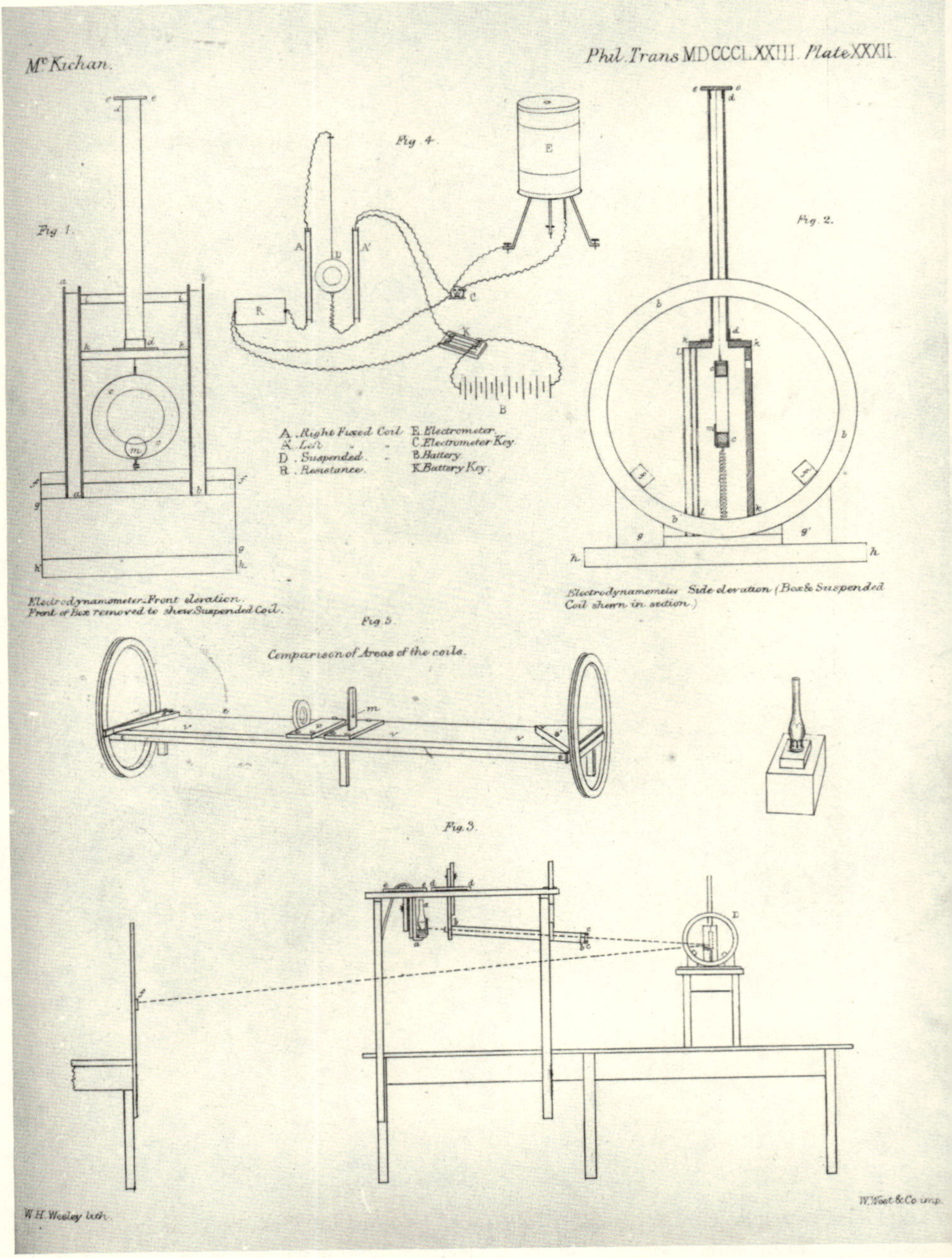

Fig. 3. McKichan's revision of the Glasgow setup: the electrodynamometer and the coil comparisons are illustrated. *Philosophical Transactions* 163 (1873), plate 32.

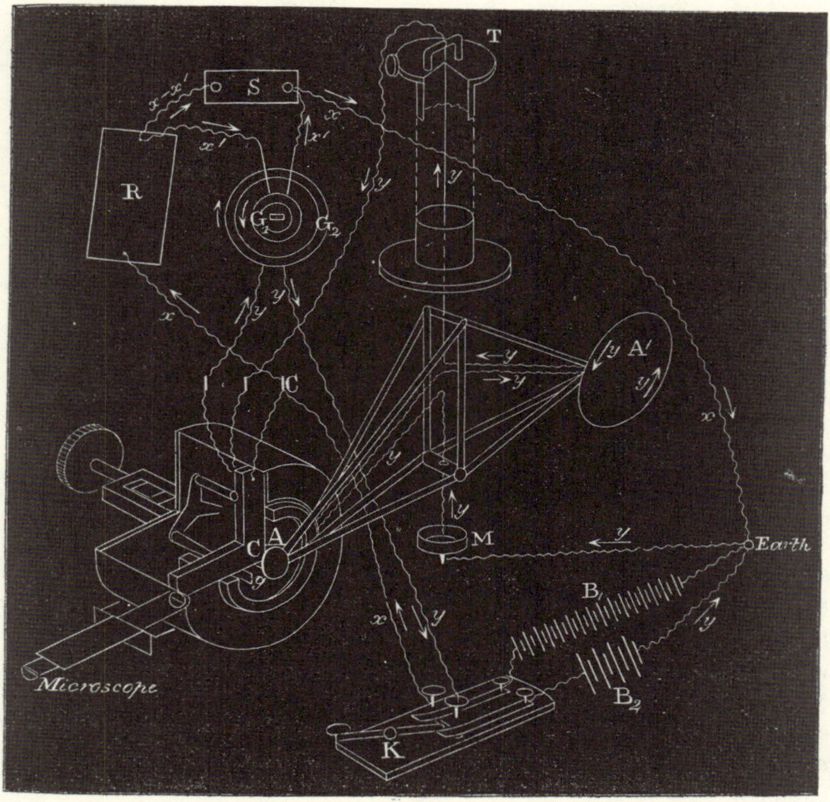

FIG. 4. Maxwell's design of an experiment to measure v in London, 1867–69. Willoughby Smith's one-megohm resistance is at R, the current balance is AA'. *British Association Reports* (1869), 437.

electrostatic and electromagnetic systems: "We require . . . a series of resistances in steps." So everything depended on the reliability of the British Association resistance standard. Writing from his retreat at Glenlair in 1868, Maxwell told Hockin to find ways of increasing accuracy. Hockin reported "the results most unsatisfactory," and especially "the box containing B.A. units did not at all agree. . . . I shall do them again. . . . If the error comes from a bad contact anywhere the result will not come out the same. This brings 'v' from 22 x 10^7 to 24.5 x 10^7 [meters/second]."[37] Eventually Maxwell managed to force the value up to 28.8 x 10^7 meters/second, still remote from Foucault's value of 29.8 x 10^7 meters/second but somewhat better than Weber's old number. He did this by following Hockin's suggestion, suppressing almost one-third of his runs, "on account of the micrometer being touched during the observation of equilibrium," and unremittingly blaming the value of the British Association resistance unit which, he noted, was almost 9 percent larger than Weber's 1862 determination and 1.2 percent smaller than the value of the resistance unit derived from James Joule's exactly contemporary work on the heating effect of currents. He could blame the B.A. unit because his best value for *v* was like those now being made elsewhere in London and Glasgow. "The method of experimenting appeared capable of considerable accuracy; but some difficulties arose from want of constancy in the batteries, from leakage of electricity &c so that many of the experiments were known to be faulty." Maxwell estimated that the "probable error" of the dozen runs which survived his scrutiny was about 1/6 percent, although the published runs range over 3 percent. He stressed that "none of the results were calculated till after the conclusion of the experiments," though correspondence with Hockin shows the two men working hard to explain away a 10 percent difference between their early runs and light speed. Needing larger batteries, possibly Grove cells, a better workforce, and much better resistance boxes, his work stalled. It could not be revived until, and may well have prompted, his move to Cambridge in 1871.[38]

The same summer at Edinburgh, in the midst of his own team's efforts to measure *v*, Thomson made his first important public remarks on Maxwell's claimed relation between *v* and light speed. He reminded the British Association of Weber's "elaborate and difficult experiments" which he described as a "*monumentum aere perennius.*" He mentioned Maxwell's recent claims that the two parameters should be "exactly equal," and then cautioned against a hasty endorsement of Maxwell's model or his results. "Weber's measurement verifies approximately this equality. . . . The most accurate possible determination of Weber's critical velocity is just now a primary object of the Association's Committee on Electric Measurement, and it is at present premature to speculate as to the closeness of the agreement between that velocity and the velocity of light."[39] Thomson, conscious of the secondment of so many of his assistants to international cable work and of the general implications of this ratio for his telegraphic model, saw these measures as testimony to the salutary effect of commercial telegraphy on the values of physics. But despite his hesitant admission that Maxwell's recent work with Hockin seemed to bring the critical velocity closer to light speed, his resonant remarks about "approximate verification" and "premature speculation" demanded an answer from within a well-equipped laboratory capable of building a new metrological monument. In Cambridge the gap between laboratory and workshop, which some of the University's mathematical coaches rather liked but which Maxwell sought to bridge, was crucial. In the very first year of laboratory operations, in spring 1874, Maxwell made his class use the Kew Observatory's magnetometer to measure the geomagnetic field strength and direction. Then he checked with the Observatory to compare his group's results and to receive instruction on the best way of setting up the great instrument. The same problems developed in his exchanges with Fleeming Jenkin about the best ways of running the British Association equipment. He asked Jenkin for the recipe for the governor he had designed in the mid-1860s, since "I must have a speedy friction governor essentially on your principle but with some few alterations to secure good balancing." Maxwell did not know where to get one. He also needed lessons on the design of the old equipment, such as the guard ring condensers which were to play a crucial role in his redetermination of the ohm and hence his new measure of *v*.[40]

It was important to reorganize social and material technologies at the Cavendish Laboratory. Maxwell joined the Society of Telegraph Engineers. In July 1873 he reported that "it is impossible to procure many of the instruments as they are not kept in stock, and have to be made to order." But standards instruments needed personal supervision during construction: "Their whole value depends on their fulfilling conditions which can as yet only be determined by trial, so that it may be some time before everything is in working order." Technology transfer and design need settings which cultivate tacit skills. These cannot be developed in long-distance, formal, communications between maker and user. In May 1874 Maxwell heard from Kew Observatory about a brass ring for his commutator which "must have been lost when the instrument was sent down—for I believe we have had occasion to move the apparatus." Two years later the great Glasgow maker James White, who built the Cavendish's absolute electrometer for £65, told Maxwell that "as you notice it is not square with the stem, it must have been by some accident put out by having the pair of springs squeezed together, this will also account for the disc being so hard against the stops." White gave Maxwell a recipe for correcting the disc suspension. He should use a pair of forceps until the image of a hair spring could be seen through the magnifying lens against the electrometer's scale. These instructions proved hard to follow. So, as Maxwell argued in 1877, the cultural and practical distance between operative and researcher had to be effaced. "The experimenter has only occasional opportunities of seeing the instrument maker, and is perhaps not fully acquainted with the resources of the workshop, so that his instructions are imperfectly understood by the workman. On the other hand the workman had no opportunities of seeing the apparatus at work, so that any improvements in construction which his practical skill might suggest are either lost or misdirected."[41]

Changing this situation was hard work. When William Garnett was hired as demonstrator in the Cavendish he soon "introduced into the Laboratory workshop a few engineer's and joiner's tools." This was not enough. A good example of the trouble is provided by the role of Becker at Elliott Brothers, the suppliers of the standard resistances. In autumn 1878 Maxwell needed some good new standards from them for his experiments on *v*. Elliotts' sent two resistance boxes for Hockin to examine. Hockin found that the temperatures at which these resistances were correct were more than 4 °C higher than the firm supposed. "We have therefore a difference which perplexes us still. Might we suggest that your standard be compared with Mr. Hockin's?" At least three resources existed for developing more effective skills: closer links with London and Glasgow engineers; recruitment from the new University engineering department, under its professor James Stuart's controversial leadership; and hiring in-house laboratory technicians. In the mid-1870s the technician Robert Fulcher from Stuart's works in Balgonie was switched to the Cavendish. Stuart's own shop began to make instruments for the lab, and this helped the new Cambridge Scientific Instruments Company expand from its status as supplier for Michael Foster's physiology laboratory into a general scientific instruments concern. Trotter recalled that in the Stuart workshop, supplied with fine lathes and blocks, "the highest accuracy was demanded throughout." But "accuracy" meant manual skill: "No micrometer vernier or 'go-and-not-go' gauge was used. But it was splendid practice." The shop began to provide the Cavendish with equipment and jobs for students.[42]

The skills of men like Fulcher were crucial for the electromagnetic standards project. Maxwell began designing new methods for redetermining the resistance unit and for assessing the value of *v*. They needed high-quality precision skills. He reckoned that a better value of *v* could be obtained by returning to a suggestion he had made in the mid-1860s, with a commutator linked to an air condenser. The current strength in the circuit could be measured in electromagnetic units using a balancing constant current, while the electrostatic value of the quantity of charge could be found from the commutator's speed and the dimensions of the condenser. This method still needed a good value for the British Association resistance standard. James Joule, Manchester doyen of precision measurement, was much concerned with this value because it was used to calibrate his own estimate of the mechanical equivalent of heat. He told Maxwell that he had discussed the project with Thomson and Balfour Stewart: "We agreed that it would be most important if you would undertake fresh experiments for redetermining the unit of resistance. I earnestly wish you would do this and then my results will sink or swim." With a range of interests in play, the laboratory was put to work. The aim was to match Maxwell's support staff to the problems of eliminating the troubles his team had had at King's in the 1860s. He devised a null method with little dependence on battery output variation and no need to calibrate the ballistic galvanometer. Surviving ballistic galvanometers from this period include one designed by Maxwell and then used in the standards project, and one obtained by the Cavendish from the Cambridge chemistry professor in early 1875. Once again, commutators played a central role. Maxwell baptized this the "Determination of *v* by wippe," and introduced it into revisions of the *Treatise*. He told the high wrangler George Chrystal that "I am glad the commutator with the crank worked well. Such a commutator is very much wanted for many experiments. Of course, there must be some kind of counter for absolute measurements of time." To ensure very fast current decay when the commutator was running, he wanted to fit a narrow radial gap filled with an insulator inside the gunmetal former which held the mutual inductance coils. Maxwell's very last lecture at Cambridge was devoted to an investigation of the shape of such a decaying current in a mutual inductance system. Ambrose Fleming's notes on this lecture were posthumously included in the next edition of the *Treatise*.[43]

Good coils and formers were needed. In summer 1877 they could still not be made in Cambridge. Maxwell ordered them from the London engineers Thomas Horn. His troubles with distant, and thus unverifiable, makers began here. Fulcher would solve them by making the coils on site. Supervision and hierarchy mattered too. In January 1878 Hockin was asked again about other labs' work, including those of Carey Foster in London, where temperature effects on resistance were used. Fleming, then a new student at St. John's College, was given the task of getting the London method working in Cambridge.[44] Chrystal was given the surveillance job. During 1878 he told Maxwell of the troubles he faced when winding coils onto the gunmetal formers. The main problem was technical support: "Owing to the dearth of assistance I could get no further than a preliminary trial which promised success." Maxwell ordered Fulcher onto the job, and even tried to persuade his notoriously inefficient laboratory assistant Pullin to help: "I think he is converted." By July Chrystal confirmed that "I propose having the coils wound by Fulcher in my immediate presence, so that I can give account of every layer if necessary." Because of this new in-house resource, replacing vulnerable long-range contacts with an immediate link between Stuart's technicians and Maxwell's wranglers, the professor was able to tell the university that at last "much of the apparatus to be used in Cambridge which has hitherto been ordered from London may in future be constructed in Cambridge and tested while in the maker's hands by those who are to use it."[45]

Chrystal's work on winding coils is a good example of the way in which the new laboratory regime bred new skills, even among the wranglers. Precision started to look like hard and disciplined labor. Second wrangler and Smith's prizeman in 1875, Chrystal was at once elected to a fellowship at Corpus Christi College and for the next two years formed part of the Cavendish workforce. The skills he learned were so important that even after his departure in autumn 1877 for the mathematics chair at St. Andrew's he remained a crucial element in Maxwell's attempt to measure the resistance unit and *v*. "I should still have a good part of the year to come to Cambridge and work," he reassured Maxwell. "The word research is becoming by the way . . . fatal—one loathes to hear it." Chrystal spent the later 1870s working with Garnett and Fulcher on winding coils. They were sent from London and then made and checked in Cambridge to be used for the experiments on *v*. "I do not know whether it is worth while now to redetermine the absolute unit of resistance, but there remains *v* and now that we have a graduated condenser the thing might I suppose be done," he told Maxwell in summer 1878.[46] So each summer Chrystal came to Cambridge and reported to Maxwell on the team's work. Maxwell told him to get Fulcher to wind the coils onto large bobbins and determine their size. "The counting of the windings is the main thing for it." Maxwell had lots of experience with such coils. He reckoned that "for accuracy of measurement and calculation it is better not to increase the depth too much because the outer layers of winding compress the inner ones and so change their electrical constants." Changing coil design was easier than changing the magnetization. High-class German experiments were simply too expensive to emulate: "Kohlrausch makes the length of his coil practically infinite: This uses more wire than is necessary."[47]

Chrystal got a cathetometer from the Kassel firm F. W. Breithaupt, equipped with a 20x telescope costing almost £40, a cynosure of precision design recommended by Maxwell. He took almost a hundred readings of the channels in which the wires were to be laid and used a robust beam compass to take each measure. He learned to keep each reading "in case of corrections turning up" and "set a man to calibrate the cathetometer by means of a fixed length which is hoisted up and down." Troubles multiplied: the iron bar which carried the cathetometer was twisted and he could not measure the bottom of the channel. Three weeks later he at last got Fulcher to start winding the coils onto the channels. The wire was bad: it was hard to insulate the channel and Chrystal tried using a paraffin-soaked ribbon between the wire and the metal bobbins. Maxwell suggested rolling a disc round the wire to get its length more accurately, but Chrystal "was not successful in getting constant results. I thought the cathetometer safer, in whose scale by the way I am disposed to believe."[48] By March 1879 this belief had strengthened into faith. Maxwell, back in Cambridge, asked Chrystal for the records of his trials so that the coils could now be fitted into the experiments on *v*. Chrystal told him that he was much more confident in the dimensions of the coil than in their coefficients of induction. The former were all carefully recorded: "I can give you the number of layers, the number of rounds in each layer and four diameters of each layer. . . . I have all the actual Cathetometer readings." On the other hand, the induction coefficients "were very rough." Chrystal had bad news about many of the resources they needed: "Somebody else must finish the absolute unit experiment." He stressed the difficulty in insulating the coils from the bobbins and the bad wire quality. His trials had also shaken "my faith in Thomson's electrometer a little. The leakage from the quadrants was so bad that I could get nothing but the roughest results." So he recommended a new engine to drive the commutator to maintain constant speed; a new electrometer from White's; and, above all, a lab assistant at the tuning fork to "regulate a large resistance."[49] Chrystal's experiences showed that troubles of long-distance calibration which had plagued the isolated laboratory could only be solved with a new pattern of workshop skill developed inside and around the Cavendish.

"Our Theory is Certainly Not Contradicted by the Results"

The solution to the problem of skill involved the integration of instrumentation and training. The induction coils, magnetometers, and commutators became emblematic of laboratory initiation. Maxwell's *Treatise* helped explain this status. In the key section of the book comparing values of v with those of light speed, he noted that the two quantities were "of the same order of magnitude," a fact known for a decade or more. "In the meantime our theory, which asserts that these two quantities are equal, and assigns a physical reason for this equality, is certainly not contradicted by the comparison of the results such as they are." But this argument from noncontradiction was scarcely compelling, hence the stress on increasing accuracy within the laboratory. The *Treatise* described and rationalized instrumentation as part of academic physics. Maxwell's brilliant analysis of ballistic galvanometry was one example. Elsewhere, chapters on instruments outlined a hierarchy of tools, ranging from "null or zero methods" such as the electroscope, through registration devices and scale readings, to "instruments so constructed that they contain within themselves the means of independently determining the true values of quantities." Maxwell here explained the doctrinal and practical significance of absolute measures. His influential program notes on the South Kensington scientific instruments exhibition of 1876 served a similar purpose. Instrument design embodied the principles of physics they were designed to illustrate and then to investigate. "There must be a prime mover or driving power, and a train of mechanism to connect the prime mover with the body to be moved." Dynamics was basic in theory and energy transmission basic in design. Electrical instruments were to be classed into sources, channels, restraints, reservoirs, overflows, regulators, indicators and, finally, scales, such as the ohm. By making instruments look like machines, Maxwell helped link workshop with laboratory practices.[50]

This machine vocabulary aided the introduction of the British Association standards into the laboratory. In the 1870s, Garnett's custom was to introduce tyros to the Cavendish by getting them to work on the Kew magnetometer, as it "afforded practice not only in reading scales and making adjustments, but also in time observations, counting the beats of a watch while observing the vibrating magnet." Maxwell reported in late 1876 that "as we get richer in apparatus, mathematical lectures give way to experimental, and the blackboard to the lamp and scale."[51] The status of the British Association machinery was also visible in the laboratory layout. The magnetic room occupied the east end of the ground floor, a site of honor, with solidly based stands and a relatively large space for ancillary equipment. These devices were central to the rites of entrance to the laboratory. The wrangler Richard Glazebrook began work at the Cavendish in 1876. He was ordered to help Chrystal by using a quadrant electrometer to supervise the constant e.m.f. of a set of Daniell cells. Glazebrook soon gave up, revealing the difference in values between the standards program and the varsity *cursus honorum:* "the measurement which I was set to do did not seem likely to afford material for a Fellowship Dissertation."[52]

This was one way in which it was hard to form a cadre of skills. There were others. The young natural scientist A. P. Trotter was set to work on the Kew magnetometer in 1878 and although this led to "the production of fairly good results," Garnett had to teach Trotter the basic principles of measuring the magnet's moment, "an abstraction to which I had not been introduced." Trotter was also taught to measure the resistance of a ballistic galvanometer by measuring the logarithmic decrement of its swings. "This experiment was good practice, but what a way of measuring the resistance of a galvanometer!" It was good practice in the principles Maxwell's regime sought to inculcate. The Cavendish professor began to develop a theoretical technology using his *Treatise* to interpret the behavior of this ballistic galvanometer. "We are learning to distinguish between shoves, kicks and jerks in the galvanometer needle," he told Jenkin. "Several men are able to rig up the various combinations out of their own heads."[53] This work is not to be interpreted as the simple-minded application of truths set out in Maxwell's principles. Theoretical and practical resources were produced together. In cases like these, the "men" were learning to see the laboratory bench electromagnetic skills as instances of more general principles of changing current and field strength. But, as his collaborators often emphasized, there was a very wide gap between visions of a trained workforce and the actual production of routine practical skills. For example, in 1874 W. M. Hicks, inspired by Maxwell's teaching, tried to use a ballistic galvanometer balanced between two coils of different sizes set at different distances to measure the speed of electromagnetic waves: "Of course, nothing came of it, but the practice was worth a great deal to me." Hicks's student trials have been cited as the sole example of any Cavendish cLaboratory attempt directly to measure the speed of electromagnetic radiation. They are rather more significant as indications of the status of the electromagnetic theory of light in Cambridge in the 1870s. The testimony of contemporary electrical engineers who worked at the Cavendish supports the impression that confidence in this theory hinged on the precision measurement program on electromagnetic units. Trotter, who later became a distinguished electrical engineer and whose brother, Coutts Trotter, was a wrangler who studied with Helmholtz and taught physics at Trinity College, simply denied any knowledge of Maxwell's *Treatise* when he was a natural scientist at the Cavendish in 1878–79. He first read the book when he inherited it from his brother later in the 1880s. Another electrical engineer who worked at the laboratory in this period, J.E.H. Gordon, published an authoritative account of the program in his 1880 treatise on electromagnetism. There he baldly stated that it was almost impossible to measure the speed of electromagnetic disturbances, and that "the indirect method of comparison of units is as certainly a measure of the velocity of disturbance, and is capable of far greater accuracy than is ever likely to be obtained by the direct method."[54]

The confidence in this "accuracy" depended on the patient multiplication of sites at which such exacting laboratory practices could be performed. These values were distributed in company with the resistance boxes which the Cavendish validated. In summer 1877 the Cambridge physicists heard from Heinrich Weber at Zurich that in a new program of trials using the heating effects of a current-carrying resistance he had shown that Wilhelm Weber and Friedrich Kohlrausch had both erred considerably in their prior resistance measures. The following May, Heinrich Weber asked Maxwell to send him a set of validated resistance standards, and Maxwell ordered Chrystal to dispatch boxes to Zurich "after being compared with the B.A. standard and the resistance determined at a standard temperature." The distribution of such units helped make the authority of the Cambridge system. Its revamped standards program, inaugurated by Lord Rayleigh and his assistants after Maxwell's death in 1879, added to this prestige with a massive increase in the finance, time, and resources devoted to the measurements and the internationalization of the Cambridge units.[55] Henry Rowland also played a major role in this process of multiplication. Between summer 1877 and spring 1878 he ran a series of new trials on the determination of an absolute resistance standard at Baltimore, winding all the coils himself, using very fine wire, and getting results which sat rather well with the revised British Association standards. He told Maxwell that his new method, using the mutual induction of two separated coils, would eliminate the intensity of earth-magnetism and require a very accurate homemade galvanometer. The superiority of Rowland's instrument shop was soon demonstrated. Maxwell counseled him on the "probable error of the resistance of the connexions," and arranged a Cambridge comparison of the Elliott's ohms which Rowland acquired. Elliott's told Rowland that "there is considerable difficulty in determining who is right." Furthermore, Rowland had some bad news for Maxwell: "My value [of resistance], when introduced into Thomson's and Maxwell's values of the ratio of the electromagnetic to the electrostatic units of electricity, caused a yet further deviation from its value as given in Maxwell's electromagnetic theory of light." Rowland could only offer a small consolation: "Experiments on this ratio have not yet attained the highest accuracy.[56]

Rowland's comments were rather striking. He did not doubt the truth of Maxwell's account, even though his own work damaged its authority. Rowland maintained the resistance program into the 1880s, putting his students to work, requisitioning $10,000 for fresh trials, and offering to pay the Cavendish to ship its coils to Baltimore. In late 1881, after the important Paris Electrical Congress which provisionally ratified the Cambridge program of resistance standards, Rayleigh told Rowland that "we are beginning operations to compare the B.A. unit with mercury and it would be a good thing if you would refer your absolute measurements also to mercury." It was hard to get the Cambridge and Johns Hopkins values to agree at a distance. Rowland could find "no error of much account" in his own trials. He needed Rayleigh's advice "to determine the reason of the discrepancy between my experiments and your own." There was much discussion about exchange of hardware with Cambridge. Rayleigh and Glazebrook doubted the worth of getting versions of the Baltimore coils in Cambridge: "We shall not be disposed to do anything more in the matter here unless and until you confirm your former value." But Rayleigh was prepared to "spare the frame and revolving rings used in the B.A. method," if Rowland would pay the costs and guarantee the coils' use. "It will be interesting to see what result you come to," Rayleigh wrote in summer 1883. He even offered Rowland a bet of £5 that the Baltimore physicist would "get a value for the B.A. unit nearer to 0.970 [ohms] than to 0.991."[57] It was the engineering network which answered these challenges. In summer 1878 W. E. Ayrton reported on trials he had performed in Japan with John Perry, using a revised form of Weber's capacitor method to lower the errors in coil methods, and giving a value for *v* of 29.8 x 10^7 meters per second. Ayrton was much concerned with "the exact commercial measurement of electrical quantities." It was within the setting of a telegraphy laboratory that this value of *v* was produced. Commercial resources warranted these electromagnetic measures. The value was confirmed the following year by Hockin in a series of tests on standard condensers to be used in cable testing. Hockin's galvanometric tests on condenser discharge were explicitly part of a project to set up "some one condenser or condensers as a provisional standard" for telegraphy: he worked out a value for *v* of about 29.9 x 10^7 meters per second.[58] The work was repeated in 1883 by the Cambridge wrangler J. J. Thomson. A year later at Baltimore William Thomson took it for granted that the most recent values of *v* and light speed were all but identical. Indeed, he now claimed that he had already established this fact three decades earlier in his first work on telegraphic signaling. But he flatly denied that this identity warranted the claim that light waves were transverse undulations in an electromagnetic ether. By the early 1880s values for *v* seemed to be converging around a mean of about 29.6 x 10^7 meters per second, matching Foucault's value of 29.84 rather well. While astronomers continued to argue about light speed, especially after the repetition of Fizeau's trials by Cornu in Paris in 1874–75, further work in the Cavendish by J. J. Thomson's team and at Cleveland by A. A. Michelson on light speed brought more secure values as new engineering techniques in interferometry and electrotechnology were deployed. Michelson's project was especially vital, linked as it was with an attempt to make light waves into universal length standards. Optical metrology was an important resource for this new electromagnetism.[59]

The identity of light speed and the ratio of the electrostatic and electromagnetic units was established over a period of three decades by a worldwide network of engineers and laboratory technicians. By the time all participants had agreed to the security of this identity, its role as the principal evidence for Maxwell's theory had already been displaced by the achievements of Heinrich Hertz. Long before this agreement a coherent program of research which took the identity for granted had been launched among Cambridge mathematicians and their colleagues. Some of the most useful values of *v* were produced by those hostile to Maxwell's electromagnetic theory. So the exactitude of this identity was neither sufficient nor necessary in recruiting believers. Instead, it became what Ludwik Fleck has called "a signal of resistance" shared by a set of collective investigators, a tenet to be made part of a communal project and not easily to be given up within that community. This claim about the collective quality of precision values is not new. In his address to the British Association in summer 1882, looking back over the history of the project to make absolute resistance standards, Lord Rayleigh sternly counseled that "the desire for great accuracy has sometimes had a prejudicial effect. . . . The comparison of estimates of uncertainty made before and after the execution of a set of measurements may sometimes be humiliating, but it is always instructive." Rayleigh was the acknowledged guru of late Victorian precision measurement. When Rayleigh announced the existence of a hitherto unknown inert gas in 1894, Glazebrook recalled that "it was sufficient for me to be told that Rayleigh had discovered a discrepancy between the densities of chemical nitrogen and that derived for the air to know that this must be the case." Rayleigh himself noted that in the case of the absolute resistance unit "discrepancies" were omnipresent. In his survey he mentioned the highly various endeavours of Weber, Kohlrausch, Thomson, Maxwell, and Rowland. "The history of science teaches only too plainly the lesson that no single method is absolutely to be relied upon." It is no coincidence that Rayleigh, Glazebrook, and their colleagues were instrumental in 1900 in establishing Britain's first national laboratory devoted to the development of precision measures. Exact measures were the product of reorganizations of working practices and of the management of a range of different workplaces. The criteria which such measures had to meet were not to be stipulated in advance, but were established in the course of the project. The label of precision attached to any measure hinged on cultures of communal trust and was a consequence of the strength of the social relations between these separate and complex institutions.[60]

Notes to Chapter Six

I am grateful for permission to cite papers held in Cambridge University Library; Imperial College Archives; Eisenhower Library, Johns Hopkins University; and Glasgow University Library. Thanks also to Peter Harman, Bruce Hunt, and Norton Wise.

1. Garnet Wolseley, *The Soldier's Pocket Book* 5th ed. (London: Macmillan, 1886), 13–14.

2. Thomson to Maxwell, 24 August 1872, Cambridge University Library MSS ADD 7655 / II / 62; Maxwell, "Dimensions" (1877), *Encyclopaedia Britannica,* 24 vols. (Edinburgh: Black, 1875–89), 7:240–41; the same passage is in Maxwell, *Treatise on Electricity and Magnetism,* 3d ed., 2 vols. (Oxford: Clarendon Press, 1891), 1:1–2.

3. Maxwell, "Dimensions of Physical Quantities," Cambridge University Library MSS ADD 7655 / V h / 4. The document must be later than 1867: it is

written on the back of an examination paper by H. Pearson, who sat the Tripos that year.

4. Fleeming Jenkin, "Is one man's gain another man's loss"? (1884) in *Papers Literary and Scientific* 2 vols. (London: Longmans, Green, 1887), 2:140–54, pp. 141–42; John Herschel, *Preliminary Discourse on the Study of Natural Philosophy* (London: Longmans, 1830), 128.

5. H. M. Collins, *Artificial Experts: Social Knowledge and Intelligent Machines* (Cambridge, Mass.: MIT Press, 1990), 22–26.

6. James Clerk Maxwell, "On Physical Lines of Force: Part 3," in W. D. Niven, ed., *Scientific Papers of James Clerk Maxwell,* 2 vols. (Cambridge; Cambridge University Press, 1890), 1:499–500.

7. Pierre Duhem, *Les Théories Electriques de J. C. Maxwell* (Paris: Hermann, 1902), 62; Joan Bromberg, "Maxwell's Displacement Current and his Theory of Light," *Archive for History of Exact Sciences* 4 (1967), 218–34, p. 227; A. F. Chalmers, "Maxwell's methodology and his application of it to electromagnetism," *Studies in History and Philosophy of Science* 4 (1973), 107–64, pp. 134–37; Salvo d'Agostino, "Weber and Maxwell on the Discovery of the Velocity of Light," in M. D. Grmek et al., eds., *On Scientific Discovery* (Dordrecht: Reidel, 1980), 281–93, p. 287; Daniel M. Siegel, *Innovation in Maxwell's Electromagnetic Theory* (Cambridge: Cambridge University Press, 1991), 136–43.

8. William Thomson, "Notes of Lectures on Molecular Dynamics and the Wave Theory of Light," in Robert Kargon and Peter Achinstein, eds., *Kelvin's Baltimore Lectures and Modern Theoretical Physics* (Cambridge Mass.: MIT Press, 1987), 12. See Crosbie Smith and Norton Wise, *Energy and Empire: a biographical study of Lord Kelvin* (Cambridge: Cambridge University Press, 1989), chap. 13; Bruce Hunt, *The Maxwellians* (Ithaca: Cornell University Press, 1991), 162–68 discusses Thomson's opposition in 1884–88 and notes his temporary conciliation with the Maxwellians.

9. William Thomson, *Popular Lectures and Addresses,* 3 vols. (London: Macmillan, 1889–94), 2:160–61(1871); 1:443 (1876) (my emphasis); I:83, 119 (1883).

10. "Report of the Committee appointed by the British Association on Standards of Electrical Resistance," *British Association Reports* (1863), 111–24, p. 124; James Clerk Maxwell and Fleeming Jenkin, "On the Elementary Relations between Electrical Measurements," ibid., 130–63, pp. 149, 153–54. The link between the BAAS committee and Maxwell's electromagnetic theory is discussed by Salvo d'Agostino, "Experiment and Theory in Maxwell's Work," *Scientia* 113 (1978), 469–80, who attributes the principal 1863 report to Maxwell. For Thomson's drafting of this report, see Silvanus P. Thompson, *Life of Lord Kelvin,* 2 vols. (London, Macmillan, 1910), 1:419.

11. For the relation between Cambridge values and the establishment of the Cavendish, see Simon Schaffer, "A manufactory of ohms: late Victorian metrology and its instrumentation," in Susan Cozzens and Robert Bud, eds., *Invisible Connexions* (Bellingham: SPIE, 1992), 23–56.

12. Maxwell, *Treatise on Electricity and Magnetism,* 2:436; Henry A. Rowland, "Research on the Absolute Unit of Electrical Resistance," *American Journal of Science* 15 (1878), 281–91, 325–36, 430–39, pp. 282–84; for Thomson's re-

turn to skepticism after 1894, see Smith and Wise, *Energy and Empire,* 488–91; David B. Wilson, *Kelvin and Stokes* (Bristol: Hilger, 1987), 168–69; Ole Knudsen, "Mathematics and Physical Reality in William Thomson's Electromagnetic Theory," in P. M. Harman, ed., *Wranglers and Physicists* (Manchester: Manchester University Press, 1985), 149–79, pp. 171–76.

13. Rowland to Maxwell, March 1876, in Nathan Reingold, *Science in Nineteenth Century America* (Chicago: Chicago University Press, 1964), 269. See also Samuel Rezneck, "An American Physicist's Year in Europe: Henry A. Rowland, 1875–1876," *American Journal of Physics* 30 (1962), 877–86, p. 885.

14. For the London experiments, see *British Association Reports* (1863), 111–24, and for the issue of units, see [Fleeming Jenkin], "Report of the Committee on Standards of Electrical Resistance," *British Association Reports* (1864), 345–49, p. 345, and Fleeming Jenkin, "Electrical Standard," *Philosophical Magazine* 29 (1865), 248. The British Association work is discussed by Bruce Hunt, "The Ohm is where the Art is: British telegraph engineers and the development of electrical standards," *Osiris*, forthcoming.

15. For the teamwork, see Ian Hopley, "Maxwell's work on electrical resistance: the determination of the absolute unit of resistance," *Annals of Science* 13 (1957), 265–72, pp. 266–68; the governor is discussed in Maxwell, *Scientific Papers*, 112; Maxwell to Thomson 1863, Glasgow University Library MSS Kelvin M13. Details of the floorboard technique and the value of π are given in Maxwell to Thomson, 31 July 1863, Glasgow University Library MSS Kelvin M14; [Thomson], "Report of the Committee on Electrical Standards" (1863), 120, and by William Garnett, Cambridge University Library MSS ADD 8385 no. 10. Data for the experiments of 1863–64 are contained in Cambridge University Library MSS ADD 7655 Vc/9 and Vc/12, especially the estimates of errors in the scales.

16. Werner Siemens, "On the question of the unit of electrical resistance," *Philosophical Magazine* 31 (1866), 328; cf. Werner Siemens, *Inventor and entrepreneur: recollections,* 2d ed. (London: Lund Humphries, 1966), 133, 165–66 and Fleeming Jenkin, "New Unit of Electrical Resistance," *Philosophical Magazine* 29 (1865), 477–86. The problem of Maxwell's treatment of error is discussed in Ian Hopley, "Maxwell's work on Electrical Resistance: the Determination of the Absolute Unit of Resistance," *Annals of Science* 13 (1957), 265–72, p. 269.

17. Friedrich Kohlrausch, "Determination of the Absolute Value of the Siemens Mercury Unit of Electrical Resistance," *Philosophical Magazine* 47 (1874), 294–309, 342–54, on pp. 295, 301 (first published November 1870); Siemens, *Inventor and entrepreneur,* 167. For Kohlrausch and Weber in Göttingen, see Christa Jungnickel and Russell McCormmach, *Intellectual Mastery of Nature: Theoretical Physics from Ohm to Einstein,* 2 vols. (Chicago: University of Chicago Press, 1986), 2:72–74. For Kohlrausch's *Leitfaden*, see David Cahan, "The Institutional Revolution in German Physics, 1865–1914," *Historical Studies in Physical Sciences* 15 (1985), 1–65, pp. 48–50, and Kathryn Olesko, *Physics as a Calling: Discipline and Practice in the Königsberg Seminar for Physics* (Ithaca: Cornell University Press, 1991), 409–12.

18. John David Miller, "Rowland and the Nature of Electric Currents," *Isis*

63 (1972), 5–27, pp. 6–9; Maxwell to Rowland, 9 July 1873 and 9 July 1874, in Reingold, *Science in Nineteenth Century America*, 265, 267.

19. Rezneck, "An American Physicist's Year in Europe," 879–82; Rowland, Diary of European Tour, Rowland MSS Box 22, pp. 57–58.

20. Henry Rowland, "Note on Kohlrausch's Determination of the Absolute Value of the Siemens Mercury Unit of Electrical Resistance," *Philosophical Magazine* 50 (1875), 161–63, p. 161.

21. For the trials at Berlin, see Miller, "Rowland and Electric Currents," 10–15; Rezneck, "An American Physicist's Year in Europe," 884; Jed Z. Buchwald, *From Maxwell to Microphysics: Aspects of Electromagnetic Theory in the Last Quarter of the Nineteenth Century* (Chicago: University of Chicago Press, 1985), 75–77. For Kohlrausch and Rowland at Würzburg, see David Cahan, "Kohlrausch and Electrolytic Conductivity," *Osiris* 5 (1989), 167–85, p. 177. For Toepler and Rowland at Graz, see Jungnickel and McCormmach, *Intellectual Mastery of Nature,* 2:66; Rowland, Travel Diary, 75–76, 87–88.

22. James Clerk Maxwell, "Introductory Lecture on Experimental Physics," in Maxwell, *Scientific Papers,* 2:241–54, pp. 242–44.

23. "Report of the Electrical Standards Committee Appointed by the Board of Trade," *British Association Reports* (1891), 154–60.

24. Maxwell to Tyndall, 23 July 1868, Imperial College London MSS, archives.

25. For Weber's work see Jungnickel and McCormmach, *Intellectual Mastery of Nature,* 1:144–46. Maxwell's criticisms are in Maxwell to Jenkin, 27 August 1863, in Lewis Campbell and William Garnett, *Life of Maxwell* (London, 1884), 252; Maxwell's later criticism of Weber's method is in "On a method of making a direct comparison of electrostatic with electromagnetic force" (1868), in *Scientific Papers,* 2:125–43, p. 136 and in *Treatise on Electricity and Magnetism,* 2:417.

26. Gustav Kirchhoff, "On the Motion or Electricity in Wires," *Philosophical Magazine* 13 (1857), 393–412. For Maxwell's design of the mechanical ether model's elasticity and density in accordance with Weber's number, see Maxwell, "On Physical Lines of Force: Part 3" (1862), in *Scientific Papers* 1:497–99. Siegel, *Innovation in Maxwell's Theory,* 129–35 states that "the ratio of medium constants . . . was set in that model to get the correct ratio." The subsequent "coincidence" with light speed is described in Maxwell to Faraday, 19 October 1861, in P. M. Harman, ed., *The Scientific Letters and Papers of James Clerk Maxwell* (Cambridge: Cambridge University Press, 1990), 1:685–86; the "nearness between the two values" is celebrated in Maxwell to Thomson, 10 December 1861, ibid., 1:695. In the early 1870s Maxwell made detailed notes on Kirchhoff's paper on electricity in wires: see Cambridge University Library MSS ADD 7655 Vn/1, p. 44 ff.

27. Maxwell to Monro, 20 (?) October 1861, in Harman, *Scientific Letters and Papers of Maxwell,* 1:690 and Monro to Maxwell, 23 October 1861, in Campbell and Garnett, *Life of Maxwell,* 245. Thomson's dispatch of resistances to Göttingen is mentioned in Thomson to Helmholtz, 30 July 1856, in Thompson, *Life of Thomson,* 322, and Weber to Thomson, 14 September 1861,

Glasgow University Library Kelvin MSS W4. His opposition to Maxwell's inference from the numerical agreement, and his claim for priority on the basis of an encyclopedia article of 1860, is discussed in Crosbie Smith and M. Norton Wise, *Energy and Empire: a Biographical Study of Lord Kelvin* (Cambridge: Cambridge University Press, 1989), 458–60.

28. [Fleeming Jenkin], "Provisional Report of the Committee appointed by the British Association on Standards of Electrical Resistance," *British Association Reports* (1862), 125–35, p. 131; [Thomson], "Report on Standards of Electrical Resistance" (1863), 121. For response to Kirchhoff and Weber's work, see Kirchhoff to Jenkin, 8 June 1862, in *British Association Reports* (1862), 151; Harman, *Scientific Letters and Papers of Maxwell,* 1:686 n. 15; Smith and Wise, *Energy and Empire,* 693.

29. Hippolyte Fizeau, "Sur une expérience rélative à la vitesse de la propagation de la lumière," *Comptes Rendus de l'Académie des Sciences* 29 (1849), 90–92; Samuel Haughton and Joseph Galbraith, *Manual of Astronomy* (London: Longman, 1855), 39. Maxwell gives the right version from the *Manual* and the wrong one from Fizeau in "On Physical Lines of Force: Part 3," 500. He gives the right value from Fizeau in Maxwell to Thomson, 10 December 1861, in Harman, *Scientific Letters and Papers of Maxwell,* 695. His use of Haughton and Galbraith at Aberdeen is mentioned in Maxwell to Litchfield, 7 February 1858, ibid., 582. For enthusiasm about Fizeau, see *Revue scientifique et industrielle* 5 (1849), 393–96.

30. Léon Foucault, "Détermination expérimentale de la vitesse de la lumière" (1862), in C. M. Gariel, ed., *Receuil des Travaux Scientifiques de Léon Foucault* (Paris: Gauthier-Villars, 1878), 216–26, p. 217. Foucault's work is discussed in its astronomical context by Leverrier in *Cosmos* 21 (1862), 357–60 and Clerke, *Astronomy in the Nineteenth Century*, 232. Clerke notes that the Sun's "sudden bound of four million miles nearer" might mistakenly "shake public faith in astronomical accuracy." Herschel's comment is in "On Light: Part 1" (1864), in *Familiar Lectures,* 219–67, p. 234. Maxwell cites the discredited astronomical figure alongside Foucault's result in "Dynamical Theory of the Electromagnetic Field," in *Scientific Papers*, 1:580.

31. Maxwell to Hockin, 7 September 1864, on the safe determination of *v*, in Campbell and Garnett, *Life of Maxwell,* 255; Maxwell asks Stokes for Foucault's work in Maxwell to Stokes, 8 May 1857, in Harman, *Scientific Letters and Papers of Maxwell,* 503. For "rough" estimates of *v*, see Thomson to Stokes, 30 October 1854 and 12 February 1855, Cambridge University Library MSS ADD 7656, K74 and K78. Commercial values of *v* are described in Wise and Smith, *Energy and Empire,* 454–56 and Bruce Hunt, "Michael Faraday, Cable Telegraphy and the Rise of Field Theory," *History of Technology* 13 (1991), 1–19, pp. 8–10.

32. Maxwell on Hockin in Maxwell, "On a Method of Making a Direct Comparison of Electrostatic with Electromagnetic Force" (1868), *Scientific Papers,* 2:125–43, p. 127; cable duty, in Hockin to Maxwell, 27 July 1868, Cambridge University Library MSS ADD 7655 / II / 31. Gassiot had been a protagonist of the London Electrical Society: see Iwan Morus, "Currents from the Underworld: elec-

tricity and the technology of display in early Victorian England," *Isis* 84 (1993), 50–69.

33. Maxwell, "A Method of Direct Comparison," 126. Compare his "Experiments on the value of *v*, the ratio of the electromagnetic to the electrostatic unit of electricity," *British Association Report* (1869), 436–38.

34. Maxwell to Thomson, 17 September 1864 and 2 February 1865, Glasgow University Library MSS Kelvin M16 and Ml8.

35. Maxwell to Thomson, 17 April 1865, Cambridge University Library MSS ADD 7655 / II / 22A. Maxwell describes the method in "A Method of Direct Comparison," pp. 130–32, 135, and mentions the time required on p. 127; see I. B. Hopley, "Maxwell's determination of the number of electrostatic units in one electromagnetic unit of electricity," *Annals of Science* 15 (1959), 91–108, pp. 91–93.

36. Dugald McKichan, "Determination of the Number of Electrostatic Units in the Electromagnetic Unit," *Philosophical Transactions* 163 (1873), 409–27, pp. 409–10, 417–18, 421; Thompson, *Life of Thomson,* 524–25; William King, "Description of Sir William Thomson's Experiments made for the Determination of *v*," *British Association Report* (1869), 434–35. For the dispatch to the Royal Society, see McKichan to Stokes, 19 April 1873, Cambridge University Library MSS ADD 7656 RS 929. For the use of these values by Maxwell, see "A Method of Direct Comparison," 136 and *Treatise on Electricity and Magnetism,* 2:436 with J. J. Thomson's additions to the third edition (1891), ibid. For further comments, see J.E.H. Gordon, *A Physical Treatise on Electricity and Magnetism*, 2 vols., 3d ed. (London: Sampson Low, Marston, Searle and Rivington, 1891), 2:228–29.

37. Maxwell, *Treatise on Electricity and Magnetism,* 2:418; Maxwell to Jenkin, 27 August 1863, in Campbell and Garnett, *Life of Maxwell,* 252; Hockin to Maxwell, 15 May 1868, Cambridge University Library MSS ADD 7655 / II / 30, published in Hopley, "Maxwell's determination," 97.

38. Maxwell, "A Method of Direct Comparison," 135–36; Maxwell, "Experiments on the value of *v*," 438.

39. William Thomson, "Presidential Address to the British Association 1871," in *Popular Lectures and Addresses,* 3 vols. (London: Macmillan, 1889–94), 2:160–61; Thompson, *Life of Kelvin,* 1026.

40. Whipple to Maxwell, 6 May 1874 and Maxwell to Jenkin, 22 July and 18 November 1874, Cambridge University Library MSS ADD 7655 /II / 77, 241–42. For Thomson on *v* and telegraphy, see Smith and Wise, *Energy and Empire*, 457–58.

41. Maxwell's report of July 1873 in Campbell and Garnett, *Life of Maxwell*, 267–68; his report of 1877 in *Cambridge University Reporter* (15 May 1877), 434; compare Maxwell on Henry Cavendish in Maxwell to Chrystal, 21 September 1877, Cambridge University Library MSS ADD 8375 f. 14. His membership of the STE is reported in Preece to Maxwell, 7 March l873, Cambridge University Library MSS ADD 7655 / V i / 12. His exchange with Kew is Whipple to Maxwell, 6 May 1874, and with Glasgow is White to Maxwell, 5 June 1876, in Cambridge University Library MSS ADD 7655 / II / 77 and 113. The supply of the

absolute electrometer is recorded in the Cavendish Laboratory account book, Cambridge University Library MSS ADD 7655 / V j / 3 (29 May 1876).

42. *History of the Cavendish Laboratory 1871–1910* (London: Longmans, Green, 1910), 35 on Garnett; Elliot Brothers to Maxwell, 25 October 1878, Cambridge University Library MSS ADD 7655 / II / 165 for resistance temperatures; T.J.N. Hilken, *Engineering at Cambridge University 1783–1965* (Cambridge: Cambridge University Press, 1967), 63–74, and M. J. Cattermole and A. F. Wolfe, *Horace Darwin's Shop* (Bristol: Adam Hilger, 1987). See Kenneth Lyall, *Electrical and Magnetic Instruments* (Cambridge: Whipple Museum of the History of Science, Catalogue 8, 1991) nos. 23–24, 451 (Wh: 1339–40, 4376). For the first Cavendish experiments on the temperature dependence of Elliots' standards, see G. Chrystal and S. A. Saunder, "Results of a Comparison of the British-Association Units of Electrical Resistance," *British Association Reports* (1876), 13–19. For Trotter on Stuart's shop, see A. P. Trotter, "Elementary Science at Cambridge, 1876–1879," Cavendish Laboratory 82.T.2 (written 1945), 34.

43. For instrument design, see Hopley, "Maxwell's determination," 99–101, and id., "Maxwell's work on electrical resistance—proposals for the redetermination of the BA unit of 1863," *Annals of Science* 14 (1958), 197–210, pp. 201–2. See Maxwell to Chrystal, 7 August 1876, Cambridge University Library MSS ADD 8375 f. 4; Maxwell, *Treatise of Electricity and Magnetism*, 396, 420–21. The ballistic galvanometers are described in Lyall, *Electric and magnetic instruments,* nos. 14 and 17 (Wh: 1318, 1334). Joule comments to Maxwell (1876), in Cambridge University Library MSS ADD 7655 / II / 123. I am very grateful to Otto Sibum for access to his work on Joule's researches.

44. For Horn, see Hopley, "Redetermination of the BA unit," 208; see Hockin to Maxwell, 14 January 1878, Cambridge University Library, MSS ADD 7655 /II / 150; Ambrose Fleming, "Some memories," in *James Clerk Maxwell: a commemoration volume* (Cambridge, 1931), 116–24, p. 119.

45. For Chrystal and Fulcher, see Hopley, "Redetermination of the BA unit," 209; Chrystal to Maxwell 9 July 1878, and Maxwell to Chrystal 9 July 1878, Cambridge University Library MSS ADD 7655 / II / 159 and MSS ADD 8375 f. 16. Maxwell's report is in *Cambridge University Reporter* (2 April 1878), 420.

46. Chrystal to Maxwell, 5 July 1877 and 9 July 1878, Cambridge University Library MSS ADD 7655 / II / 134, 159. Maxwell discusses the St. Andrew's appointment in Maxwell to Chrystal, 12 July 1877, Cambridge University Library MSS ADD 8375 no. 11.

47. Maxwell to Garnett, 11 July 1876, ibid., no. 3; Maxwell to Chrystal, 30 December 1876, ibid., no. 6.

48. Maxwell to Chrystal, 9 July 1878, Cambridge University Library MSS ADD 8375 no. 16; Chrystal to Maxwell, early July 1878 and 29 July 1878, Cambridge University Library MSS ADD 7655 / II / 161–62. The Breithaupt cathetometer is described in Cavendish laboratory accounts book, Cambridge University Library MSS ADD 7655/ V j /3 (2 August 1877).

49. Chrystal to Maxwell, 6 March 1879, Cambridge University Library MSS ADD 7655 / II / 183.

50. Maxwell, *Treatise on Electricity and Magnetism,* 1:326–27, 2:436; program notes (1876) in Maxwell, *Scientific Papers,* 2:505–27, discussed in Peter

Galison, *How experiments end* (Chicago: Chicago University Press, 1987), 23–27.

51. *History of the Cavendish,* 34–35; Maxwell to Campbell, Christmas 1876, in Campbell and Garnett, *Life of Maxwell,* 304–5.

52. R. T. Glazebrook, "Early days at the Cavendish Laboratory" in *James Clerk Maxwell: a commemoration volume* (Cambridge: Cambridge University Press, 1931), 130–41, pp. 135–37.

53. Trotter, "Elementary Science at Cambridge," 20; Maxwell, *Treatise on Electricity and Magnetism,* 2:382–91; Maxwell to Jenkin, 22 July and 18 November 1874, Cambridge University Library MSS ADD 7655 / II / 241 and 242. Compare the same remarks in Maxwell to Rowland, 9 July 1874, in Reingold, *Science in Nineteenth Century America,* 268.

54. *History of Cavendish Laboratory,* 19. Hicks's work is discussed in Thomas K. Simpson, "Maxwell and the Direct Experimental Test of his Electromagnetic Theory," *Isis* 57 (1966), 411–32, pp. 425–26. Trotter reminisces in "Elementary Science at Cambridge," 32. For "accuracy" of the units' ratio, see Gordon, *Physical Treatise on Electricity and Magnetism,* 2:224. For Gordon's work at the Cavendish, see Maxwell to Gordon (? 1879) Cambridge University Library MSS ADD 7655 / II / 217; J.E.H. Gordon, On the Determination of Verdet's Constants in Absolute Units," *Philosophical Transactions* 167 (1877), 1–34.

55. Heinnch Weber, "Electromagnetic and Calometric Absolute Measurements," *Philosophical Magazine* 5 (1878), 30–43, 126–39, 189–97, on p. 32; Weber to Maxwell 19 May 1878, Cambridge University Library MSS ADD 7655/ II / 154; Maxwell to Chrystal, 17 July 1878, Cambridge University Library MSS ADD 8375 no. 18. For Rayleigh's program, see Robert Strutt, *Life of John Strutt, Third Baron Rayleigh,* 2d ed. (Madison: University of Wisconsin Press, 1968), 109–20.

56. Henry A. Rowland, "Research on the Absolute Unit of Electrical Resistance," *American Journal of Science* 15 (1878), 28l-91, 325–36, 430–39, pp. 282, 284; Rowland to Maxwell, [summer 1877] and [summer 1879], Cambridge University Library MSS ADD 7655 / II / 105, 192; Maxwell to Rowland, 22 April 1879, in Johns Hopkins University, Milton S. Eisenhower Library Special Collections, Rowland MSS 6. See Rowland Notebook "on ratio of electric units 1878–1880," Box 25, Rowland MSS 6.

57. Rowland to Rayleigh, 23 June 1883, in Imperial College London MSS Rayleigh Correspondence; Rayleigh to Rowland, 2 April 1992 and 20 August 1883, in Johns Hopkins University, Milton S. Eisenhower Library Special Collections, Rowland MSS 6. See Rowland Notebook on resistance comparison, December 1881–January 1882, Box 25, Rowland MSS 6.

58. W. E. Ayrton, *Practical Electricity* 6th ed. (London: Cassell, 1894), iii; Graeme Gooday, "Teaching Telegraphy and Electrotechnics in the Physics Laboratory: William Ayrton and the Creation of an Academic Space for Electrical Engineering in Britain 1873–1884," *History of Technology* 13 (1991), 73–111, p. 93; C. Hockin, "Note on the Capacity of a certain Condenser and on the value of *V*," *British Association Reports* (1879), 285–90.

59. William Thomson, "Lectures on Molecular Dynamics" in Kargon and

Achinstein, *Kelvin's Baltimore Lectures,* 42: Thomson refers to his 1854 exchange with Stokes (see above, n. 30) and to "Velocity of Electricity" (1860), amplified in *Baltimore Lectures on Molecular Dynamics and the Wave Theory of Light* (London: Clay, 1904), 688–94. For Cornu and light speed, see A. Cornu, "Determination of the velocity of light and of the Sun's parallax," *Nature* 11 (4 February 1875), 274–76, and H. de Kericuff, "Sur la vitesse de la lumière et la parallaxe du soleil," *Les Mondes* 36 (1875), 372–74. For Michelson's interferometry as applied technology, see D. H. Stapleton, "The context of science: the community of industry and higher education in Cleveland in the 1880s," in Stanley Goldberg and Roger H. Steuwer, eds., *The Michelson Era in American Science 1870–1930* (New York: American Institute of Physics, 1988), 13–22.

60. Lord Rayleigh, "Address to the Mathematical and Physical Sciences Section," *British Association Reports* (1882), 437–41, p. 438; Glazebrook to Strutt, 2 November 1924, in Strutt, *Life of Rayleigh*, 419. For facts as signals of resistance, see Ludwik Fleck, *Genesis and Development of a Scientific Fact* (Chicago: Chicago University Press, 1979), 101–2.

SEVEN

PRECISION AND TRUST:

EARLY VICTORIAN INSURANCE AND THE POLITICS OF CALCULATION

Theodore M. Porter

[T]here is no manufactory for actuaries where
you can get one made to order.
(Edward Ryley, 1853)[1]

HENRY James Brooke, fellow of the Royal Society, collaborated in founding the commercially successful London Life Assurance Association. His obituary notice in the *Proceedings of the Royal Society*, reprinted by the *Assurance Magazine* of the Institute of Actuaries, remarked on the "tone of extreme precision by which all his subsequent acts and observations were characterized." Subsequent to what? Did such a man become a precisian by way of a university education in mathematics, or owing to the demands of actuarial practice? Not according to the obituarist. Brooke's behavior reflected "the precise habit of thought and expression which the active study of the law must necessarily induce."[2]

We tend to associate precision with measurement, machinery, and mathematics. But it is also a cultural value. In educational and even research laboratories, exact measurement began very early to mean intense discipline tending to self-effacement.[3] This was never simply a cognitive matter, nor a monogamous relationship of man to machine, but a strategy for organizing a heterogeneous work force, and a response to bureaucratic pressures.[4] Those pressures are all the more prominent when we shift our attention to the role of precision outside the laboratory, especially where the objects of quantification are human and the purposes administrative or political.[5]

None other of the human sciences acquired the discipline of mathematics so early as did the business of the actuary. We might thus expect to see in the eighteenth- or nineteenth-century actuary the prototype of the modern quantitative expert, one whose expertise derives from mastery of a powerful set of techniques rather than from skill and judgment attained through long experience. And it is true that by the early nineteenth century the best life insurance offices depended on extensive calculations to set their rates. But the processing of numbers by actuaries is not the place to look for an ideology, or even a practice, of faith in precision. The ability to calculate precisely was a minimal requirement for the novice actuary. These calculational skills were to be applied to the preparation of life tables and the determination of premiums. Actuaries were unanimous in recognizing the importance of reliable statistical records. But they did not believe in the possibility of precise measurement, of reducing their work to calculational routines. Precision was not gained through the determined efforts of actuaries, but imposed on them by members of Parliament and other regulatory authorities in pursuit of political and administrative ends. British actuaries thought of themselves as an elite—as gentlemen whose integrity and judgment had earned the public trust. The pursuit of precision meant the denial of that trust, in the name of democratic openness and public scrutiny. It was not achieved simply through technique, but as the outcome of a process of political struggle and cultural self-definition.

Select Lives

Statistics and calculation helped to make life insurance reputable. Traditionally it had not always been easy to distinguish from gambling, especially when the life insured was not one's own. Probability calculations permitted Equitable Assurance to offer policy contracts extending over the whole life of the insured for relatively modest premiums. The history is a touching one, a triumph of rational calculation over the presumed lessons of undisciplined and hence false experience. Thus did a mutual life insurance company with no initial capital combine unprecedentedly low rates with rock-solid security. Indeed, its actuaries were too conservative. For half a century after its founding in 1762, Richard Price and then William Morgan held out against the clamor of members for the share of the steadily accumulating profits to which they were entitled.

This hoard of cash, a tribute to actuarial caution, should caution us too. We must be wary of understanding the triumph of Equitable Assurance as a straightforward vindication of calculation and statistics. Its actuaries chose the most conservative life table from several available. They assumed a low interest rate on their investments. After performing the calculations they added a percentage for an extra measure of safety. Finally, and most interestingly, they did not make life insurance available indiscriminately. Their tables covered, at least in theory, the general population. But they insured only "select lives," persons of temperate habits and good health.[6] This seemed a necessary precaution, since otherwise they quite reasonably anticipated that people would contract for life insurance mainly when their health was deteriorating. Insuring bad lives would lead to higher payouts than the mortality tables predicted. Restricting new insurance contracts to healthy, middle-class people achieved the opposite. Death rates for those persons who had recently entered into an insurance contract were almost infallibly lower than the population average, according to tables prepared by the companies.

Insurance, it should be understood, was a business of managing risk, not merely of compensating customers who were badly treated by it. Fire insurance companies were able to intervene most actively, by keeping fire engines to protect the property of their clients. The practice of insurance against sickness was more ingenious, and more meddlesome. William Sanders, explaining to a Parliamentary select committee on friendly societies in 1849 how he ran the Birmingham General Provident and Benevolent Institution, observed that reliable tables were important for insurance against sickness, but that appropriate rules were absolutely crucial, to limit claims to the genuinely sick. "I would rather trust a society with moderate tables with good rules, than a high one with bad rules."

> SIR H. HALFORD (of the committee): The stringency of the rules consists in the smallness of the payments?
>
> SANDERS: Of course . . . ; we do not allow our members to insure such an amount in sickness, as, looking at their circumstances and income, would prove a temptation to fraud.
>
> HALFORD: You do not refer to any strict supervision as to the reality of the sickness?
>
> SANDERS: That is inquired into, of course; we pay nothing but upon a surgeon's certificate. In addition to that, the parties are visited by ordinary members, and those visits are weekly reported to the secretary.
>
> HALFORD: Of course you interdict work during sickness?
>
> SANDERS: Our rules on that point are more stringent than most.[7]

Such strictness and surveillance were essential to the operation of a large institution. Other friendly societies, especially in Scotland, maintained low rates by acting in effect as charities, only paying out when sickness was attended by genuine need.[8]

The selection of lives in life insurance was also a form of intervention. Its skillful exercise was crucial for the health of the companies. The Pelican, a subsidiary of the successful Phoenix fire insurance company, was largely unprofitable through most of the nineteenth century. Clive Trebilcock, author of an excellent company history, explains that the Pelican "simply was not proficient at selecting which lives to insure."[9] It seems they insured too many dissolute aristocrats (table 1), while other companies enlisted the sober, middle classes.[10]

Age (x).	Completed the age x.	Existing 31st Dec., 1855, between the Ages x and $x+1$.	Died between the Ages x and $x+1$.	Number exposed to the Risk from the Age x to $x+1$.	Probability of Dying in the Year.	Probability of Surviving the Year.	Mean Duration of Life.
0	2534	31	197	2518·5	·07821	·92179	52·00
1	2358	38	38	2339·0	·01625	·98375	55·37
2	2326	27	20	2312·5	·00865	·99135	55·25
3	2328	42	9	2307·0	·00390	·99610	54·73
4	2315	42	10	2294·0	·00436	·99564	53·93
5	2302	27	11	2288·5	·00481	·99519	53·16
6	2293	45	4	2270·5	·00176	·99824	52·42
7	2276	34	6	2259·0	·00266	·99734	51·51
8	2271	49	10	2246·5	·00445	·99555	50·66
9	2250	30	8	2235·0	·00358	·99642	49·88
10	2249	42	9	2228·0	·00404	·99596	49·04
11	2238	36	13	2220·0	·00586	·99414	48·23
12	2230	26	8	2217·0	·00361	·99639	47·52
13	2231	28	11	2217·0	·00496	·99504	46·68
14	2227	36	5	2209·0	·00226	·99774	45·91
15	2225	34	9	2208·0	·00408	·99592	45·02
16	2223	33	16	2207·5	·00725	·99275	44·21
17	2211	30	14	2196·0	·00638	·99362	43·53
18	2197	27	16	2183·5	·00733	·99267	42·82
19	2188	36	17	2170·0	·00783	·99217	42·13
20	2174	48	18	2150·0	·00837	·99163	41·46
21	2135	42	35	2114·0	·01656	·98344	40·80
22	2089	32	20	2073·0	·00965	·99035	40·47
23	2077	31	24	2061·5	·01163	·98837	39·87
24	2053	36	18	2035·0	·00885	·99115	39·33
25	2035	31	23	2019·5	·01139	·98861	38·67
26	2010	38	25	1991·5	·01255	·98745	38·10
27	1985	24	15	1973·0	·00760	·99240	37·58
28	1978	40	17	1958·0	·00868	·99132	36·86
29	1949	27	18	1935·5	·00930	·99070	36·18
30	1938	36	24	1920·0	·01250	·98750	35·51
31	1899	35	11	1881·5	·00585	·99415	34·96
32	1882	30	19	1867·0	·01018	·98982	34·17
33	1867	34	14	1850·0	·00757	·99243	33·51
34	1851	31	12	1835·5	·00654	·99346	32·75
35	1835	26	10	1822·0	·00549	·99451	31·97
36	1815	33	25	1798·5	·01390	·98610	31·15
37	1773	49	11	1748·5	·00629	·99371	30·58
38	1738	31	17	1722·5	·00987	·99013	29·77
39	1720	28	15	1706·0	·00879	·99121	29·08
40	1697	30	23	1682·0	·01367	·98633	28·33
41	1657	36	16	1639·0	·00976	·99024	27·71
42	1626	26	22	1613·0	·01364	·98636	26·99
43	1604	37	17	1585·5	·01072	·98928	26·34
44	1565	38	10	1546·0	·00647	·99353	25·63
45	1537	39	22	1517·5	·01450	·98550	24·80
46	1489	26	22	1476·0	·01491	·98509	24·16

TABLE 1. On the Rate of Mortality amongst the Families of the Peerage: Males. English actuaries did not rely on a few general tables for the whole population, but calculated endlessly to reflect experience with a great variety of subpopulations, and also to provide information suitable for different types of insurance contracts. From *Assurance Magazine* 9 (1860–61), 319.

On this account the Pelican suffered higher mortality than other companies. The key importance of proper selection was universally recognized. The Anglo-Bengalee Disinterested Loan and Life Assurance Company, created by Charles Dickens in his novel of 1843–44, *Martin Chuzzlewit*, advertised to Dickens's readers its irresponsibility by admitting lives indiscriminately. Dr. Jobling, the company doctor, received a commission on every policy.[11]

The selection of lives was foremost among the concerns of the committee set up in 1848 to consider the advisability of forming an actuarial society. The committee called for occasional meetings "to establish uniformity in dealing with certain points which are of constant occurrence in Assurance Offices but in respect of which there is much diversity of practice." The first item on their list was: "To determine the propriety or otherwise of paying fees to the medical referees of parties whose lives are proposed for assurance."[12] The medical referees themselves embodied a makeshift solution to a difficult problem of trust. A sound company would take care that medical as well as financial expertise was represented on its board. The customary practice among life insurance companies in the early decades of the industry was to require a personal appearance of every applicant before the assembled directors. There an inspection would take place, and a decision would be reached about whether this was indeed a select life. But sometimes this was grossly inconvenient, especially if the applicant lived far from London. Charles Babbage reported in his study of insurance institutions in 1826 that most companies were willing to dispense with this visit for a certain percent. How much this ought to be, he added disapprovingly, had never been calculated.[13]

Clearly the companies would in any event require some information about the lives they were considering. The most convenient source of advice was their agents in other cities, who had solicited the business in the first place. But the agents might have no medical expertise, and in any case it was dangerous to place too much trust in persons working on commission. Trebilcock shows that in the case of fire insurance, at least, the poor judgment or cupidity of some agents caused Phoenix Assurance huge early losses in St. Thomas and then Liverpool.[14] The Pelican appointed a medical adviser to its Board in 1828, and tried to keep checks on the quality and credentials of its doctors. But their attention to medical matters was only fitful. The large number of canceled policies attests to the frequency of mistakes. The Board was generally more interested in investments than in actuarial or medical work. Perhaps this was why it suffered such high mortality.[15] Royal Exchange Assurance was more successful with its actuaries and medical examiners, and hence also with its life insurance business. It appointed a medical adviser fourteen years later than the Pelican, in 1842. It did not require a medical certificate of applicants for insurance until 1838. We should probably read this not as a sign of indifference, but rather of intense personal interest, an unwillingness to delegate to others these crucial decisions about the quality of lives.[16]

Four actuaries called before a Parliamentary Select Committee on Joint Stock Companies in 1843 described the identification of quality lives in some detail. First the candidate was asked if he had had "certain named diseases." He was asked for reference to his "medical attendant, and to some private friend who is acquainted with his habits of life and general state of health." Letters of inquiry were then sent to the friend as well as the doctor, and the candidate himself required to appear "before either the directors at the insurance office, or some medical officer they may appoint, or both." The committee chairman, Richard Lalor Sheil, was not convinced that an appearance before the board could accomplish any useful end. "I consider it very useful," replied Charles Ansell. "But the main reliance is placed upon the medical report, is it not?" he was asked. "I am not prepared to say that; indeed, I know cases in which the directors are bold enough to differ *in toto* from the medical officer, and accept lives which their medical man rejected, and sometimes the contrary." Another actuary, Griffith Davies, interjected that the directors almost never accept a life the medical officer has rejected, but often reject applicants the medical officer has approved. "There is another advantage," continued Ansell, "which is sometimes derived from men of the world seeing the lives which are proposed for assurance; and that is, that men's habits are frequently indicated by their appearance; and it leads often to inquiries as to the parties' habits of life," such as use of spirits.[17] Life insurance was not for the loose or disreputable.

Since the companies were not yet very large or bureaucratic by midcentury, actuaries too were involved in selection of lives, and occasional bits of advice on this matter were printed in the journal of the institute of actuaries, the *Assurance Magazine*. In 1859–60 it published a collection of medical maxims for identifying bad lives. "The practised eye of the medical examiner will at once detect the advanced drunkard in the characteristic bloated countenance," and reject his application for life insurance. An attack, however slight, of apoplexy "renders a life quite ineligible," and no respectable company would seriously consider "a gouty person who is a free liver and of sedentary habits."[18]

Especially revealing is a short paper on "Corpulence in Connection with Life Assurance" in its first volume, in 1850. The author, a Dr. Chambers, assured his readers that "the bulk of the frame is pretty accurately shown by the height, and consequently the quantity of fat which ought to be attached to that frame may easily be calculated. Its deviations from the normal proportion may, therefore, easily be arrived at. . . ." A weight significantly "above or below the standard" indicates higher than ordinary risk. Chambers attached to his paper a table of normal weights according to height, thus furnishing a splendid instance of the statistical definition and valorization of normality.[19] The purpose of the exercise was not in the first instance to attach moral worth and blame to individuals, though. It was to create objective medical facts where there was little opportunity to form relations of personal trust.

> If, for example, a proposal be sent from the country, backed merely with the opinion of a referee whom we do not know, that "no signs of disease are discoverable," and that "the proposer has a robust appearance," our knowledge of the tendencies of his constitution is small indeed. But if to this it be added that he is five feet eight inches high, and weighs eleven stone, we feel a certain degree of safety in accepting him. . . . I may add, too, it facilitates much the explanations of the reasons for a refusal or acceptance, which the directors will sometimes require from their medical advisors; for it depends on a reasoning comprehensible to all, and capable of reduction to figures . . . and can avoid the vagueness of a mere negative opinion.[20]

For similar reasons, life insurance examinations provided an important incentive for the use of advanced medical technology toward the end of the nineteenth century.[21]

Despite such efforts, the acceptance and rejection of lives had not by 1860 been reduced to a system, and actuaries regarded this as prominent among the uncertainties that demanded actuarial judgment rather than slavish adherence to formulas. Indeed, divergent professional ideals were inseparable from the mid-nineteenth-century debate about the existence of "laws" of sickness and mortality. William Farr, head of the statistical department of the General Register Office, compared his life tables to Halley's remarkable prediction of the periodic return of the comet. This was in line with the prevailing statistical rhetoric, that natural laws governed human affairs with the same inflexible rigor they displayed in astronomy and mechanics. Significantly, as John Eyler points out, the life tables that Farr's office prepared were used by few insurance companies. His tables covered the population at large, while the companies sold almost all their insurance to the more prosperous classes. We can recognize also in Farr's universalist claims an expansive program for life insurance, which he wanted the government to provide for working people as well as the prosperous middle classes.[22]

This commitment to laws of mortality reached an extreme in the claims of the self-styled discoverer Thomas Rowe Edmonds, who prepared a set of tables "founded upon the discovery of the law which, in my belief, governs the mortality, according to age, of all nations and classes of men, from the earliest infancy to extreme old age." He spoke of "immutable quantities which, like the law of gravitation, form part of the foundations of the universe."[23] Similar, though milder, rhetoric was deployed often by actuaries to defend the soundness of life insurance. Ultimately, though, most of them grounded insurance in the skillful practices of respectable gentlemen rather than timeless laws of nature. A system of life contingency calculation based on fixed mortality schedules and fixed interest rates, argued Edwin James Farren, is at best suitable for the infancy of life insurance. The neophyte actuary, trained in this "exquisite logic," must be surprised when he confronts the real world, marked by the inescapable variability of the "so-called law of mortality." The "assumption of absolutism" in this domain is no longer useful.[24]

Arthur Bailey and Archibald Day argued that no general law of mortality was yet known, and that if one is someday discovered "it will represent the law that really prevails among the living, moving, thinking men that inhabit the earth, much in the same way that the statue of the Apollo Belvedere represents their bodily form. Such a law will never supersede, in our pursuits at least, the exercise of . . . careful judgment and sound discrimination." Actuaries must calculate on the basis of "dependent risks," which apply only within a given company at a given time. The tables apply to lives actively selected, whose quality may determine the viability of the company. William Lance of Lloyd's explained, "It is well known to actuaries that the success of a Life Assurance Office does not necessarily follow from accepting assurances at rates determinable upon tables of mortality, but according to the judgment with which lives are selected, and the premiums improved at interest."[25] Tim Alborn suggests that Augustus De Morgan's subjective interpretation of probability reflected the actuary's preference for expert judgment over mere mechanical calculation.[26]

Ian Hacking offers a striking instance of the creation of a statistical law in *The Taming of Chance*. Testifying before a select committee concerned with friendly societies in March 1825, John Finlaison declared that sickness was not subject to any known law of nature. By April, the committee had browbeat him into confessing that, as the committed summarized his testimony, sickness "may be reduced to an almost certain law." But, as Hacking notes, Finlaison's own words were far more cautious, and in this respect he was at one with most of his fellow actuaries. In 1853, H. Tompkins reviewed the reports and tables Finlaison had produced in connection with this issue, and asked why the quinquennial returns of friendly societies had not been mined to give genuinely reliable tables of sickness. The council of the Institute of Actuaries was moved by Tompkins's temerity to publish a rare anonymous editorial. "The notion that there is a 'fixed' rate of mortality and a 'fixed' rate of sickness is evidently untenable. There is reason to believe that these rates differ in every association, not widely perhaps, but characteristically." For that reason friendly societies of working men could never manage their own affairs, but need the assistance of well-established institutions and middle-class professionals. "All experience seems to prove that a high order of intelligence, and integrity strengthened by a somewhat conspicuous position, are alone fitted to administer successfully affairs of the kind in question. . . ."[27]

Practised Computers

Since actuaries furnished such abundant denials that their profession could be reduced to the performance of calculations, it is important to ask just what role mathematics did play in the British insurance industry. Notwithstanding this insistence on expert judgment, life insurance was a thoroughly quantitative business. The amount of space in the *Assurance Magazine* devoted to probability mathematics and to the presentation of statistical tables makes this clear. Preparation of life tables and rate structures was the principal business of the actuary. Mathematical reasoning was undoubtedly important. What it could not provide was precise measurement. Because of the essential heterogeneity of company practices and of the populations insured, no mere process of collecting and tabulating results from the population at large could yield numbers that would be valid for any particular company. Mathematical solutions to insurance problems in the *Assurance Magazine* often provoked skeptical responses, which were duly published. A mathematical piece on the value of an inheritance in perpetuity conditional on the earlier death of some elder brothers was answered twice in the next issue. The problem could be solved mathematically by presupposing the validity of the tables, the critics conceded. "That would not be the view taken by an actuary before whom the case came in the ordinary way for an opinion." A proper solution would depend partly on factors ignored by the tables, such as the states of health of the various parties. Hence, "actuaries would, in stating a value, be guided more by their own judgment than by any tabular or mathematical value, which can only, without great trouble, be approximate."[28]

The reliance of actuaries on quantitative analysis is most obvious not where calculation was used, but where it wasn't. While actuaries held devoutly to the view that the variability of phenomena made judgment indispensable, none believed that insurance could be managed by judgment alone. Only through extensive calculations could it be demonstrated that an insurance company or friendly society with large reserves might face insolvency when its members had become older. Laymen and especially working people without much knowledge of mathematics could rarely be convinced.[29] Actuaries urged new companies to rely on tables from similar firms until they had acquired sufficient experience of their own. New types of policies, such as insurance on the lives of older men with older wives that would only take effect "in the event of issue," inspired efforts to construct tables providing the appropriate risks.[30] It remained common in mid-century for the more conservative companies to require insured persons traveling abroad to surrender their policies, because nobody knew how much the risk was increased in India, Africa, or the Caribbean. Others imposed somewhat arbitrary, and generous, surcharges. At the same time, actuaries worked diligently to gather information on the experience of Europeans abroad, so that insurance contracts could be maintained on military officers and colonial administrators without increasing the risk to the company. Their inquiries produced some of the best evidence we now have concerning European mortality in tropical colonies.[31]

The unease of actuaries in the absence of systematically gathered quantitative data on risks is evident also in their writings on fire and marine insurance. Samuel Brown, one of the more vigorous champions of mathematical probability as a tool of insurance, complained of the failure of the companies to gather up and share their experience with risks to buildings and shipping. Others argued that losses in these categories display regularities from year to year comparable to those that govern human mortality.[32] Underwriters involved in marine and fire insurance, though, were less easily convinced. Buildings and technologies changed too rapidly for past results to be generalized into the future. J. M. McCandlish wrote on fire insurance in the *Encyclopaedia Britannica*: "The slightest observation reveals an endless diversity in the risks undertaken, and even if an absolute law could be reckoned on, the risks would require careful and accurate classification before the law could be deduced. But, in point of fact, the risks are always changing."[33] Clearly the companies relied on experience, but they used it in a more secretive and informal way than did the life insurers. In the absence of regulatory interference, there was insufficient incentive to systematize and rationalize that experience. When losses of a certain type or in a particular city became noticeable, the companies colluded to raise prices and restore profitability.[34]

Life insurance companies could not deal with risk so casually. They offered long-term, whole life contracts, in which premiums greatly exceeded the value of risk for several decades, but then fell considerably below it. Rates could not be readily adjusted if experience showed that they were too low. Sound instinct or seasoned judgment by itself provided little guidance to the pricing of so complex a contract. Calculation, if not definitive, was at least approximate. Cautious, responsible companies normally based their calculation on low interest rates and conservative life tables, then added a percentage for safety. They might also make adjustments for the expected quality of lives or for unusual investment opportunities. By 1850 most companies were operated at least in part on mutual principles, which meant that some of the profits were returned to the insured. The measure of profitability was anything but self-evident, and an actuary testifying before a parliamentary select committee in 1853 admitted that what is called a return of nine-tenths of the profit may in fact be no more than one half.[35] The point here is that reliance on calculation was by no means inconsistent with the exercise of discretion, provided that nobody pretended to perfect quantitative precision. In life insurance, judgment entered not as a fundamental alternative to calculation, but as a set of strategies for setting up the computation and then adjusting its results. Determination of premiums and bonuses was a matter of skill and experience, never of blind mechanism.

Where calculation was enlisted in settling affairs between companies or adjusting contracts between parties, there was reason for greater rigidity. Charles Jellicoe published, for example, a "solution" to the problem of apportioning property between a tenant for life and the "remainder man." Since the solution depended on which party wanted the contract dissolved, it could not eliminate the need for "judgment and discretion," and yet Jellicoe took some trouble to define the provenance of the rules in a way that would be legally defensible. He also contributed a paper showing that the rules governing the amalgamation of two insurance companies "are susceptible of the nicest and most accurate adjustment," so that "the rights of all may be most scrupulously maintained." Another topic requiring precision was the distribution of liabilities when a single accident was covered by policies issued by more than one company.[36]

Also, it is clear that precise calculation was assigned moral value, especially for younger actuaries. The Institute of Actuaries defended its written examination, mostly mathematical, not on the ground that this was the best way to demonstrate competence, but because nothing was better suited than an examination to keep this incipient profession free of "admixture with all 'baser matter.'"[37] A paper in 1860 on the construction of mortality tables defended the worth of precise and correct calculations. The author conceded that various simplifying approximations could give values "as near the *truth* as the values correctly deduced would be." But we should aspire to make our conclusions fully consistent with our premises. "To this very proper regard for logical consistency, which is the foundation of mathematical science, we owe the construction of tables of annuities certain to five or six decimal places; for it cannot be pretended that *any* assumed rate of interest represents the value of money so exactly as to render such extreme accuracy at all necessary to the abstract justice of the case."[38]

It was also argued, paradoxically, that precise calculation helped to form a mature judgment. In 1854, Peter Gray explained that the construction of life tables from survivorship data can be accomplished according to at least two distinct methods. The "logarithmic method," has the disadvantage of yielding "no more than seven figures" in its results. Seven figures may or may not be sufficiently precise. "On this point different computers will entertain different views." It was an advantage of his other method, "construction in numbers," that it permitted the calculation of an arbitrary number of digits. He admitted that such calculations would quickly go beyond the reliability of the data. But he considered that the act of computation had its own value. While "practised computers" might find his discussion excessively minute, "younger members" would confer a great benefit upon themselves as well as the profession by calculating tables according to this more precise method. "They would find that they had acquired such an intimate acquaintance with the structure and properties of the tables, that they could apply them to practical purposes with a facility and confidence which without this preparation long experience alone could have imparted."[39]

Henry Porter explained the value of mathematical study for actuaries mainly in moral terms. Apart from its direct usefulness, it "has the effect of promoting and improving our powers of judgment, of creating in us care and caution, and of indirectly producing those very qualities for which, I believe, actuaries are noted." Those qualities, he conceded, were not universally admired, and insurance directors had been known to ridicule "the prominence, phrenologically speaking" of their actuary's "organ of caution." Certainly Porter was not prepared to view insurance as a preeminently technical discipline. "I believe that it is now generally considered, that a very abstruse mathematical knowledge is not absolutely requisite for the general business of an actuary." Switching to (faint) praise, he spoke against the opinion of some, "that it is rather detrimental than otherwise, as we know of many instances which prove that the highest scientific knowledge and perfect business habits are not necessarily incompatible." Mathematics, he determined is essential for actuaries, even if "the prosperity of some Companies has been sacrificed to the closet meditation of the profound theorist." No such ambivalence attached to Greek and Latin, which Porter deemed necessary as aids to the mastery of the relevant technical vocabulary, or to physiology, which would aid the actuary in judging doubtful lives and deciding on an appropriate supplement to the premium.[40]

It is necessary to add that Porter began his lecture by calling mathematics "the foundation of all actuarial knowledge." Statistics, meaning numerical data, provide "the very foundation on which the superstructure of life assurance is raised." Still, "it is not sufficient at once to adopt the numerical result arrived at by calculation." Mere calculation can lead to absurdities. A senior wrangler, "without experience, would be helpless in a Life Office." Men of experience recognize the crucial importance of "*judgment*" in actuarial practice. Porter's lecture was a paean to "judgment and experience," which "cannot be taught," but only acquired through an apprenticeship, as in all professions. He considered that the license to practice should only be conferred after a "searching examination, by gentlemen of undeniable attainments" in this refined art.[41]

By coincidence, Porter himself had been in the first cohort of successful candidates to undergo such an examination, under the auspices of the new Institute of Actuaries. If self-interest is to be invoked to explain his pronouncements, that interest was at least a collective one, widely shared among English actuaries. Edwin James Farren deemed it "a well-known feature of the advancement of learning, that the more knowledge becomes prevalent, the less positive do opinions become." Actuaries had too much experience to believe any more in "abstract truths" and "fundamental axioms" from which all practical conclusions could be derived.[42] This sensibility that knowledge is local, and that even general rules are useless except to those who understand the conditions under which they should be applied, was especially valued as a defense against bureaucratic intervention. It was expressed most persistently in response to an 1853 inquiry into "assurance associations" by a select committee of the House of Commons.

A Select Committee Seeks Exact Rules

Charles Dickens chose life insurance as the exemplar of fraudulent enterprise in *Martin Chuzzlewit*. The Anglo-Bengalee Disinterested Loan and Life Assurance Company disguised its rapaciousness with a veneer of solidity and trust. That was in 1843. The novel drew on stories of shady insurance operations taken as evidence in 1841 and 1843 by the Select Committee on Joint Stock Companies. In ensuing years, speculative investment in life insurance reached new heights. Despite attempts to regulate insurance through the Joint Stock Companies' Act of 1844, Parliament had almost no control over this activity. The Select Committee on Assurance Associations, which carried out its inquiries in 1853, was able to learn from Francis Whitmarsh, Registrar of Joint Stock Companies, how many companies had been provisionally or completely registered in the intervening years. It seemed that many, probably more than a hundred, had already failed, though the bureaucratic machinery was not adequate to offer any certainty. The word on the street was not encouraging. James Wilson had written on the problem for the *Economist* newspaper, and was as well informed as anyone about business practices in life insurance. As chairman of the select committee, he looked to actuarial precision to provide readily understandable information so that people would be able to learn for themselves which companies were solid.[43] Nearly all testimony to the committee came from professional actuaries, most representing the older and more respectable companies rather than the new, possibly shady ones. These actuaries were polite but unmoveable. Precision is not attainable through actuarial methods. A sound company depends on judgment and discretion. Actuaries are gentlemen of character and discernment. Trust us.

Wilson set forth his analysis plainly in the form of leading questions. Life insurance is a long-term proposition, and the insured need some basis for confidence that the company will still be there when, at their death, a claim at last comes due. Premiums depend on age of admission to a company, but are set at a rate that is to remain fixed over the life of the insured. Death rates for those recently insured are relatively low, both because of their youth and because they are select lives. Hence a new company will accumulate a lot of capital in its early years. If it is responsible, it will set much of this aside in anticipation of an increasing rate of claims twenty or thirty years later. But many companies seemed not to be responsible. They paid their officers and directors high salaries. They expended vast sums on advertising to bring in new lives and new revenue. And many did not exude solidity. A certain Augustus Collingridge had in just a few years set up and closed down several companies. Among the companies investigated by William S. D. Pateman were several of doubtful reliability. The Victoria Life Office "consisted of one room, over a milliner's shop, in New Oxford-street, containing only two chairs, a broken table, and a large number of prospectuses, printed on foreign note paper." The Universal Life and Fire Insurance Company was lodged in a room in a very small house, "which the parties never occupied, some person calling there for the letters; no one had called for letters during the last six or seven days, in consequence of the police having been making inquiries" respecting the proprietor.[44]

John Finlaison, the government actuary, was among the more cooperative witnesses. Insurance mathematics, he explained, "is extremely simple: there is no difficulty whatever in determining the actual condition of an office at any given time. . . ." An insurance company is solvent if it has resources to pay the present value of all outstanding insurance contracts. Would it be possible, then, to display all this in standard, published accounts? Wilson inquired. No, because actuarial judgment is involved both in predicting the rate of interest at which assets will grow and in choosing a life table. His own practice was to calculate interest at 3 1/2 percent and to use a life table he had prepared, whose validity he knew from personal experience. Of course if the companies do not exercise due care in admitting lives, the tables will not predict actual mortality. Also, the value of fixed assets will vary as interest rates fluctuate, and only long experience prepares one to appraise them. Wilson soon picked up the drift of Finlaison's arguments. "If I understand you rightly, you are of the opinion that so much depends upon the discretion and good management of the office, apart from anything that can be shown on paper, that you would not be disposed to place much confidence in any check that would be obtained by accounts?" Finlaison agreed. He also worried that publication of accounts would lead to invidious comparisons which would work to the disadvantage of new companies. Even where there was reason to suspect insolvency, the appropriate remedy was a discreet, confidential inquiry by an independent actuary working with the company's regular actuary, not a loud public investigation. Insurance fraud was very rare, he explained. A respectable board of directors guarantees the integrity of the office. Moreover, actuaries calculate conservatively, by adding a margin for safety and profit on top of calculated premiums.[45]

The record of testimony before the select committee provided little comfort to those who hoped that accounts could be rendered uniform and precise. The witnesses were nearly unanimous. It is impossible to judge companies against any single set of life tables, because they are run according to different principles and insure lives of varying quality. Income and assets cannot be fixed from a single interest rate, because investments are so diverse. Some actuaries are optimistic, others pessimistic, and there is no legislating away these differences. Edward Ryley, a decidedly unfriendly witness, put the argument most bluntly. Actuaries disagree. "If you appoint a Government actuary, you must necessarily select him from existing actuaries; there is no manufactory for actuaries where you can get one made to order." Uniform rules of calculation, imposed by the state, might yield "uniform error."[46] Charles Ansell, testifying before another select committee a decade earlier, argued similarly, then expressed his fear that the office of government actuary would fall to "some gentlemen of high mathematical talents, recently removed from one of our Universities, but without any experience whatever, though of great mathematical reputation." This "would not qualify him in any way whatever for expressing a sound opinion on a practical point like that of the premiums in a life assurance."[47]

Wilson and his fellows on the 1853 committee did not simply capitulate before this barrage of expert testimony. Can we not apply "some general average rule," he asked Ansell, "so as to work out the result upon some given general principle?" Ansell said no. How about using John Finlaison's life table, applying a uniform interest rate of 3 1/2 percent, and adding 10 percent for contingencies, in order to determine whether a company has adequate resources to meet its expected liabilities? There would be great difficulties and valid objections, replied Ansell. All these things depend on the particular circumstances of each company. And the best calculated scale of premiums will avail them nothing if a company has been careless in selecting its lives.[48]

The actuaries were by no means contemptuous of the search for general principles. Often, their responses to Wilson's initial queries were favorable. Without fail, though, they added qualifications and stipulations that undermined the principles. Many believed in general maxims; none would concede the possibility of precise, standardizable rules. Samuel Ingall, for one, had a ready reply when asked for a general test of the solvency of a company. It should have one half the premiums received on existing policies in hand, as capital. This was more informative than the usual answer to Wilson's question, that a company should have funds to buy off its policies. Ingall further declared that during its first twenty years a company should accumulate each year, on average, 1 percent of its potential liabilities. But soon he admitted that these maxims only applied if business was approximately steady, and moreover that actuaries disagreed about how the profits of a company should be calculated. This reservation sounded rather crucial to Wilson, who then asked how one could test the accuracy of the returns. Ingall replied: "I think the best security is the character of the parties giving them."[49]

This sentiment was repeated like a refrain by the actuaries. Although actuarial principles are well established, said Ansell, there remains much uncertainty "in applying known principles to different scales or premiums." It takes as long as two years for an actuary to assess the economic health of a company, argued James John Downes. Even to understand a company report requires minute knowledge, so that the actuary of an office inevitably understands its position better than any other person, however clever. Since we must depend on the skill and integrity of the actuary to prepare the data, he might as well be trusted to make the final calculation. Actuaries are "gentlemen of character," reported William Farr, and the government should leave the preparation of accounts to them. No quantitative measure of solvency can be adequate, insisted Francis Neison. A considerable expenditure of funds for publicity when a company is founded may be the best way to secure its future. There is always "a special knowledge beyond the accounts, not appearing in the books of the institution." The success of a company depends ultimately on skilled management.[50]

These were brave claims from a profession so new. The Institute of Actuaries had been founded only in 1848. The office of the actuary had been recognized in legislation as early as 1818, in an act permitting friendly societies to be enrolled by the government and to receive certain benefits if their life tables were evaluated by a committee including at least "two professional actuaries, or persons skilled in arithmetical calculations."[51] But William Morgan of the Equitable complained in 1824 that some people "call themselves actuaries, who are nothing but schoolmasters and accountants." Worse, a clergyman and magistrate at Southwell remarked that it was difficult to put this law into effect, since "how to define who is an actuary exceeded my power."[52] Even in 1843, the actuary John Tidd Pratt testified to another select committee that mere schoolmasters were often asked to certify tables because "no one knows who an actuary is."[53] The main object of the new institute of actuaries was to secure some of the recognition befitting a profession. This entailed more than a demonstration of technical competence, and the oft-expressed disdain for mere calculation must be understood in part as a strategy of legitimation.[54]

In any event, British actuaries had no faith in quantification according to a rigid protocol. Argued Downes: "We may have a person who is a good theoretical actuary, who may apply formulas with very great facility, but who has had no experience in the working of companies, and, therefore, he would not be able to apply those forms and theorems usefully to the business of a life insurance company." Charles Jellicoe, testifying for the Institute of Actuaries, explained that subtle shades of difference can be decisive on such issues as the margin to be set aside for future risks. "Then you do not differ in point of principle, but you differ in the mode of applying that principle?" he was asked. True. Indeed, an actuary could make a company's books look good by overvaluing assets and undervaluing risks. But this would be fraud, said Wilson. Perhaps, Jellicoe replied, but the company could always argue "that the lives were particularly good ones, that the mortality, therefore, would be small," and the like. For such reasons, you cannot effectively legislate a minimal guarantee fund; it is a matter of detail in each office. Parliament should simply "do the best they can to get persons whose judgment and discretion enable them to do the duty properly," by authorizing the Institute of Actuaries to grant a license or diploma to those who meet its standards.[55]

This was very far from what the select committee wanted to hear. Wilson aimed to clear up the problem of proliferating, possibly insolvent life insurance companies with a minimum of government intervention. The committee never disputed the truth of what it was told again and again: that government meddling in the business of insurance would cause far more trouble than it would solve. Wilson hoped that the mildest intervention would suffice. His ambition depended only on the possibility of making insurance readily interpretable, of standardizing the calculations sufficiently that independent mathematicians could certify their validity, and so that potential clients could judge the companies for themselves from a few crucial numbers. But the actuaries resisted his suggestions, in the name of judgment, of subtle shades of meaning and innumerable points of detail. The government sought public knowledge, while the actuaries denied its possibility. The government sought a foundation for faith in numbers, while the actuaries demanded trust in their judgment as gentlemen and professionals.

Such actuarial skepticism was heard whenever someone on the select committee suggested that the government might tell the companies how to keep their books. None of the actuaries would countenance such interference. The most sympathetic view of the committee's suggestions was expressed by a renegade, the Scottish actuary William Thomas Thomson, who assented to the proposition that a single form of balance sheet would be appropriate for all companies. He added that in any event the companies could easily maintain their existing form of accounts and with modest effort translate them into a standard form for public purposes. He even spoke favorably of American-style regulation, as enacted recently in New York. None of this appealed in the slightest to the other expert witnesses. When Thomas Rowe Edmonds was asked if Mr. Thomson's forms would provide the kind of information the public needed, he answered sharply: "Not in the least."[56]

To reject quantification according to a mandated protocol was not quite the same as denying that useful information could be conveyed by numbers. Standardization was what the actuaries opposed. Actuarial work involved a kind of precision, but the numbers remained personal and local so long as the object to be measured had not been clearly defined by public law. This would necessarily entail a large measure of centralized control, which they could not approve. They did not on this account refuse to report quantitative information. Without exception, the actuaries claimed to regard clear, accurate reporting as good business practice. Hence there was no need for new legislation. One strong opponent of regulation went so far as to lay the blame for unintelligible reports on government interference. Companies naturally become secretive when surrounded by "lynx-eyed" inspectors, "searching out any kind of seeming irregularity in these accounts, and making the most of it, to the prejudice of the institution."[57]

Wilson and the committee had greeted warmly the idea that a public officer, an auditor or actuary, was just what was needed to assure that the accounts were clear and accurate. The only actuary who favored anything like this was Thomson, the Scot, who however wanted independent auditors rather than government officers to certify the accounts. Either possibility seemed far too meddlesome to the others. Public investigation could be justified, they argued, only where there is reasonable suspicion of fraud. Routine public involvement in the regulation of insurance would weaken the self-reliance of citizens, and thus aggravate the very problem the government had set out to solve.

Still, many allowed that the government might reasonably require publication of financial records. This, the actuaries argued, should take the form of a straightforward factual presentation. It should not include summary numbers standing for assets and liabilities, which would imply that solvency could be determined by anyone capable of noticing which was the greater. Permitting interpretive accounts in public records would make the government "a medium for advertising all sorts of opinions and fallacious valuations, . . . publishers of puffs."[58] By indicating simply the quantity and nature of each holding, or presenting their own figures along with an explanation of the principles by which they were calculated, the companies would provide potential customers all they needed to evaluate the security of a policy. Not that everyman could interpret this information for himself. That would presuppose the impossible—a highly standardized document, with everything reduced to a few categories and expressed in the same terms. Rather, customers could learn of the security of a company by consulting an actuary, much as they might consult an attorney about matters of law. The public could not learn much from a balance sheet, said Thomson, but it would permit professional actuaries to form "a very decided opinion concerning the solidity of the company." Would it be a correct opinion? asked Wilson. "I should say if it is decided it must be correct, so far as depends on the individuals' judgment."[59]

The underlying issues emerge here very clearly. The validation of expertise meant that standardization was unnecessary. If the key entities at issue resisted precise measurement, the government was left with a choice between trust and intrusive regulation to fill the gap. To be sure, the preferred solution of the less militantly antigovernment actuaries required exact quantitative reports, but this was basically a matter of a modest opening of their books, not of precise measurement of anything. Support for uniformity came from the government, not from mathematical actuaries. There's no need for precision when you have a profession.

Precision and Political Culture

There is a politics of precision. It does not operate only in the domain of government, bureaucracy, and regulation. Precision in science, too, must be understood as a form of knowledge sanctioned by a community, a response to pressures arising both within and outside of this community. At the same time, it is important to recognize that bureaucracies are by no means unflagging supporters of precise public knowledge. In the later Victorian period the British civil service became more hierarchical, mainly in consequence of the Northcote-Trevelyan reforms. A hierarchical political order has less need of the objectivity that comes with precision, though the appeal of precision is likely to be felt wherever trust and deference are threatened.[60]

This is not the only predisposing cause of social quantification. In a large-scale industrialized society, numbers provide an appropriately impersonal way of learning about people in alien settings. Some institutions in Victorian Britain continued to function mainly through intimate knowledge and face-to-face contact. Life insurance did not. The statistical investigations of poverty, crime, education, sanitation, and employment of working people that became almost an obsession in Britain beginning around 1830 reflected the greater distances between classes in an industrializing society. The quantifying impulse of middle-class reformers is nicely caricatured by some Glasgow studies of nutrition, designed to minimize the cost per unit of energy in meals served at the Night Asylum for the Houseless.[61] Criminal justice, as Martin Wiener shows, provides a striking illustration of institutional reconstruction in an emerging "society of strangers." Judicial discretion, based on close personal knowledge, was the heart and soul of law in eighteenth-century England. This discretion came at the price of great arbitrariness, and was no bar to cruelty. It was narrowed in the early nineteenth century. Perhaps the best emblem of precision for academic intellectuals in the age of Foucault is Jeremy Bentham's rotary whipping machine, which was to chasten criminals by administering a preset number of strokes with unvarying force.[62]

Still, quantification need not entail the sacrifice of discretion. Charles Booth's largely quantitative study of the poor in London was based on years of close contact with the objects of his investigation, and freely combined the results of "personal observation" with general statistics. He "never pretended to greater precision than his sources permitted," writes Gertrude Himmelfarb. He did not scruple at using moral language, describing people and classes as "degraded," "shiftless," or "honourable."[63] The possibility of quantifying people presupposes a position of strength in relation to those people, but the abandonment of judgment in favor of mechanical precision generally reflects political weakness or exposure.

In 1853, English actuaries proved strong enough to resist the effort to rein in their judgment by standardizing their practices. The actuaries sought to shore up the flagging trust in individuals by establishing a credible profession. This was by no means an unmixed success,[64] but it sufficed for a time to hold off the rather weak forces of bureaucratic and legal intervention. Still these were always a threat. The rhetoric of precision came to be valued by actuaries as a second line of defense, in accordance with the logic of Wilson's hope that standardized reporting would render more intrusive government regulation otiose. As was explained in the ninth edition of the *Encyclopaedia Britannica* (1881): "The business of life assurance being founded on well-ascertained natural laws, and on principles of finance which in their broad aspect are of the simplest description, there exists no necessity for frequent close scrutiny of the affairs of an assurance office. . . ."[65]

I thank Ayval Ramati for research assistance on this paper, and John Carson, Lorraine Daston, and Norton Wise for comments on an earlier draft.

Notes to Chapter Seven

1. Testimony of Edward Ryley in *Report from the Select Committee on Assurance Associations*, in British Parliamentary Papers, 1852–53, vol. 21, 243–53, p. 246. Hereafter this volume will be cited as *SCAA*.

2. "A Sketch of the Life of Henry James Brooke," *Assurance Magazine*, 7 (1857–58), 286–88, p. 286.

3. Graeme Gooday, "Precision Measurement and the Genesis of Physics Teaching Laboratories in Victorian Britain," *British Journal for the History of Science*, 23 (1990), 25–51; also id., "The Morals of Energy Metering," this volume; Kathryn M. Olesko, *Physics as a Calling: Discipline and Practice in the Königsberg Seminar for Physics* (Ithaca: Cornell University Press, 1991); Zeno Swijtink, "The Objectification of Observation: Measurement and Statistical Methods in the Nineteenth Century," Lorenz Krüger et al., eds., *The Probabilistic Revolution, volume 1: Ideas in History* (Cambridge: MIT Press, 1987), 261–86; Lorraine Daston, "The Moral Economy of Science," Critical Problems in the History of Science and Technology conference, Madison, Wisconsin, 1991.

4. Simon Schaffer, "Astronomers Mark Time: Discipline and the Personal Equation," *Science in Context* 2 (1988), 115–45; John Heilbron, "The Measure of Enlightenment," in Tore Frängsmyr, John Heilbron, and Robin Rider, eds., *The Quantifying Spirit in the Eighteenth Century* (Berkeley: University of California Press, 1990), 207–42; Heilbron, "A Mathematician's Mutiny, with Morals," in *World Changes: Thomas Kuhn and the Nature of Science*, ed. Paul Horwich (Cambridge, Mass., 1993), 81–129.

5. Theodore M. Porter, "Quantification and the Accounting Ideal in Science," *Social Studies of Science* 22 (1992), 633–51; Porter, "Objectivity as Standardization: The Rhetoric of Impersonality in Measurement, Statistics, and Cost-Benefit Analysis," *Annals of Scholarship* 9 (1992), 19–59; Porter, *Trust in Numbers: The Pursuit of Objectivity in Science and Public Life* (Princeton: Princeton University Press, 1995).

6. Lorraine Daston, *Classical Probability in the Enlightenment* (Princeton: Princeton University Press, 1988), 174–82.

7. *Report from the Select Committee on the Friendly Society Bill*, *Parliamentary Papers*, 1849, XIV, 1, testimony by William Sanders, pp. 43–56.

8. Ibid., testimony of Francis G. P. Neison, pp. 1–2. The low reported rates of

sickness in Scotland were discussed also in the select committee hearings of 1824. Ian Hacking, *The Taming of Chance* (Cambridge: Cambridge University Press, 1990), 51–52, infers that in this case Scottish mortality really was lower than the English tabulators could imagine. But Neison's seems the more likely explanation.

9. Clive Trebilcock, *Phoenix Assurance and the Development of British Insurance, volume 1, 1782–1870* (Cambridge: Cambridge University Press, 1985), 605.

10. Robin Pearson, "Thrift or Dissipation: The Business of Life Insurance in the Early Nineteenth Century," *Economic History Review* [2] 43 (1990), 236–54, suggests that much life insurance was not purchased to provide security for dependents, but rather to secure debts, at the insistence of creditors.

11. Charles Dickens, *Martin Chuzzlewit* (New York: Penguin, 1968), chap. 27, pp. 509–10. Dickens also let his readers know that there was no paid-up capital, that the company put false numbers in its advertisements, and that its premiums were too low. The shadiness of the enterprise was of course most glaringly evident from the character of its officers.

enmmonds, *The Institute of Actuaries, 1848–1948* (Cambridge: Cambridge University Press, 1948), 11.

13. Charles Babbage, *A Comparative View of the Various Institutions for the Assurance of Lives* (1826; reprinted New York: Augustus M. Kelley, 1967), 125.

14. Trebilcock, *Phoenix Assurance* (n. 9), 211–12, 419, 552.

15. Ibid., 607–8.

16. Barry Supple, *Royal Exchange Assurance: A History of British Assurance, 1720–1970* (Cambridge: Cambridge University Press, 1970), 176–77, 99.

17. *First Report of the Select Committee on Joint Stock Companies, Parliamentary Papers*, House of Commons, 1844, vol. 7, 1, pp. 147–48.

18. Stephen H. Ward, "Treatise on the Medical Estimate of Life for Life Assurance," *Assurance Magazine* 8 (1858–60), 248–63, 329–43, pp. 252, 338, 336. This was reprinted from the *American Life Assurance Magazine*.

19. See Hacking, *Taming of Chance* (n. 8).

20. Dr. Chambers, "Corpulence in Connection with Life Assurance," *Assurance Magazine* 1 (1850–51), 87–89, p. 88.

21. At least in the United States; see Audrey B. Davis, "Life Insurance and the Physical Examination: A Chapter in the Rise of American Medical Technology," *Bulletin of the History of Medicine* 55 (1981), 392–406.

22. John Eyler, *Victorian Social Medicine: The Ideas and Methods of William Farr* (Baltimore: Johns Hopkins University Press, 1979), 84; Theodore M. Porter, *The Rise of Statistical Thinking, 1820–1900* (Princeton: Princeton University Press, 1986).

23. Thomas Rowe Edmonds, "On the Discovery of the Law of Human Mortality," *Assurance Magazine* 9 (1860–61), 170–84, p. 170; Edmonds, "On the Laws of Mortality and Sickness of the Labouring Classes of England," ibid., 5 (1854–55), 127–45, p. 142.

24. Edwin James Farren, "On the Improvement of Life Contingency Calculation," *Assurance Magazine* 5 (1854–55), 185–96; 8 (1858–60), 121–27, pp. 185–87, 121.

25. Arthur Bailey and Archibald Day, "On the Rate of Mortality amongst the Families of the Peerage," *Assurance Magazine* 9 (1860–61), 305–26, p. 318. William Lance, "Paper upon Marine Insurance," *Assurance Magazine* 2 (1851–52), 362–76, p. 364.

26. Timothy Lee Alborn, *The Other Economists: Science and Commercial Culture in Victorian England*, Ph.D. dissertation, Harvard University, 1991, pp. 121–22.

27. Hacking, *Taming of Chance* (n. 8), 47; "Editorial note," *Assurance Magazine* 3 (1852–53), 15–17, pp. 15, 17.

28. Two anonymous letters, "Solution of Problem Proposed by Juvenis," and "The Same Subject," in *Assurance Magazine* 12 (1864–66), 301–2.

29. *Select Committee on Friendly Societies*, 1849 (n. 7), testimony of Francis G. P. Neison, 1–26, p. 8.

30. Charles Jellicoe, "On the Rates of Mortality prevailing . . . in the Eagle Insurance Company," *Assurance Magazine* 4 (1853–54), 199–215; Archibald Day, "On the Determination of the Rates of Premiums for Assuring against Issue," ibid., 8 (1858–60), 127–38.

31. Philip Curtin, *Death by Migration: Europe's Encounter with the Tropical World in the Nineteenth Century* (Cambridge: Cambridge University Press, 1989). An example of such inquiry is Charles Jellicoe, "On the Rate of Premiums to be charged for Assurances on the Lives of Military Officers Serving in Bengal," *Assurance Magazine* 1 (1850–51), no. 3, 166–78. Trebilcock, *Phoenix Assurance* (n. 9), 552–65 discusses the surcharges imposed by the Pelican.

32. Samuel Brown, "On the Fires in London during the 17 Years from 1833 to 1849," *Assurance Magazine* 1, no. 2 (1851), 31–62; Lance, "Marine Insurance" (n. 25), 362. Still, it was well known among the professionals that there will be fluctuations from mean values. Probability arguments were used to argue that a friendly society should have at least 150 to 300 members to be reasonably safe. Charles Babbage explained this to a select committee in *Report from the Select Committee on the Laws respecting Friendly Societies*, *Parliamentary Papers*, House of Commons, 1826–27, III, 869, pp. 28–33. The committee members undercut him rather effectively by asking about the effects of epidemic diseases, a case to which the standard probability calculus did not apply.

33. J. M. McCandlish, "Fire Insurance," *Encyclopaedia Britannica*, 9th ed., 1881, vol. 13, p. 163.

34. Trebilcock, *Phoenix Assurance* (n. 9), 355, 419, 446.

35. Testimony of Francis G. P. Neison, *SCAA*, 189–207, p. 204.

36. Charles Jellicoe, "On the Valuation of Property held for Life and in Reversion," *Assurance Magazine* 6 (1855–57), 61–66; Jellicoe, "On the Principles which should govern Assurance Companies in Amalgamating," ibid., 7 (1857–58), 255; Richard Atkins, "On the Settlement of Losses by Fire under Average Policies," ibid., 8 (1858–60), 1 ff.

37. "Proceedings of the Institute of Actuaries of Great Britain and Ireland," *Assurance Magazine* 1 (1850–51), no. 1, 103–12, p. 110.

38. William Matthew Makeham, "On the Law of Mortality and the Construction of Annuity Tables," *Assurance Magazine* 8 (1858–60), 301–30, pp. 301–2.

39. Peter Gray, "On the Construction of Survivorship Assurance Tables," *Assurance Magazine* 5 (1854–55), 107–26, pp. 125–26.

40. H. W. Porter, "On some points connected with the Education of an Actuary," *Assurance Magazine* 4 (1853–54), 108–18, pp. 108–11.

41. Ibid., 108, 112, 117, 116.

42. Edwin James Farren, "On the Reliability of Data, when tested by the conclusions to which they lead," *Assurance Magazine* 3 (1852–53), 204–9, p. 204.

43. Alborn, *Other Economists* (n. 26), 236.

44. Testimony of William S. D. Pateman in *SCAA*, 262–93, p. 282.

45. Testimony of John Finlaison in *SCAA*, 49–64.

46. Ryley in *SCAA*, 246. See also, inter alia, George Taylor's testimony on p. 30, Charles Ansell, p. 70, and James John Downes, p. 105.

47. *Select Committee on Joint Stock Companies* (n. 17), 49.

48. *SCAA*, testimony of Charles Ansell, 64–85, pp. 69, 74, 82.

49. Samuel Ingall in *SCAA*, 156–71, pp. 158–59, 165.

50. Ansell, p. 81; Downes, pp. 105, 107, 108; Neison, p. 197; Farr, p. 303, all in *SCAA*.

51. See *Report of the Select Committee on the Laws Respecting Friendly Societies*, *Parliamentary Papers*, House of Commons, 1824 IV, 322, p. 18, referring to 59th Geo. 3. c. 128. The committee report observes that this provision was deleted before passage, but several witnesses spoke as if the measure were in effect.

52. Testimony in ibid. by William Morgan, 49–53, p. 52 and Thomas John Becher, 27–36, p. 30. Becher determined finally that an actuary was a mathematician, someone who could solve a "question in fluxions" and "make . . . calculations algebraically as well as arithmetically."

53. *Select Committee on Joint Stock Companies* (n. 17), 81.

54. On the unsuitability of calculational and measurement routines for the formation of an elite see also Simon Schaffer, "'Accurate Measurement is an English Science,'" this volume.

55. Downes, p. 108 and Jellicoe, pp. 188, 184 in *SCAA*.

56. Thomson, pp. 85–104 and Edmonds, p. 138, both in *SCAA*.

57. Francis G. P. Neison in *SCAA*, 189–207, p. 196.

58. John Adams Higham in *SCAA*, 208–23, pp., 213, 220.

59. Thomson in *SCAA*, 97. See Alborn, *The Other Economists* (n. 26), 239.

60. Porter, *Trust in Numbers* (n. 5). On elitism in the British civil service, see Peter Gowan, "The Origins of the Administrative Elite," *New Left Review* 61 (1987), 4–34; Harold Perkin, *The Rise of Professional Society: England since 1880* (London: Routledge, 1989); Roy McLeod, ed., *Government and Expertise: Specialists, Administrators, and Professionals, 1860–1919* (Cambridge: Cambridge University Press, 1989).

61. M. Norton Wise, "Work and Waste: Political Economy and Natural Philosophy in Nineteenth-Century Britain (III)," *History of Science* 18 (1990), 221–61, pp. 222–24.

62. Martin J. Wiener, *Reconstructing the Criminal: Culture, Law, and Policy in England, 1830–1914* (Cambridge: Cambridge University Press, 1990), 105.

63. Gertrude Himmelfarb, *Poverty and Compassion: The Moral Imagination of the Late Victorians* (New York: Alfred M. Knopf, 1991), 97–98, 105, 116.

64. See Simmonds, *Institute of Actuaries* (n. 12), on their problems. One sad episode deserves to be singled out: in 1859 they learned that an assistant secretary had embezzled most of their funds, and it took a voluntary subscription to restore them to solvency.

65. G. M. Low, "Life insurance," *Encyclopaedia Britannica*, 9th ed., vol. 13 (1881), 175.

EIGHT

THE IMAGES OF PRECISION:

HELMHOLTZ AND THE GRAPHICAL METHOD IN PHYSIOLOGY

Frederic L. Holmes and Kathryn M. Olesko

IN the experimental sciences great discoveries are not necessarily coextensive with great experiments. Landmark conclusions are often reached through the accumulation of evidence from converging investigations in which no single experiment is decisive, or memorable in method or execution. There is, however, a select group of historic experiments whose intrinsic qualities—of novelty, ingenuity, precision, elegance, beauty, or some combination of these—cause them to stand out as distinctly as do the discoveries with which they are associated. Famous examples are Lavoisier's use of the special properties of mercury calx to demonstrate the composition of the air by analysis and synthesis; Mendel's experiments on peas, and more recently the Meselson-Stahl semiconservative DNA replication experiment. Such experiments are sometimes labeled "classical," and they often play a pedagogical role in textbooks or in classroom and laboratory demonstrations long after the research forefront has moved beyond them.

To such a select circle belongs each of the four sets of experiments in physiology published by the young Hermann Helmholtz between 1845 and 1852. The first of these demonstrated that there is a chemical change associated with the contraction of an isolated muscle. The second showed that the isolated muscle produces heat in its contractions. The third and fourth measured, by two independent methods, the velocity of the nerve impulse. The succession of these experiments displays also a progressively more sophisticated realization by Helmholtz of the requirements for exact measurements in scientific experiments, and of the diverse manifestations of precision appropriate to different aims and circumstances.

In a previous paper on Helmholtz's physiological experiments we have traced, in narrative form, the development of these investigations. Here we shall focus on the second, graphical method by which he measured the velocity of the nerve impulse. A comparison of this method with the first method, using the ballistic galvanometer to detect minute time intervals, can elucidate not only Helmholtz's evolving experimental style, but some of the multiple forms of precision—both quantitative and qualitative—that were emerging in mid-nineteenth-century scientific practice.

Quantification and precision measurement are often assumed to have transformed only the physical sciences during the nineteenth century, and to have reached the biological sciences during the twentieth century. Although many prominent areas of biology did remain qualitative during the nineteenth century, one can, on closer examination, find strong movements toward quantification of those problems that appeared amenable to measurement. One reason for this trend was that, by the 1840s, the two physical sciences in which quantitative methods had become most pervasive—physics and chemistry—offered methods more effectively applicable to physiological phenomena than had ever before been available. We can distinguish clearly two general investigative streams. One of them derived from the rapid earlier growth of organic chemistry, giving rise to the nascent subfield of physiological chemistry. The other derived from the new methods of experimental physics oriented particularly around the recently discovered electromagnetic phenomena. The standards and problems of precision measurement applied to physiological problems reflected those prevailing in the parent fields from which these methods were transferred. Helmholtz participated in both movements. The analytical methods he applied in his study of the chemical changes in muscle action were those developed by Berzelius in animal chemistry. In the study of animal heat and the velocity of the nerve impulse, he employed current methods of physical measurement. In both domains he pressed the demands for precision and accuracy well beyond the standards existing in the fields he entered.

To understand the methods Helmholtz chose for measuring the nerve impulse, it is necessary to stress that he had originally devised both of the methods in question for another purpose—to analyze the time-course of muscle contractions. His objective changed during the course of his investigation as he realized, perhaps initially by accident, that he could adapt a method intended to measure the minute time intervals between successive stages of a muscle contraction to measure the time intervals required for a stimulus to reach the muscle along different lengths of its attached motor nerve. Helmholtz never returned explicitly to his original purpose. Nevertheless, the results of the graphical method that he developed were taken by his contemporaries to be as powerful for the further study of muscle contraction as they were for the demonstration of the propagation of the nerve impulse. These two purposes of the same experimental procedure invoked two different meanings of precision. For the velocity of the nerve impulse, the need was to establish a numerical result reliable enough to guarantee the reality of the phenomenon. For the study of muscle contraction, the salient requirement was not to attain numbers, but mechanically recorded curves that could represent the time-course of the muscle contraction with sufficient resolution to reveal qualitative characteristics of the process.

The Graphical Method and Exactitude in Physiology

Helmholtz applied to his investigation of the mechanical characteristics of the individual muscle contraction a graphical method devised in 1847 by Carl Ludwig for another physiological purpose. By an arrangement that transmitted changes in arterial blood pressure measured with a manometer, through a mechanical linkage, to a stylus that moved up and down on a drum revolving horizontally around a vertical axis, Ludwig was able to "obtain curves whose height comprises an expression for the blood pressure, and whose width comprises a determination of the time."[1] Similarly transmitting the respiratory motions of the thorax through a second stylus onto the same drum, he could record simultaneously the motions of inspiration and expiration. By comparing these curves, obtained from dogs and horses, he could detect the effects of changes in the amplitude and frequency of the respiration on the amplitude and rhythm of the systolic and diastolic phases of the heartbeat, and from these results draw some conclusions about the mechanism through which the respiration modifies the arterial pressure.[2]

Several historians have noted that Ludwig's kymographic method led "immediately to the study and analysis of a wide range of physiological events that had previously been inaccessible to researchers." To later physiologists, the introduction of the kymograph even seemed to mark "the birth of modern physiology."[3] The method, the mechanically recorded graphs, and the mode of analysis that they made possible became so commonplace in later physiology, however, that the question has not been fully explored, why Ludwig wished to make the particular kind of measurements that his method permitted: that is, the continuous variation of a quantitative value over very short time intervals.

It is beyond the topic of this paper to examine comprehensively the state of experimental physiology during the 1840s that induced its practitioners to begin posing questions that required answers in such a form; but a suggestive glimpse of the situation can be gleaned from the well-known "Vorrede" published at about the same time Ludwig was inventing his graphical method, by his close friend Emil duBois-Reymond. During a period in which experimental physiology was itself emerging in Germany as a discipline organized in teaching-research laboratories, the perceived need for greater rigor took various forms, including the desire to make the subject an "exact" science in the image of the physical sciences. To some, such as Purkynje's former student, Gabriel Gustav Valentin, this aim expressed itself in measuring as many physiological quantities as possible.[4] To Theodor Fechner it meant seeking mathematical formulae that could express biological forms and functions to whatever degree of approximation was required.[5] DuBois-Reymond viewed some of these efforts as premature and excessive. In the present condition of physiology he thought it seldom possible to derive a mathematical expression for the causal connections between observable phenomena, or to measure all the variables that such an expression would entail. What *was* practical, was to observe "all the possible values" of a given effect when each condition known to influence the effect was varied.

> The dependence of the effect upon each condition is now presented in the form of a curve, whose exact law, to be sure, remains unknown, but whose general character one will most often be able to trace. It will almost always be possible to determine whether the function grows or diminishes with the variable investigated. In other cases one may be able to discover distinctive points of the curves, the sense of its bending with respect to the abscissa, whether it approaches asymptotically a constant value, etc.
>
> In my opinion, in most parts of physiology, in the present condition of things the application of mathematical analysis cannot reach further.

Even this much, duBois-Reymond believed, can be sufficient to test particular causal explanations, especially when the curves produced display salient features.[6]

DuBois-Reymond and Ludwig were forming at this same time a group, including also Helmholtz and Ernst Brücke, who shared similar methodological assumptions concerning experimental physiology. It is, therefore, likely that Ludwig had such views in mind when he conceived his graphical method to investigate the influence of respiratory motions on arterial blood pressure. In any event, his method, and the experiments for which he initially devised it, manifested in practice the principles duBois-Reymond advocated for establishing functional relationships between physiological processes.

Nor is it fortuitous that it was on the particular field of the study of the circulation that Ludwig brought this method to bear. As one of the phenomena of physiology most susceptible to mechanical analysis, in particular as a complex example of fluid dynamics, the circulation was in the early nineteenth century a favorite subject for quantitative investigation, and it attracted particularly those, such as Poisseuille in France and E. H. Weber in Germany, who wished to apply physico-mathematical analysis to biological problems.[7] Moreover, ever since William Harvey, those who wished to study the heartbeat and the associated arterial pulse had found the rapidity of the phenomena an obstacle to their observations. In Ludwig's laboratory, his "young friend Gerau" had begun the study of the influence of respiratory movements on the circulation, but had concluded after a long series of fruitless attempts "that without a more exact [method of] simultaneous time determinations" "a result could not be reached."[8] It was that difficulty that immediately prompted Ludwig to invent the kymograph and the graphical method.

Alfred Volkmann, a well-known physiologist at the University of Halle, commented in 1851 that

> After Ludwig invented an instrument that permits one to represent the variable forces of the heart through curves, it immediately suggested making other motive forces also visualizable [*anschaulich*] and measurable with the help of this instrument. It cannot be surprising, therefore, that Herr Helmholtz and I came independently of one another to the idea of applying the Ludwig kymograph to the reproduction of curves which can bring the contraction of a stimulated muscle and its subsequent extension into view [*zur Anschauung*].[9]

Although Volkmann and Helmholtz may have come to the same basic idea, they implemented it, as we shall see, in ways that manifested very different understandings of the goals and requirements for precision in experimental physiology.

Helmholtz, like Ludwig himself, was induced to adopt a graphical method because he wished to study a phenomenon whose rapidity, like that of the heartbeat, made it impossible to analyze without a method that could enable him to follow the variations in a measured quantity during minute time intervals.

As part of the broad program in which he had been engaged since 1849, to study the interrelated physical and chemical processes associated in muscle action, Helmholtz turned, in 1848, to an investigation of "the mechanical work they are able to perform." To do so, he needed to examine "The processes occurring in the simple contraction of a muscle."[10] Those processes took place so quickly, however, that Eduard Weber, whose comprehensive work on muscle action served as Helmholtz's point of departure, had regarded the movements of contraction and relaxation as essentially instantaneous.[11] This was the first time that Helmholtz had confronted the problem of measurements over very short time intervals, because in his previous studies of the chemical changes and heat production in the muscle action he had stimulated one of a pair of muscles to prolonged tetanic contractions in order to maximize their cumulative effects. The recent availability of Ludwig's graphical method now made it possible for Helmholtz to treat a process which appeared to ordinary observation to begin and end almost in the same moment, as one spread out across a measurable period of time.

The apparatus that Helmholtz initially constructed for his graphical experiments is nowhere described in detail, but it appears to have combined a simplified version of Ludwig's revolving drum and recording stylus with an apparatus similar to one Weber had used to measure the height to which a suspended isolated frog muscle raised a weight hanging from it. His conceptual starting point was Weber's interpretation of the muscle as like "an elastic band of variable elasticity." In its excited state the muscle is, according to this view, in greater tension at the same length. When attached to a given weight it will begin to lift it as soon as its internal tension exceeds the weight. Helmholtz's own hypothesis, departing from that of Weber, was that these changes in the tension and the length of the muscle occupy "spaces of time" (*Zeiträume*). The immediate question he posed for his experiments was, therefore, "In what spaces of time and stages does the energy of the muscle rise and fall after an instantaneous stimulus?" The stimulus he provided by means of the momentary current induced through opening and closing of a Neeffian induction coil.[12]

The curves that the stylus traced on the smoked surface of Helmholtz's revolving drum gave "the heights to which the weight was raised during the spaces of time measured along the abscissa," because on this moving surface "time differences reappear [*wieder erscheinen*] as spatial differences." These curves were so tiny that Helmholtz had to observe them through a microscope and copy them by hand on a larger scale. The single such curve that he later published is reproduced in figure 1.

These results did not satisfy Helmholtz. It appeared to him that the

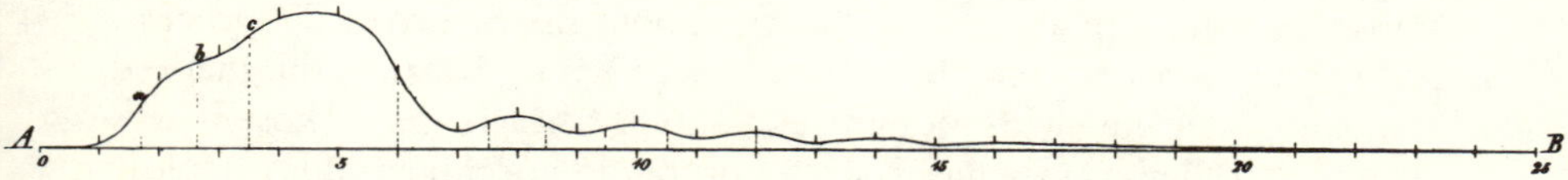

FIG. 1. Helmholtz's graphical representation of the time course of muscle action. (Ref. 10, endplate)

friction in the muscle and the apparatus, which no amount of refinement could eliminate, prevented him from "perfecting" the method sufficiently to answer his question. When he returned to the problem a year later, therefore, he abandoned his graphical apparatus for a different, more precise method that had in the meantime occurred to him for measuring minute time intervals. To what degree Helmholtz felt at the time he was pursuing his first method that it provided too little "exactitude," and to what degree he looked back on it that way only after he had available to him the second method, is impossible to ascertain from the sparse record of his experimental activity during this time. In any case, even after applying the second method with success, he still considered the first method valuable enough to include in his publication on the subject, because "it affords an overview [*Überblick*] of the course of the contraction that can be reached more quickly and through a simpler process of reasoning."[13]

If we examine more closely what the first method revealed to Helmholtz, it turns out to be quite a lot. In the most general terms, it proved that "the energy of the muscle does not fully develop at the moment of an instantaneous stimulus, but only gradually increases, most often only after the stimulus has ceased, reaches a maximum, and again disappears." By analyzing the form of his curves more closely, he was able to interpret in greater detail the "stages" of the process. Noting that there were alternating sections of the curve in which it was concave upward and concave downward, he interpreted these as phases in which the elastic tension of the muscle was greater than and less than the weight attached to it, and the points of inflection as moments at which the two forces were in equilibrium.[14]

If we look again at Helmholtz's initial question, we can see that this method enabled him to define phases of the muscle contraction in a manner that fit Weber's mechanical interpretation of the muscle as a variable elastic band, and to order these phases along the spaces of time over which the contraction and relaxation of the muscle took place. Already with these first efforts, he had fulfilled duBois-Reymond's criteria for the application of mathematical analysis to physiological phenomena: to establish through the salient characteristics of a curve expressing the values of an independent and a dependent variable, general functional relationships that can serve as a test of theories about the nature of the process. Helmholtz seems to have been dissatisfied mainly because the method could not determine with enough "exactitude" the lengths of those spaces of time that the several phases of the contraction occupied. It is not clear whether he felt a need for greater accuracy in order to answer more refined questions concerning the nature of a muscle contraction, or whether he strove for more precise measures of the time intervals as an end in itself.

Qualitative Precision and the Images of Persuasion

In our previous essay we have described the "entirely new" method with which Helmholtz began, during the fall of 1849, to measure the time required during muscle contractions,[15] and will restrict ourselves here to brief supplementary comments. The method, adopted from Pouillet, posed, as Helmholtz remarked, in principle "no limit on the smallness of the parts of time whose measurement is made possible in this way." The method relied on the fact that "the time during which a galvanic current of known intensity" passes through the windings of a sensitive multiplicator [galvanometer] "can be calculated exactly by the changed motions" the current imparts to the magnetic needle of the instrument. Helmholtz arranged his apparatus so that the onset of this current coincided in time exactly with the stimulus imparted to the muscle or its attached nerve, and the cessation exactly with the time at which the muscle lifted a given weight off of its resting place. By this method he was able to measure with unprecedented precision "the time that passes from the moment of the stimulus to the point at which the elastic force of the muscle reaches a definite value" measured by the "over-burden" of the weight attached to it.[16]

Although the new method could measure with extreme accuracy the length of time of a single phase of the muscle action defined in this manner, it was not able to do so for any other of the phases into which a complete contraction might be divided. By performing a series of experiments with different weights, however, "one learns in what spaces of time following one another the different grades of the energy [of the muscle] develop." That is, from a series of measurements with different weights attached to the muscle, Helmholtz could measure the time required for the tension in the muscle to equal the "opposing forces" of each of these weights. By plotting these points and connecting the points with lines, he constructed curves showing, not the actual motion of the muscle in any given contraction, but the way in which the tension in the muscle *would have* increased over time if it had been stimulated at constant length. The curve was similar in form to the curve that one would obtain by connecting the points of inflection of the curves recorded graphically. Therefore, the results of the new method appeared to Helmholtz "not to contradict" those of his first method. They revealed, moreover, as the first method could not, "that a time passes after the stimulus, before the overall energy of the muscle begins to climb."[17]

Helmholtz described his "preliminary" graphical method afterward as possessing "less exactitude" (*geringere Genauigkeit*) than his second "more perfect" method.[18] As applied to the problem of muscle contraction, however, this comparison was somewhat inaccurate. The two methods complemented each other, in that each revealed something about the nature of the muscle contraction that the other could not. Had he stuck to his original subject of investigation, the superiority of his new method would not have been evident: in part because the more precise measurement of time intervals was less central to this purpose than was the clear, semiquantitative characterization of muscle action as a function of spaces of time not too small to be measured graphically. His newfound ability to measure time intervals with greater precision than his task required led him, however, by steps that we cannot reconstruct in fine detail, to recognize that he could employ the same method to measure the difference in the time required for a stimulus applied to a distant portion of the nerve attached to the muscle to reach the muscle, than for a stimulus applied to the nerve close to the muscle. Here the capacity to measure limitlessly small intervals would be crucial, because his mentor, Johannes Müller, had attempted to determine the times that elapsed between a nerve stimulus and a muscular response in frogs, and found it to be "infinitely small and unmeasurable."[19]

In January, 1850, Helmholtz reported to the *Akademie der Wissenschaften* in Berlin that he had found the length of time for a stimulus to traverse a section of nerve 50–60 millimeters long was "0.0014 to 0.0020 of a second."[20] The velocity of propagation of the nerve impulse implied in these numbers was surprisingly slow. Nevertheless, its detection had required precise measurements of time intervals far more minute than had ever before been attempted in experimental physiology.

Helmholtz's initial report drew praise from the physiologists who were most important to him, especially Johannes Müller and Carl Ludwig. Yet, as his friend duBois-Reymond cautioned him, there was also considerable skepticism. In our previous paper we have discussed the way in which Helmholtz responded to doubts about the validity of his results. Analyzing the data from his earliest trials by the method of least squares so prominent in some approaches to precision measurement, especially at Königsberg, where he was then teaching, Helmholtz established the probable errors in his measurements of time, and expressed the outcome in the form "the velocity of propagation: 24.6 ± 2.0 meters per second." Helmholtz did not analyze his data in this way in order to arrive at an *exact* value of the nerve propagation velocity—he did not produce an average result from the varying results of the several sets of trials he had conducted—but to assess the reliability of his data. The analysis based on least squares sustained his confidence that he had measured a real phenomenon rather than an artifact of his method.[21]

When he wrote up his results in a long paper submitted to the *Physikalische Gesellschaft* in Berlin in July, 1850, Helmholtz was already well aware, in part from criticisms by duBois-Reymond, that the "description" of his method for measuring the velocity of the nerve impulse "presents . . . difficulties." He presented the data from each of his individual experiments in complicated tables, with extended discussions, and even provided for those readers not familiar with the method of least squares an elementary illustration of its meaning.[22] He was also cognizant that to grasp his results one needed to understand general methods of physico-mathematical analysis and the physical laws underlying the operation of the multiplicator, skills that were foreign to many contemporary physiologists. For such reasons he decided to return to the graphical method despite the "lesser exactitude" that his prior efforts with it had yielded. In his paper he stated his intentions:

> I shall remark, finally, that the method used in our preliminary experiments to trace temporal processes, which was not useable for the investigation of the muscle contraction, can probably serve for an easier and quicker exposition [*Darlegung*] of our results concerning the velocity of transmission in the nerve than the method hitherto pursued. One needs only to arrange the mechanism so that two contraction curves corresponding to two stimuli from different places on the nerve must completely coincide if the propagation velocity were infinitely large. Since the latter is not the case, they will, in fact, be separated, and the difference along the abscissa between every two corresponding points of the curves will correspond to the propagation time. If my means permit it, I reserve for myself [the opportunity] to carry out the experiment in this manner.[23]

During the fall of 1850 Helmholtz spent much time, as he wrote his father in December, on "the construction of another apparatus [for time measurements] and on the necessary preparation."[24] In characteristic style, Helmholtz designed a highly effective, elegant apparatus. He carefully considered the factors necessary to reduce friction in the mechanical linkage connecting the isolated muscle to a point tracing a line on a drum rotating as nearly as possible at constant speed, and to ensure that the stylus would transform the shortening or lengthening of the muscle into a purely vertical line. The most critical feature of the instrument was a mechanical device that automatically initiated successive nerve stimuli at exactly the same point along the abscissa of the revolving drum, fulfilling the preconditions he had set out in his anticipatory description of the method.[25]

In January, 1851, Helmholtz resumed his work on the nerve impulse, after a "fortunate interruption" imposed by his remarkable invention of the ophthalmoscope. He worked in two directions. In order to satisfy himself that "when an induction current exerts its physiological effect I do not ascribe to the effects of the nerves time differences that belong to electrical currents," he developed a mathematical theory of "the duration and course of electrical currents induced through variation in currents." He concluded that in comparison to the time intervals he was measuring, the induced currents were instantaneous. At the same time, he began to produce with his new graphic apparatus tracings (*Zeichnungen*) of the contractions of frog muscles stimulated successively from two different places on their attached nerves. His early results were already persuasive enough to him so that, while awaiting new spring frogs in April, he sent "the older tracings to [Alexander von] Humboldt, in order to impress him through their appearance [*durch Augenschein*]."[26]

During the summer vacation of 1851, Helmholtz toured physiological laboratories in Germany, Switzerland, and Austria, ostensibly to obtain information and advice that would help him to organize the teaching of physiology at Königsberg, but evidently also to make contacts and to publicize his own investigations. "Everywhere," he wrote his wife Olga in August, "I demonstrate my frog curves."[27] Along the way he had opportunities to compare the persuasive power of his two methods for demonstrating the velocity of the nerve impulse.

In Göttingen, where physiologists of the older school, in particular Rudolph Wagner, were in close contact with the physicist Wilhelm Weber, and younger men such as the anatomist Gustav Bergmann and the "mathematical optician Listing" were more familiar with physical reasoning, Helmholtz found that his demonstration by the Pouillet method had already converted the group; but he could not be certain that all of them understood it. "It was pleasant for me to see," he wrote Olga, "that they have worked their way into my rather difficult nerve work and agree with it, or at least, as it seems, because of Weber's judgment they have sufficient trust in my physical knowledge to believe in my results."[28]

Traveling next to Marburg, where he encountered the physiologist Hermann Nasse, Helmholtz experienced the more immediate impact his graphical method could have, even on someone not attuned to his physicalist approach:

> I had to wait a long time for Professor Nasse, because he gave a lecture until 6. He belongs to the school that prefers to explain life in the most mystical way possible. He therefore stands in direct opposition to us. He has delivered much industrious and worthy work, but nothing significant. Consequently he received me at first only politely, avoided scientific subjects, and finally expressed doubt about the matter of my nerve propagation. Then I paraded my frog curves in front of him. He grasped these more readily than the earlier [method], and they made the matter extremely plausible to him. Now he became completely different . . . and we parted as the best of friends.[29]

Helmholtz stayed for several days with Ludwig in Zurich, where they talked about "every possible physiological and physical subject."[30] There was no need to persuade Ludwig with the curves, because he had been fully convinced of the validity of the measurements by the earlier method, but Ludwig undoubtedly gave Helmholtz helpful advice on the further refinement of his graphical methods. As soon as he had returned to Königsberg, Helmholtz continued his experiments on the nerve impulse.[31]

In early 1852 he completed them and published his results in a second article entitled "Measurements of the Velocity of Propagation of a Stimulus in the Nerves."

In the introduction to this paper Helmholtz defined differently than he had in his first paper the advantages and disadvantages of his two methods. The Pouillet method he now affirmed, "offers the best guarantees, where it is a question of the reliable execution of exact measurements," but has the great drawback that the "specified result can only be extracted through an extended and tedious series of experiments." The graphical method, on the other hand, "allows a much simpler and more easily executed proof of the propagation time in the nerves."[32] The main difference between this comparison and the one he had earlier made in anticipation of his efforts to develop the second method was that he had then focused on the ease with which results once gained can be persuasively presented, whereas here his attention fixed on the ease with which one could employ the two methods to attain further results. The two perspectives merge in an assessment of persuasive power, if we assume that the effects of results presented to other investigators are most persuasive to those who repeat and extend them in their own laboratories.

Helmholtz devoted most of this paper to the description of his apparatus, the experimental method, and evaluations of possible sources of instrumental errors.[33] He reproduced several of the curves recorded in his experiments: one representing an optimal experiment in which the excitability of the muscle did not change between the two successive stimuli of the nerve, and two others showing that even if the excitability decreased somewhat the curve representing the stimulus from a longer distance in the nerve was displaced to the right of the curve representing the closer stimulus. Direct visual inspection of the curves carried the burden of Helmholtz's new demonstration. "The greatest advantage of the method," he asserted, "is that in each individual tracing of the two curves belonging together, one can immediately recognize from their form" whether or not the muscles worked in exactly the same way in both cases.[34] The power of the curves to produce this immediate recognition is as obvious to us as it was to those to whom Helmholtz himself showed his curve. Figure 2 is the example in which the two curves show identical contractions:

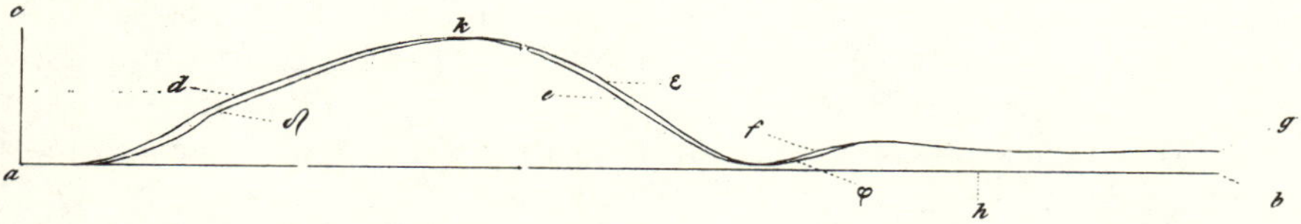

FIG. 2. Comparative displacement of time course of muscle action (identical contractions). (Ref. 25, endplate)

Figure 3 illustrates how a muscle "worked" slightly differently after the second stimulation than after the first.

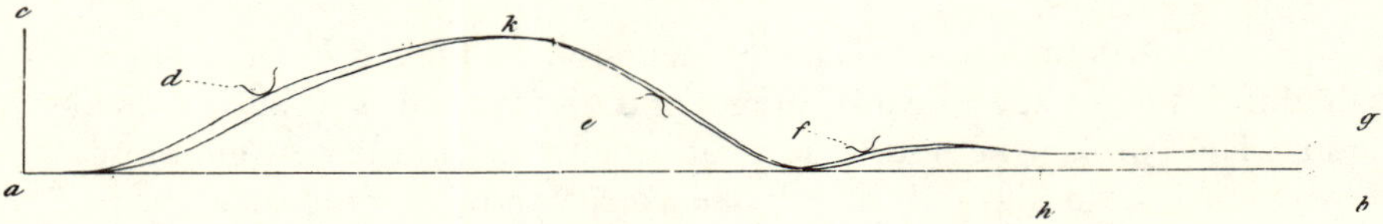

FIG. 3. Comparative displacement of time course of muscle action (slightly greater excitability in second contraction). (Ref. 25, endplate)

When all phases of the muscle contraction appeared exactly alike, the only plausible interpretation of the displacement to the right of the one stimulated from a longer distance in the nerve is the delay required for that stimulus to be propagated along that additional length of the nerve.

Concerning the "absolute value of the propagation velocity," Helmholtz was almost casual. He determined it simply by measuring the horizontal distance between any two corresponding points on the curves. In contrast to the first paper, in which he had supplied the complete data from each of his successful experiments, he merely stated that the horizontal distance between points on the best curves shown on his published plate was "about 1mm." The length of the nerve between the two points of stimulation being 53mm, "it follows [that] the velocity of propagation is 27.25 meters per second." The only evaluation of the accuracy of this result he made was to note that "the most probable value deriving from the earlier experiments was 26.4 meters."[35] The "horizontal distance between the two curves," Helmholtz acknowledged, "cannot be measured with very great exactitude."[36] The approximate value the measurement yielded was, however, good enough to allow him to employ his graphical method for the advantages that it possessed over the Pouillet method. Helmholtz was adopting a pragmatic standard, defending a less precise method as adequate in practice, one that permitted him to exploit the ease of execution and interpretation that the method offered. Implicit in his presentation of the method was that standardization of the procedure would allow him and other investigators to explore the velocity of the nerve impulse under varied physiological conditions, and to obtain numerical values good enough for comparative purposes. His approach illustrates well what Steve Woolgar, following other ethnomethodologists, terms "practical reasoning" in science. In the course of their daily work scientists regularly assess the connections between specific research results and underlying realities as "good enough," or "practically adequate," even though such connections can, in principle, always be called into question.[37]

Following Helmholtz's own evaluation, we have in our previous discussion of these two methods described his move from the Pouillet to the graphical method as one away from complicated techniques of precision measurement to a quantitatively less rigorous method possessing the greater demonstrative power of direct visual evidence. As we have seen, his principal motivation for developing the graphical method when he had already proven his case to his own satisfaction through the "more perfect" Pouillet method was to gain more demonstrative power. If, however, we extend the meaning of precision to encompass not only exact quantitative measurement, but also "strict expression" and "exact definition" in a qualitative sense, then we can view Helmholtz's move also as a shift from one form of precision to another. Among the criticisms that had been made of his first presentation of his results with the Pouillet method was that he had not strictly eliminated the possibility that the differences he observed in the time between the stimulus and the attainment in the muscle of a tension sufficient to lift the weight were due to differences in the time of the first phase of the muscle action, rather than in the time required for the stimulus to reach the muscle.[38] The graphical method eliminated such a possibility more effectively than had any of his subsequent efforts to deal with the problem with the Pouillet method, by showing "at a glance" that the two actions of the muscle appeared identical in all phases. In this qualitative sense, therefore, the graphical method was more "exact" than the quantitatively more accurate Pouillet method.

This viewpoint becomes stronger if we treat Helmholtz's paper not only as the study of the velocity of the nerve impulse that he intended it to be, but also as a continuation of the study of muscle contraction itself that he had begun in 1848. Despite his own opinion that the graphical method was not "usable" for this purpose, he discovered with it, in passing, remarkable features of the changes in the contraction as the excitability of a muscle decreases. At first all the vertical coordinates were proportionately diminished, while the length and form of the curve remained unchanged. With further loss of excitability, however, the duration of the contraction changed, "and indeed, what one would perhaps not have been able to suspect, it becomes not shorter, but longer." Again his curves gave clear visual expression to those modifications (fig. 4).

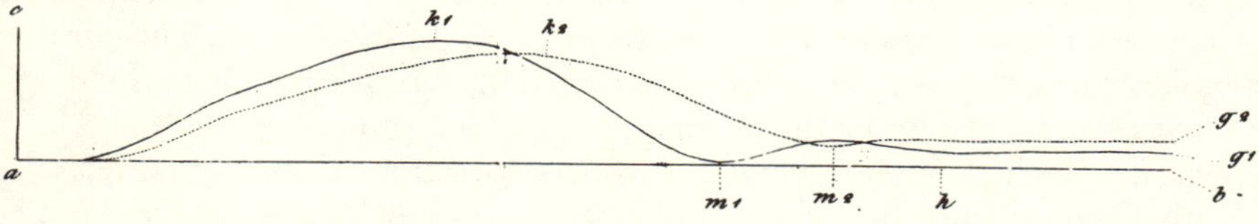

FIG. 4. Comparative displacement of time course of muscle action (decreased excitability in second contraction). (Ref. 25, endplate)

This striking by-product of Helmholtz's use of the graphical method revealed that it offered as much promise for the future study of muscle contractions as for the measurement of the nerve impulse. Later contemporaries, in fact, looked on his paper as "a classical work" in the investigation of "the temporal course" of a muscle contraction. His apparatus became known as a myograph, and "many authors" adopted it, or minor modifications of it, "to carry the analysis of the contraction process somewhat further than Helmholtz's investigation had done."[39] In this domain the precision of his method was directed, not at exact quantitative measurement, but at the accurate reproduction of functional relationships between the height of the contraction, the time, and other variables such as the degree of excitability of the muscle. Exact definition made visible the qualitative characteristics of the contraction process that had previously been hidden from observation.

Helmholtz's drive for qualitative precision was expressed not only in his experimental design and procedures, but also in the verbal and pictorial language he used to convey to others exactly what he had done. After he had communicated his discovery of the velocity of the nerve stimulus to the Berlin Academy of Sciences in a succinct "preliminary report," duBois-Reymond complained to him that he had "presented the matter with such boundless obscurity" that his paper could serve only as a "short introduction" to induce others "to rediscover" his method.[40]

To the modern reader this alleged obscurity is not self-evident. Helmholtz's spare summary of his method and results is what we are accustomed to find in preliminary reports, for which one expects a fuller description to follow. Nor does it appear that those who heard the paper, communicated by Helmholtz's mentor Johannes Müller, had as much trouble with it as duBois-Reymond claimed. Nevertheless, Helmholtz took the criticism to heart. He lamented that

> The composition of such a note is very difficult, because it is intended mainly to be only a brief indication [*Andeutung*] for specialists, to which they must invent for themselves further details, and it is much more risky to judge at what level of knowledge one should aim than is the case in formulating the matter in extenso, where one assumes the least possible knowledge.[41]

Helmholtz's comment nicely illustrates the extent to which "strict expression" and "exact description" are matters of social interaction. The question of whether his account of his experiments was precise or murky depended not only upon the words he employed, but upon the context within which they were read. His concise description could be precise only to those who could fill in from their own knowledge of the relevant phenomena and instruments the details he did not supply.

In his extended papers Helmholtz took great pains to eliminate obscurity. In both his 1850 and 1852 papers on the velocity of the nerve impulse he described his apparatus and his protocol in meticulous detail. He combined drawings of the apparatus, in the execution of which he was careful to ensure that they depicted the essential features with maximum clarity, and verbal descriptions closely keyed to the drawings. Text and drawings together enabled the attentive reader to conjure a more precise mental image of the instruments and of their operation in Helmholtz's experiments than could be achieved through either the visual or the verbal mode alone.[42]

A Comparison to Ideal Representations of Data

Alfred Volkmann had applied Ludwig's kymograph to the study of the temporal course of muscle contractions independently of Helmholtz, even though he began somewhat later. When Helmholtz reported in his paper of 1850 that friction posed an insuperable barrier to accurate results with this method, Volkmann decided to stay with the use of the kymograph and attempt to solve the friction problem. He proceeded by substituting for the pointed stylus a bristle made of a human hair, and claimed to have succeeded so far that the friction between it and the darkened paper covering the revolving cylinder no longer retarded significantly "the movement of the bristle tracing the curves." Volkmann presented his method at a meeting of the Leipzig Scientific Society in January 1851; that is, just as Helmholtz was beginning to perform similar experiments with his improved graphical apparatus.[43]

At the time he took up this problem, Volkmann was fifty years old—a scientific generation older than Helmholtz. Trained during the 1820s in anatomy and physiology, Volkmann acquired a reputation as an outstanding teacher and investigator, the latter mainly for his studies of the nervous system. During the 1840s, Volkmann joined the movement toward a more quantitative, physically oriented physiology, by embarking on experimental studies of the dynamics of the circulation.[44] Admiring the methods of the "1847" group, including Helmholtz and duBois-Reymond, Volkmann sought contact with these younger reformers. His own knowledge of the physical concepts and mathematical methods they brought to physiology was limited, but he was determined himself to subject the phenomena of muscle contraction to a physical-mathematical analysis.

At the time Volkmann discussed his work at Leipzig in January, he had not yet surmounted to his satisfaction the "major difficulty" for his experiments, how to produce "stimuli of the shortest duration and of known and constant strength." His results so far were, therefore, "rather inadequate." Nevertheless they sufficed to persuade him that "the lines traced by the muscle on the ascending side of the curve are true parabolas," whereas "a glance shows immediately" that on the descending side "the first half is upwardly convex, and the second half is downwardly convex."[45]

Unlike Helmholtz, Volkmann did not display any examples of actual curves traced by his apparatus, but depicted the ideal form that such curves would take (fig. 5).[46]

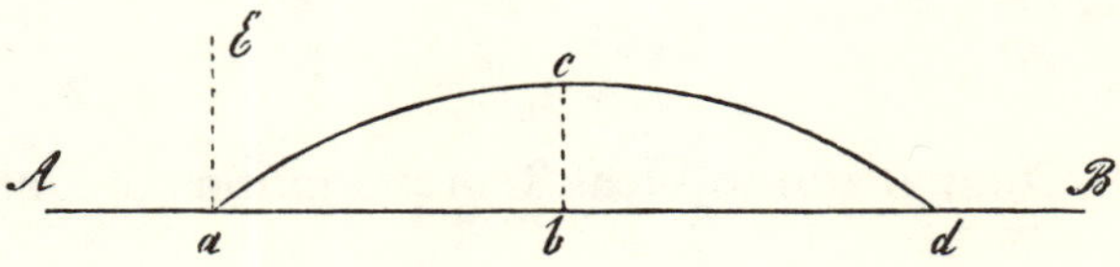

FIG. 5. Volkmann's ideal curve representing muscle action.

By doing so, and by labeling the parameters and variables algebraically, Volkmann imparted to his presentation an aura of mathematical abstraction, in contrast to the empirical basis of the curves Helmholtz presented.

By the time he gave a second presentation to the Scientific Society, in June 1851, Volkmann had made considerably more progress. Now he regarded neither the ascending nor the descending curves as "strict parabolas," because the beginning of the curve was downwardly convex. Nevertheless, he concluded, by comparing the observed values for the height of the curves at successive points along the abscissa with the values calculated from the mathematical equation for a parabola, if one first took care to locate the apex of the curve, "the differences between the observed and calculated values of x [the ordinate] fall within the observational errors." Volkmann did not support this evaluation with a formal analysis of the probable errors, but did include tables of calculated and observed values and their differences, such as in table 1.[47] "After it has now been proven that the curve traced by the muscle clearly approaches a parabola," Volkmann went on, "the question arises, how one must think of the force of the muscle which can give rise to such a line?" His answer was that the force of the muscle must be exerted suddenly, at the beginning of the contraction, like exploding gunpowder. The rest of the motion represents a "constantly decelerating ascent," which "finds the law of its expression in the form of the parabola."[48]

Volkmann thus attempted what duBois-Reymond had asserted was seldom possible in the current state of physiology: to reduce an observed functional relationship to a mathematical expression upon which one could then base a causal explanation of the phenomena. Volkmann's procedure contrasts with that of Helmholtz, who had been content to con-

Abscissen.	Ordinaten.	*y*.	*x*.	*x*' berechnet.	Differenz.
1 Mill.	0,4 Mill.	1 Mill.	0,05 Mill.	0,05 Mill.	0,00 Mill.
2 »	0,8 »	2 »	0,1 »	0,14 »	+ 0,04 »
3 »	1,4 »	3 »	0,25 »	0,27 »	+ 0,02 »
4 »	1,9 »	4 »	0,40 »	0,45 »	+ 0,05 »
5 »	2,5 »	5 »	0,60 »	0,66 »	+ 0,06 »
6 »	3,1 »	6 »	0,9 »	0,92 »	+ 0,02 »
7 »	3,6 »	7 »	1,25 »	1,22 »	— 0,03 »
8 »	4,0 »	8 »	1,6 »	1,57 »	— 0,03 »
9 Mill.	4,4 Mill.	9 Mill.	2,0 Mill.	1,95 Mill.	0,05 Mill.
10 »	4,75 »	10 »	2,4 »	2,39 »	— 0,01 »
11 »	5,1 »	11 »	2,9 »	2,85 »	— 0,05 »
12 »	5,4 »	12 »	3,5 »	3,36 »	— 0,14 »
13 »	5,6 »	13 »	4,1 »	3,92 »	— 0,09 »
14 »	5,75 »	14 »	4,6 »	4,52 »	— 0,08 »
15 »	5,9 »	15 »	5,2 »	5,16 »	— 0,04 »
16 »	5,95 »	16 »	5,6 »	5,83 »	+ 0,23 »
17 »	60				

TABLE 1. Volkmann's observations of the contractions of a frog's thigh muscle (data-based representation of Fig. 5). (Ref. 47, pp. 56–57)

nect the explanatory concept of the muscle as an elastic band with the salient qualitative features of his empirical curves.

In 1845 Karl Wunderlich, one of the leading promoters in Germany of a new "physiological medicine," wrote that "in medical judgments, as in judgments about all human situations and complexities, mathematical certainty can only be attained in isolated cases; more often a calculation of probabilities alone is admissible." [49] His view might have served as an admonition to Volkmann, who retained an older-style belief that the ideal parabola his measured ordinates "approached" lay closer to the truth than did his empirical values.

Both duBois-Reymond and Helmholtz appreciated Volkmann as an industrious, competent experimentalist, but neither was impressed with his talent for theoretical interpretation of his work. During the Easter holiday, while Volkmann was probably composing the above arguments, duBois-Reymond visited him at Volkmann's request in Halle. Afterward duBois-Reymond wrote Helmholtz: "I would have had a good time, if I had been able, in the course of eight days to make it comprehensible to Volkmann that the measure of the force is the change in the velocity. He is occupied with an investigation similar to yours, and wants to analyze the phenomena mechanically, and as you can see from the above, he lacks the ABC's."[50] At the same time, duBois-Reymond sent one of Volkmann's tracings, on the assumption that his method of recording on smoked paper with a bristle might be useful to Helmholtz.

Helmholtz replied in June, thanking him for Volkmann's curves. His own method, using a steel point that traced along a smoked glass cylinder, he believed, caused "even less friction than Volkmann's hair. Nevertheless, Volkmann's method is simpler, cheaper, and can provide longer tracings." Concerning duBois-Reymond's comments about Volkmann, Helmholtz wrote that they did "not surprise me. The same lack of knowledge is the worm that has undermined his laborious work on hemodynamics. So much industry, such good will, an achievable goal in sight, and nearly total lack of success because of the absence of simple mechanical principles!"[51]

DuBois-Reymond and Helmholtz displayed in their attitudes toward Volkmann some of the natural impatience of youth with their elders trained in a bygone era. Nevertheless, we can agree that Volkmann's knowledge of the physical concepts and mathematical methods he deployed in his study of muscle physiology was limited in comparison to their own understanding of the physical and mathematical foundations they applied to physiological questions. That Helmholtz could achieve a "classical work" in a field of experimental physiology in which the experienced Volkmann could make only a minor contribution, was as much due to the precision with which Helmholtz reasoned as to the precision of his instruments and his measurements. Moreover, the contrast between the restrained inferences concerning the nature of muscle action that Helmholtz drew from his curves, and the ambitious yet vague conceptions of the muscular force that Volkmann drew from experimental curves that may not have been notably inferior in quality to those of Helmholtz, suggest that the attainment of precision involves also circumspect judgments about the limits inherent in such data.

The Power of Demonstration

We return now to reflect further on the relative power of demonstration, or persuasion, manifested by the two methods with which Helmholtz measured the velocity of the nerve impulse. An initial caveat is that all of our evidence for the persuasiveness of the graphical method is drawn from Helmholtz himself. Consequently, it is difficult to distinguish between (1) the relative persuasive value of the two methods for Helmholtz himself; (2) his anticipations about their persuasive power for others; (3) his perceptions of their persuasive effects on others whom he directly encountered; and (4) the actual experiences of others in being persuaded or unpersuaded by either method. Finally, we should note that to separate the effects of the two methods may be artificial. For some people it may not have been one or the other method that clinched the case, but their convergent results. Helmholtz himself implied in his comparison at the end of his second paper, of the numerical values arrived at by the two methods, that their near agreement was the strongest argument for the validity of both methods.

If we nevertheless assume that Helmholtz's opinion of the superior demonstrative capacities of his graphical method was realistic, and his experiences concerning the responses to his two methods on his tour in the summer of 1851 were in some sense paradigmatic for the responses of other scientists whose acceptance of his conclusion he did not personally witness, then it is profitable to examine more closely the particular power of those "curves." From his own statement, Helmholtz appeared to have felt that the principal advantage of the method was that it offered an "easier and quicker exposition." The exposition was, however, not merely easier and quicker, but also in a visual mode. The other method relied upon logical calculations and analysis derived from measured numbers. The time necessary for a stimulus to traverse a section of nerve is arrived at in the first method from the difference between two numbers calculated from a series of measurements of distances calibrated as proportional to the times. In the second method this "space of time" can be "seen" as the horizontal displacement between two curves. In an article entitled "Why are Graphs so Central in Science?" Roger Krohn argues that it is because they "reemploy our powers of visual perception and pattern recognition, which are so very much greater than our ability to find meaningful relations directly in numerical lists."[52] Recent studies in experimental cognitive science help to explain such efficacy by supporting the view that humans can process information directly in images, without first reducing the visual patterns they perceive to propositional forms.[53]

The relative efficacies of different human cognitive abilities are, however, not necessarily constant. The ease with which we recognize patterns in graphical curves may be enhanced by their ubiquity in scientific presentations, particularly in the textbooks through which basic scientific skills are nurtured. At the time Helmholtz published his muscle curves, graphical representation was, according to Laura Tilling, beginning to become more popular, but was still uncommon even in the physical sciences. In physiology it was so rare that the physiologists Helmholtz hoped to persuade could not have been accustomed to assimilate information in that way. Nevertheless, the spreading usage of graphical methods of recording and presenting data at just this time suggests that Helmholtz was attuned to contemporary trends that would in the future continually reinforce the expository power of his graphical method for determining the velocity of the nerve impulse.

At several points we have put stress on the visual language that

Helmholtz, and also Volkmann, used to describe the relation between their curves and the phenomena whose properties the curves traced (sometimes noting the original German in which the visual overtones are stronger than in translation). For Volkmann the curves brought the muscle contraction "*zur Anschauung*." For Helmholtz they provided "*ein Überblick*" of the course of the contraction. Were these curves merely tracings from which one could *infer* the character of processes, or did those who viewed them see through them the processes themselves?

Dudley Shapere has argued convincingly that the specialized information with which scientists become deeply familiar extends the meaning of *observation* for them far beyond what can be perceived directly by the eye. Scientists speak seriously about "seeing" events that to the uninitiated appear only to be detected through indirect means. Helmholtz's frog muscle curves do not literally look like the contracting frog muscle hung within the apparatus that produced them. Yet the connection between the two can become so close that to the practiced observer the curve may become a more precise way of "viewing" the same event that takes place too quickly for her to observe in detail when she watches the real muscle.[54] This transformation can take place with special ease when it involves the spatial representation of temporal processes, because the ordinary language with which we describe the passage of time is embedded in spatial metaphors.[55] We have emphasized this pervasive correspondence as it applies to Helmholtz's experiments by translating the word "*Zeiträume*" that he frequently used, as "spaces of time." In describing the relation between the curves and the contractions, Helmholtz wrote that "time differences reappear as spatial differences"—not that the latter "represent" the former. The boundary between representations and images is indistinct. Even though graphical muscle curves are not literal pictures of contracting muscles, their ability to depict with precision the temporal shape of the contraction justifies the title we have chosen for this paper.

Notes to Chapter Eight

1. C. Ludwig, "Beiträge zur Kenntniss des Einflusses der Respirationsbewegungen auf den Blutlauf im Aortensysteme," *Archiv für Anatomie, Physiologie, und wissenschaftliche Medizin* (1847), 244.

2. Ibid., 242–67.

3. See Merriley Borell, "Instrumentation and The Rise of Modern Physiology," *Science and Technology Studies* (1987) 5:53–62.

4. Gabriel Gustav Valentin, *Lehrbuch der Physiologie des Menschen* (Braunschweig: Vieweg, 1844).

5. [Theodor] Fechner, "Die mathematische Behandlung organischer Gestalten

und Processe," *Berichte über die Verhandlungen der Königlich Sächsischen Gesellschaft der Wissenschaften zu Leipzig. Mathematisch-Physische Classe*, 1849, pp. 50–64.

6. Emil duBois-Reymond, *Untersuchungen über thierische Elektricität* (Berlin: G. Reimer, 1848), xxvi–xxxiii.

7. See, for example, E. H. Weber, "Ueber die Anwendung der Wellenlehre auf die Lehre vom Kreislauf des Blutes und insbesondere auf die Pulslehre, *BVKSGW* (1850), 164–204.

8. Ludwig, "Einflusses der Respirationsbewegungen," 243.

9. Alfred Wilhelm Volkmann, "Ueber das Zustandekommen der Muskelcontractionen im Verlaufe der Zeit," *BVKSGW* (1851), 1.

10. H. Helmholtz, "Messungen über den zeitlichen Verlauf der Zuckung animalischer Muskeln und die Fortpflanzungsgeschwindigkeit der Reizung in den Nerven, *AAPwM* (1850), 276–77.

11. Eduard Weber, "Muskelbewegung," in *Handwörterbuch der Physiologie mit Rücksicht auf physiologische Pathologie*, ed. Rudolph Wagner (Braunschweig, 1842–53), 3:219.

12. Helmholtz, "Messungen," 278–81.

13. Ibid., 279–83.

14. Ibid., 283–84.

15. Kathryn M. Olesko and Frederic L. Holmes, "Experiment, Quantification, and Discovery: Helmholtz's Early Physiological Researches, 1843–50," in *Hermann von Helmholtz and the Foundations of Nineteenth Century Science*, ed. David Cahan (Berkeley, 1993), 50–108.

16. Helmholtz, "Messungen," 284–85, 289.

17. Ibid., 285, 307–9.

18. Ibid., 280.

19. Johannes Müller, *Handbuch der Physiologie des Menschen*, 4th ed. rev. (Coblenz, 1844), 1:58.

20. Helmholtz, "Vorläufiger Bericht über die Fortpflanzungsgeschwindigkeit der Nervenreizung," *AAPWM* (1850), 71.

21. Olesko and Holmes, "Experiment," 83–108.

22. Helmholtz, "Messungen," 280, 328–58.

23. Ibid., 358.

24. Herman Helmholtz to Ferdinand Helmholtz, December 17, 1850, *Letters of Hermann von Helmholtz to His Parents 1837–1855*, ed. David Cahan (Stuttgart, 1991).

25. H. Helmholtz, "Messungen über Fortpflanzungsgeschwindigkeit der Reizung in den Nerven. Zweite Reihe," *AAPwM* (1852), 201–11.

26. Leo Koenigsberger, *Hermann von Helmholtz* (Braunschweig, 1902) 1:143–47; Helmholtz to duBois-Reymond, April 11, 1851, *Dokumente einer Freundschaft: Briefwechsel zwischen Hermann von Helmholtz und Emil duBois-Reymond, 1846–1894*, ed. Christa Kirsten (Berlin, 1986), 111.

27. Koenigsberger, *Helmholtz*, 1:149.

28. Ibid.

29. H. Helmholtz to Olga Helmholtz, August 10, 1851, Houghton Library, Cambridge, Mass. Undoubtedly Helmholtz did not hand his frog curves to Nasse

in silence. At a minimum he would have had to explain how they were produced. In the absence of direct evidence concerning the accompanying dialogue, we can only imagine what degree of verbal persuasion Helmholtz may have attached to the visual evidence. If we wished to fit this case into the framework of Bruno Latour's argument that "inscriptions" (of which both the frog curves and the numerical tables by which Helmholtz presented the results of his first method would be subtypes) are means "to mobilize allies" and to raise "the cost of dissent," then we could treat this encounter as a maneuver by Helmholtz to "corner" Nasse by adding "one *more* inscription" to "swing the balance of power in his favor." The outcome would then be a "victory" for Helmholtz in his effort to dominate an agonistic field. See Bruno Latour, "Drawing Things Together," in *Representation in Scientific Practice*, ed. Michael Lynch and Steve Woolgar (Cambridge, Mass., 1990), 19–68, esp. 23–24, 41–42, 44. If we can trust the tone of Helmholtz's account, however, Nasse did not behave as a vanquished dissenter. He had not been compelled to assent, but persuaded; and he responded with the pleasure of someone who has come to understand what he did not previously grasp.

30. Helmholtz to Olga Helmholtz, August 22, 1851.

31. Koenigsberger, *Helmholtz*, 1:158–59.

32. Helmholtz, "Messungen: zweite Reihe," 199.

33. Ibid., 200–211.

34. Ibid., 215.

35. Ibid., 215–16.

36. Ibid., 215.

37. Steve Woolgar, "Time and Documents in Researcher Interaction: Some Ways of Making out what is Happening in Experimental Science," *Representation in Scientific Practice*, 123–24.

38. E. duBois-Reymond to Helmholtz, March 19, 1850, *Dokumente einer Freundschaft*, 92. DuBois-Reymond stated only that "you have not been able to eliminate the time that elapses during the processes in the muscle." For his comment to contain a valid objection to Helmholtz's conclusions he must have meant something like the above.

39. L. Hermann, "Allgemeine Muskelphysik," in *Handbuch der Physiologie*, ed. L. Hermann, vol. 1, part I. (Leipzig, 1879), 23, 35.

40. E. duBois-Reymond to Helmholtz, March 19, 1850, *Dokumente einer Freundschaft*, 92.

41. Helmholtz to duBois-Reymond, April 22, 1850, ibid., 96–97.

42. Superficially this situation may appear to lend itself to the interpretation Shapin and Schaffer have made of Robert Boyle's detailed, illustrated descriptions of his air pump and of the experiments performed with it. According to them, Boyle's purpose was to make his readers "virtual witnesses," to induce them to accept that the experiments were performed as claimed, and to "compel assent" to the "matters of fact" that had been established. See Steven Shapin and Simon Schaffer, *Leviathan and the Air-Pump* (Princeton: Princeton University Press, 1985). Such a view, however cogent it may be for Boyle and experimental philosophy in seventeenth-century England, should not be generalized incautiously to experimentation in later periods. Within the organized structures of nineteenth-

century German experimental science, few would doubt that experiments performed by a capable student of the leading physiologist of the time were carried out as claimed. The apparatus and procedures were so much more complex than those associated with Boyle's air pump, that it is more natural to assume that the primary purpose of Helmholtz's meticulous descriptions was to facilitate understanding than that it was to compel assent.

43. Volkmann, "Zustandekommen der Muskelcontractionen," 1–5.

44. "Volkmann, Alfred Wilhelm," in *Biographisches Lexikon der Hervorragenden Ärzte*, ed. Gurlt et al. (Berlin, 1934), 5:797.

45. Volkmann, "Zustandekommen der Muskelcontractionen," 4–5.

46. Ibid., 2.

47. Alfred Wilhelm Volkmann, "Ueber die Kraft, welche in einem gereizten Muskel des animalen Lebens thätig ist," *BVKSGW* (1851), 54–57.

48. Ibid., 57–61.

49. Karl Wunderlich, "Das Verhältniss der physiologischen Medicin zur ärztlichen Praxis," *Archiv für physiologische Heilkunde* 4 (1845), 13.

50. E. duBois-Reymond to H. Helmholtz, May 16, 1851, *Dokumente einer Freundschaft*, 113.

51. H. Helmholtz to E. duBois-Reymond, June 12, 1851, ibid., 115–16.

52. Roger Krohn, "Why are Graphs so Central in Science," *Biology and Philosophy* 6 (1991), 188.

53. See Stephen Michael Kosslyn, *Image and Mind* (Cambridge, Mass., 1980).

54. Dudley Shapere, "The Concept of Observation in Science and Philosophy," *Philosophy of Science* 49 (1982), 485–525. Latour takes the more extreme view that the original object of enquiry disappears from view after the inscriptions have been produced. "Scientists start seeing something once they stop looking at nature and look exclusively and obsessively at prints and flat inscriptions." Latour, "Drawing," 39. The possibility that the scientist may form a mental image of the event, which combines elements of the inscription and that which she believes the inscription to have traced, Latour eliminates by ruling "'mentalist' explanations" out of consideration. Ibid., 21.

55. George Lakoff and Mark Johnson, *Metaphors We Live By* (Chicago, 1980), 30–32, 41–43.

NINE

PRECISION: AGENT OF UNITY AND PRODUCT OF AGREEMENT

PART II—THE AGE OF STEAM AND TELEGRAPHY

M. Norton Wise

Standards, Commerce, and the State

FROM MANY SOURCES we learn that standards of measurement have long been regarded as one of the foundations of justice. Olesko gives examples from American, German, and British commentators. Schaffer quotes Maxwell: "The man of business requires these standards for the sake of justice, the man of science requires them for the sake of truth, and it is the business of the state to see that our . . . measures are maintained uniform." Justice, truth, uniformity, and the state—Maxwell echoes a theme of Enlightenment that Alder gives more broadly from Condorcet: "As truth, reason, justice, the rights of man, the interest of property, liberty, [and] security are all the same everywhere, ought not all the provinces of a state, or even all states, have the same criminal laws, the same civil laws, the same laws of commerce, etc. . .?" The difference between these two quotations is that by 1870, neither Maxwell nor Olesko's informants encountered much skepticism in their call for uniformity in measurement. Rapidly expanding national and international commerce, carried especially on those new media of long-distance uniformity, the railroad and the telegraph, had made standards a necessity of life for engineers, manufacturers, and traders alike. And more than anything else, it was these new conditions of industry and commerce (along with political and military power) that drove precision measurements. The so-called "second industrial revolution" of science-based industries, especially chemical and electrical industries, was underway.

But even before the great takeoff period, European states had prepared the way with reform of the weights and measures on which easy movements of goods, money, and people depended. Olesko's discussion of standards reform in the German states makes this point unambiguously.[1]

She cites the two people most famous for establishing new levels of precision in the 1830s: Gauss, who carried the title of *Hofrat* (advisor to the court, or privy councilor) and worked on weights and measures at the command of the Hannoverian government; and Bessel, who played a similar official role for Prussia. Of course Gauss continued this work with Wilhelm Weber in the Magnetic Union, developing new mathematical techniques along with instruments of unprecedented sensitivity and portability to make observations at stations around the world, thereby to map the magnetic field over the surface of the earth, largely for navigational purposes. Weber then continued the work in electromagnetism. His scientific reputation is usually associated with the fundamental law of electromagnetic force that he established. But his contemporaries knew him equally for his precision measurements and his unified and rationalized system of absolute units.[2] His remarkable electrodynamometer naturally became a locus of authority for electrical standards in Germany, especially when the debate over establishing an international standard of resistance developed in the late 1860s between British and German telegraphic interests, as Schaffer describes. All parties recognized that to control the resistance unit was to exert considerable control over international telegraph networks.

Equally interesting is Olesko's third prominent measurer of the 1830s and 1840s, the Berlin physicist H. W. Dove, who published prolifically in meteorology and optics, edited the *Reportorium der Physik* (1837–45), and directed the Meteorological Institute of Prussia (from 1848). Olesko discusses his *On Measure and Measuring* (1833), which he developed while teaching not only at the University of Berlin, but at the *Gewerbeinstitut*, or technical institute for trade and industry. From the early 1840s he taught also at the military academy (where he and his family lived) and the Artillery and Engineering School (*Vereinigte Artillerie- und Ingenieur-Schule*). Dove's life and work at these multiple institutions in the ambitious Prussian capital epitomize the network of interdependence between bureaucratic, scientific, commercial, and military interests in standards. Significantly, he enjoyed from the late 1820s the support of that incomparably well-placed patron of science, Alexander von Humboldt and of Humboldt's friend P.C.W. Beuth through the *Gewerbeinstitut*.[3]

The *Gewerbeinstitut* provides a nice vignette of the context for precision in Berlin. It was established and run by Beuth at the Ministry of Trade to promote modern manufacturing methods and materials, especially using steam engines and associated machinery, and to educate a new breed of sophisticated young managers and entrepreneurs who could establish Prussia's independence from British suppliers of machines and materials.[4] One early student was August Borsig, who became famous in the 1840s for founding the first world-class engine works in Berlin. By his death in 1854 Borsig's works had supplied five hundred locomotives to power Prussia's fast-growing railroad system. His accomplishment depended critically on a degree of precision in casting and machining not available in Prussia before about 1840.[5]

Borsig is not one of the many students of the technical institute who studied practical physics under Dove, having left the school (without his degree) in 1825, well before Dove arrived. Rather, Borsig's great success in engine building and iron work (including architectural monuments like the Domes of the Nicholai Kirche in Potsdam and the King's Palace in Berlin) symbolized the aims of the *Gewerbeinstitut* during Dove's tenure there, when Borsig provided both a role model and apprenticeships for its students. He also associated closely with Beuth, Humboldt, and other Berlin notables who set the tone for the culture of industrial and scientific progress that rewarded Dove's expertise in physics and measurement.[6]

A similar scene of cultural motivation surrounded the rest of the Berlin physics community who shared Dove's company in the 1840s. With respect to steam engines and the pursuit of efficiency, the young Rudolph Clausius comes immediately to mind. Clausius was teaching physics as Dove's successor at the Artillery and Engineering School when he wrote his great theoretical work enunciating the Second Law of Thermodynamics in 1850. Both there and in his later teaching of technical physics at the Zurich Polytechnique, he stressed the latest technologies of heat, electricity, and magnetism.[7]

Hermann Helmholtz was a second young Berlin scientist who acquired his skills in instrumentation and measurement in part from interaction with Dove and, like his friend Clausius, from working in the private laboratory of another physicist at Berlin University, Gustav Magnus, who held equally many outside teaching posts, noteably at the Artillery and Engineering School from 1832 to 1840. Dove and Magnus provided the immediate professional setting for the founding by their students of the Berlin Physical Society in 1845, including Helmholtz.[8] Helmholtz's associations with the industrial dynamism of Berlin are best known for electrotechnology and the telegraph, where his friendship with that other renowned entrepreneur, Werner Siemens, has long been discussed. Timothy Lenoir, in a paper presented at the workshop, argues that Helmholtz depended on telegraph technology for both instrumentation and analogical models in his work on the sensations of color and tone.[9] No doubt Siemens learned as much about precision from his scientific friends as they did from him. It is the network of associations within which they all worked that seems most significant here. Siemens was a student at the Artillery and Engineering School from 1835 to 1838 while Magnus was teaching there. And of course it was Siemens who took the lead in assembling the resources for the preeminent institution of precision and unified standards in the new Second Reich, the *Physikalisch-Technische Reichsanstalt*, as discussed by David Cahan.[10]

Similar associations with commerce, industry, and national standards emerge wherever one looks at precision in science. In Britain, the leading light of measuring physics in the 1840s and 1850s was William Thomson, who first learned the importance of absolute standards of measurement through the Philosophical Society of Glasgow, an institution that served simultaneously the academic, commercial, and engineering interests of a group of people who regarded themselves as the progressive scientific managers of the city's future. He learned firsthand some of the rigors of measurements of the highest precision from Victor Regnault in Paris, who was then carrying out his extraordinary studies on high pressure steam, paid for by the French government in the interest of perfecting steam engines. Thomson had gone to Paris after graduating from Cambridge in 1845 explicitly to acquire credentials in experimental natural philosophy, since such skills were unavailable at Cambridge and remained so until Maxwell got the Cavendish laboratory operating in 1874. It was with Regnault too that Thomson learned the problems of standardizing temperature measurements and began to plan his own program for absolute thermometry and electrometry, which soon involved establishing a teaching laboratory for natural philosophy at Glasgow (the first in Britain).[11]

This program developed along two axes which intersected in energy physics, much as they did for Helmholtz, Thomson's closest German correspondent. The first axis united steam engines, theoretical thermodynamics, and heat measurements, pursued with his engineering brother James, the scientific brewer James Joule, and the engineer W.J.M. Rankine, who would become a Glasgow professor in 1852. The second axis similarly united electromagnetic theory and electrical measurements around a newfound devotion to submarine telegraphy in the mid-1850s—which promised to meet Thomson's financial as well as scientific ambitions from the design and marketing of instruments. This pursuit brought him to the center of the scientific-commercial network that constituted the British telegraph industry, in whose interests he promoted the British standards project with its system of absolute values.[12]

This was a very difficult network to manage, as Schaffer shows J. C. Maxwell discovering in his attempts to stabilize the ohm. Maxwell did not command either the material or human resources he required at Kings College, London. He required something like a national standards laboratory, which to some degree is what he made of the Cavendish. One of the morals of the story is that precision is expensive. We learn this lesson not only from Schaffer and Olesko for the nineteenth century but from Rusnock and Alder for the eighteenth. In the twentieth century, precision is one of the identifying marks of "Big Science." Two billion dollars spent on the defective Hubble space telescope have made the costs all too public. Less dramatic is the "Gravity Probe B" project, costing perhaps one-tenth as much, which will send on a satellite a little ball of 1 1/2 inch diameter and unprecedented sphericity, spinning at 10,000 rpm in ultrahigh vacuum, to test the general theory of relativity.[13] These monetary costs are one reason precision is regularly associated with large institutions whose interests are critical enough and whose resources are large enough to make it worthwhile to bear the financial burden: state agencies, international corporations, consortia of various kinds.

Reliability

But money is not enough. Precision requires standards as much as standards require precision; and standards are social creations. They require a material and moral culture of precision. In Cambridge, Schaffer shows, standardizing the ohm ultimately required introducing something of the culture of the machine shop into the Cavendish Laboratory and redirecting the education of gentlemanly wranglers to teach them the value of disciplined physical work, first under Maxwell's guidance and then more thoroughly under Lord Rayleigh's. Only under a rigorous regimen could the materials themselves be disciplined. Significantly, Maxwell succeeded by incorporating the resources of his fellow Scot, the engineering professor James Stuart—who had already violated Cambridge norms by entering commerce with the Cambridge Scientific Instrument Co.—and by hiring Stuart's Scottish mechanic Robert Fulcher. The new culture is reminiscent of Thomson's Glasgow laboratory and his close association with the instrument maker James White, who realized many of Thomson's instruments in commercial form. But Maxwell's scientific interests were not directly commercial, nor were Rayleigh's, although both were pleased to contribute to the nation's welfare. Landed gentry themselves, they fit the laboratory into Cambridge by identifying it with uniquely executed standards measurements—rejecting the image of a "manufactory of ohms"—and with establishing the electromagnetic theory of light.[14] Thus the culture of the workshop entered the laboratory in an ironically transformed role, that of oversight and discovery, a proper role for Cambridge wranglers.

Schaffer thus makes us aware that precision measurements in Cambridge involved a gentlemanly culture interestingly different from the more industrial middle-class of Glasgow, even though relying on many common characteristics. Ted Porter develops a closely related theme in his account of the opposition encountered by the government's Select

Committee when they sought to discover a system of precise measures of financial security for use in evaluating life insurance companies. The Committee heard that there could be no "manufactory of actuaries," essentially because the profession required the capacities of gentlemen. Security in the evaluation of risk, the actuaries contended, inhered in trustworthy, discerning individuals, not mechanical instruments, individuals who possessed experience, judgment, and positions of some importance. Only the Scot William Thomas Thomson thought it reasonable to establish a uniform balance sheet and government regulation. Thus the government, in attempting to standardize actuarial practice around a set of rules and instruments, faced problems similar to Maxwell's in establishing standards measurements at Cambridge. The technologies could only be born and function in a transformed cultural setting. Nurturing the new laboratory culture, however, proved easier than altering the traditions of actuaries.

The point is not, as Porter explains, that actuaries did not use precise mathematical methods in calculating their tables of annuities and lives; they did, to six decimal places. But such rigorous quantitative methods served primarily to discipline qualitative decisions, not to replace them. For the young especially, it instilled the moral virtues of responsibility, logical consistency, and due caution. It served the same function in the training of actuaries that the Mathematical Tripos served in a Cambridge liberal education, to discipline the minds of future lawyers, clergymen, and other elites. Like them, before occupying a position of any authority, actuaries had to acquire the all-important quality of judgment, a defining character trait of gentlemen. Discerning judgment was irreplaceable, the witnesses insisted, because no uniformity existed in the policies of insurance companies, in the people they insured, or the earnings to be expected on investments. No universal natural laws governed life or finance, as they did electromagnetism, despite the immense efforts of social statisticians and political economists to discover them. This is much the same argument that Joseph Banks had made in the 1760s when he attacked simultaneously the position of mathematics in natural philosophy and the gentlemanly status of mathematicians in the Royal Society. Algebra had then epitomized the evil of crank-turning methods. Mechanical tools could never replace experienced observers with a discerning judgment. Priestley made a complementary argument against Lavoisier's pretension to mathematical precision and demonstration in chemistry. Such programs violated at once the integrity of nature and the dignity of man.

Clearly gentlemanly values could pose serious obstacles to the values of precision, because uniformity of materials, methods, instruments, and people could seem to violate the very identity of a gentleman. Ironically, government regulators found themselves in a compromised position.

They put their own social status at risk by attempting to impose national standards on the actuarial profession. New mores had to attend new precision. Maxwell helped to invent them at the Cavendish, not by making it into a common machine shop, a manufactory of ohms, but by establishing it as an authority over all such manufactories. Insurance regulators ultimately succeeded, Porter suggests, only as the political power of the bureaucracy increased and as the rhetoric of precision acquired irresistible popular appeal.

How then did national standards and precision measurements acquire credibility in the German states? Olesko too argues that material means were not enough. A culture of precision was required, an "ethos of exactitude."[15] The sensibilities of this culture, however, differ in significant ways from the British examples. She focuses on the prominent role of least squares in the analysis of error. In Britain it is difficult to find the technique used in research outside of observational astronomy. Maxwell seems never to have used it in analyzing his data to measure the ohm, which immediately evoked criticism of the reliability of the B.A. unit from his German competitors.[16] It may be useful to speculate on the meaning of this difference in sensibilities.

As in Britain, the question of the truth of a measurement in Germany was whom do you trust. But the means of establishing trust, among the German commentators of the 1830s and 1840s that Olesko evaluates, had less to do with establishing gentlemanly status than with exposing possible errors to scrutiny. The trustworthiness of a precision measurement thus inhered in the open discussion of procedures for eliminating all knowable errors and especially for evaluating the remaining uncertainty. Such subjection of error to scrutiny certified the integrity of the experimenter together with that of the instruments. In this context, the least squares technique for extracting the most probable value from a set of repeated measurements made the necessary uncertainty and variability of the results explicit while minimizing its effect. It simultaneously revealed and validated the investigator's conduct.

A precision measurement for Dove, Gauss, Bessel, J. J. Littrow, Gotthilf Hagen, and others was one whose errors had been thoroughly analyzed. Their appeal was not "trust us" to choose the best answer, but "trust our procedure." Siemens's challenge to Maxwell's B.A. team at King's College to explain how they extracted a probable error of only 0.1 percent from sets of trials whose mean values differed among themselves by 1.4 percent epitomizes the difference in sensibility. Maxwell did not feel called upon to account for his method, however valid it may have been.

Two avenues for further investigation of this difference suggest themselves. The first is the difference in university culture. German universities were administered by their respective state governments. Professors were appointed, promoted, and paid by the state and one of their primary duties was overseeing state examinations for entering various arms of the government and the school system. Their lives, therefore, developed much more in tune with the needs of bureaucracy, particularly uniform evaluation procedures, than was the case in Britain, where universities were largely self-governing, private institutions and professors were not civil servants. At the same time, university students in the Germanies came predominantly from middle-class families, especially members of the civil service, while Cambridge and Oxford served more commonly as finishing schools for gentry and as vehicles for gentrification among the middle class.[17] (In this respect, the Scottish universities resemble the German more than the two elite English ones.) Given the very strong association between precision standards and bureaucracy, therefore, we might expect German investigators to place a higher value on uniform procedures for error analysis.

A second feature is more difficult to get at. It is the sense that the laws of nature themselves were sometimes ascribed a limited certainty, allowing a range of variability in the results they produced. Olesko suggests something of this kind in the works of the Berlin writer on probability for technicians, Gotthilf Hagen, and of J.A.W. Roeber at the *Gewerbeschule*. Their views may be interpreted in various ways: as uncertainty about possible human knowledge of the laws, as uncontrollable complexity of experimental circumstances, as fundamental uncertainty in nature, or all of the above. But deterministic mechanics was not their prototype for natural law. It is worth noting in addition that it was Wilhelm Lexis, an economist trained originally as a physicist, who introduced a measure of the width of a statistical distribution, the standard deviation, as a fundamental characteristic of the distribution, not of eliminable error but of essential diversity. He did so in 1877 as a critique of the preoccupation with mean values that had characterized the search for deterministic laws of organic systems, whether biological, psychological, or sociological, analogous to those of mechanics. This form of critique became a recognizable genre among a fairly wide network of commentators on social and psychological statistics and appears to have been important for the emergence of acausal quantum mechanics in central Europe.[18]

Both of these subjects, the culture of the universities and the character of natural laws, deserve a great deal more study in relation to the technique of least squares and the meaning of precision. They require, however, more of the detailed research on particular techniques and communities that the papers in this volume have explored.

Such studies reveal a fascinating diversity in the resources and motivations of local cultures. They therefore exacerbate rather than solve the problem of how cross-cultural consensus on standards ever emerges. Stressing the hard work involved, Schaffer shows how Maxwell's Cavendish results ultimately came into line with those of his various German competitors as well as with Thomson in Glasgow, and Rowland in Baltimore. Each of these labs developed their own local culture of precision, using techniques that differed significantly from the others. They typically employed different materials and measuring instruments, followed different protocols, employed different conceptions of an ideal experiment, and disagreed about the fundamental laws of electromagnetism. Transcontinental and transoceanic telegraph cables provided a constant source of motivation for establishing agreement. So too did the ubiquitous number v for the ratio of electrostatic to electromagnetic units, which appeared in every electromagnetic theory, with close relationship to the velocity of light, so that everyone required that number. But again, motivation, money, and local agreement were not enough. The various laboratories reached consensus only through travel and exchange.

Rowland traveled to Cambridge, Göttingen, and Würzburg; Thomson and Rayleigh went to Baltimore; movement between London, Cambridge, Glasgow, and Edinburgh was constant; Ayrton and Perry took their engineering practices to Japan and brought them back again in a perfected form.[19] When the people could not go they sent their material representatives: wires, coils, resistance boxes, electrometers, and dynamometers. All went out and came back with new numbers attached, sometimes agreeing, usually not. Gradually responding to critique and adjustment throughout this complex network, the numbers began to converge. Under considerable pressure from Maxwell and fellow proponents of the electromagnetic theory of light, the numbers also began to agree with new values for the velocity of light from Michelson and J. J. Thomson. It is not too much to say that the numerical values of the ohm and of the units ratio were constituted by this mutually adjusted network. Once established, the numbers could travel more or less freely through the network and wherever its authority reached. Their ability to travel made them objective, giving them a robust independence from the local values of the diverse laboratories and workshops on whose agreement they depended. But this objectivity was of course intersubjectivity, and it required that agreement be constantly maintained. That is one of the reasons that standards laboratories came into existence in the late nineteenth century. It is still their task not only to maintain the stability of master standards, itself a difficult business, but continually to redistribute those standards through the entire network of laboratories that depend on them for precision measurements. Travel remains the heart of the project, only it becomes ever more expensive to maintain the traveling capacity of numbers of ever higher precision and ever more complex interrelations.[20]

Qualitative Precision

Numbers are not the only locus of precision. Porter tells us of a Mr. H. J. Brooke, FRS, noted for the "tone of extreme precision" that informed his every act, which he imbibed from his active study of the law. Schaffer notes likewise that Rowland learned from Maxwell to distinguish experiments of illustration from those of research, the former designed to render vague concepts precise, especially in training students, and the latter to yield accurate numerical measurements. Holmes and Olesko make a similar complementarity the basis of their study of Helmholtz's graphical methods for specifying the temporal progress of muscle action and nerve impulses. They argue that Helmholtz's graphic images, because of the immediate clarity with which they expressed his claims, were more widely convincing than his more complex numerical researches. Their concern is with the qualitative precision of curves.

This topic is of particular interest historically because the so-called "graphical method" only became prominent around mid-century. We have as yet no very definitive account of why it emerged then or of why it spread so explosively, but one source is certain: indicator diagrams for analyzing power production in steam engines. Figure 1 shows its value for adjusting engines and gauging their relative horsepower.[21]

Although Boulton and Watt had been using the technique since John Southern invented it in 1796, the indicator diagram remained one of the best-kept trade secrets of the century. Its use became well known only during the 1830s. In its standard form, a stylus connected mechanically to a pressure gauge draws the diagram onto a sheet of paper mounted on a vertical drum, which makes one revolution for each cycle of the piston. The vertical motion of the stylus thus registers the continually changing pressure inside the cylinder as its horizontal position on the rotating drum registers the volume inside the cylinder. So the area under the curve gives a relative measure of the work done by the engine during one cycle (the integral of pressure times differential change in volume) as well as a clear image of the process.

The first theoretical use of such curves to explain the production of work in a steam engine was published by Émile Clapeyron, a mining engineer who designed locomotives in the course of building two railroad lines out of Paris in the 1830s. In the course of this work he apparently learned to use indicator diagrams, which he then employed in his 1834 representation of Sadi Carnot's almost unknown work on the motive power of heat (fig. 2).[22]

Clapeyron's graphic image has since become universally known as a Carnot diagram. It was crucial to the development of thermodynamics

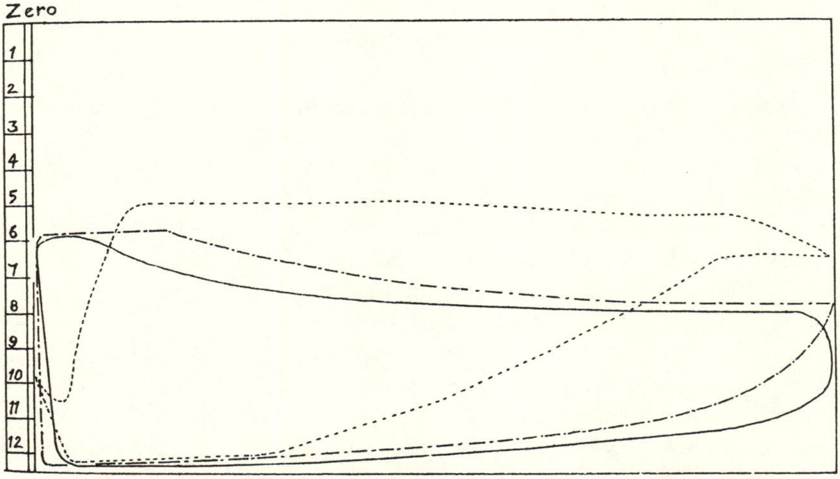

Fig. 1. A redrawn indicator diagram from 1803 showing differences obtained with different valve settings and timing. Reproduced from Hills and Pacey (n. 21); original from Portfolio 1381, Boulton and Watt Collection, Birmingham Reference Library.

both by Clausius in Berlin and by Thomson and Rankine in Glasgow. Rankine called the indicator diagram a "diagram of energy," adding what was becoming a standard refrain about graphical representation: "The principles of the expansive action of heat are capable of being presented to the mind more clearly by the aid of diagrams of energy than by means of words and algebraical symbols alone." By the 1860s every modern textbook on heat and heat engines, whether in natural philosophy or engineering, used such diagrams to explain the production of work as governed by the new laws of energy conservation and dissipation that Joule, Helmholtz, Clausius, and Thomson had enunciated.[23]

Because indicator diagrams were perceived to provide such a powerful tool for conceptual analysis of the production of work in engines they moved readily to many other systems of work production.[24] Helmholz's ingenious instruments for graphical registration of muscle and nerve action, as described by Holmes and Olesko, quite literally recapitulate the function of the indicator. Following their lead, Robert Brain and I have analyzed elsewhere Helmholtz's adaptation of engineering mechanics, dynamometers, and self-registering indicators in his early physical and physiological work, beginning with his access to the French engineering tradition through Dove and Magnus and with his famous paper of 1847 on energy conservation.[25] Briefly, in an attempt to investigate the "mechanical work" of muscles, Helmholtz designed an instrument which

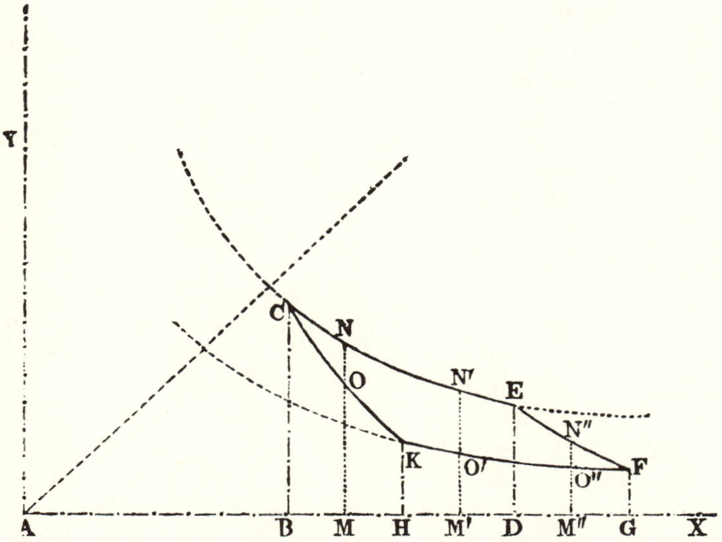

FIG. 2. Clapeyron's diagram of the "Carnot cycle," with pressure on the vertical axis and volume on the horizontal axis. (Ref. 23)

substituted for cylinder pressure the length of a muscle under constant lifting force, and which registered its temporal development through the motion of a stylus on a rotating drum. Thus he registered total work done (force times extension) on the vertical axis as a function of the time elapsed after a stimulus on the horizontal axis. Correspondingly, he spoke of his curves in terms of the stages of development of "energy" in the muscle.

Conceptually speaking, Helmholtz's adaptation of engine indicators to physiological motive forces is extremely simple. They are of particular interest because they move across immense differences in pressure and time scales. It is easy enough to say that the work of engines replaces that of muscles and therefore both should be subject to the same sort of analysis. It is quite another to compare a fifty-horsepower engine running at fifteen revolutions per minute with a frog muscle developing a few ergs of energy almost instantaneously. Part of Helmholtz's genius, as Olesko and Holmes describe it, lay in his ability to design the extraordinarily sensitive mechanisms that could realize the conceptual analogy, rendering it quite precise in terms of the time development of the muscle's energy. This remarkable skill at producing qualitative precision appears in the way he turned his technique for muscle energies to recording the propagation velocity along nerves. Here Helmholtz seems to have adapted another feature of the standard indicator, which recorded successive cycles from the same point on the drum, so that they could be directly compared visually while adjusting the engine. Helmholtz established the same sort of comparison, and thus the difference in time for the stimulus to propagate different distances along the nerve, by initiating the trace simultaneously with the stimulus. Differences in distance along the horizontal axis prior to the response of the muscle thus measured differences in propagation time along different lengths of a nerve. But again, there is no comparison between the time scale of the engine and that of the nerve impulse.

Helmholtz's activities display the complexity of the interactions through which conceptual precision has been extended into ever new areas, through people with overlapping domains of interest, transformations of instruments, and analogical extension of theories. The papers of Olesko and Holmes indicate that to grasp the significance of precision in Helmholtz's graphs is to explore the entire network in Berlin and beyond that supported and motivated his activities. Following that network for steam engines alone, we move immediately from Clausius to Magnus and Dove, Beuth and the Ministry of Trade, technical and military institutes, entrepreneurial engineers like Borsig, and to more distant engineers and scientists in France and Britain. Stripped of this cultural foundation the curve of energy of a frog muscle would be a pale and lifeless creation indeed.

Notes to Chapter Nine

1. For German measuring physics in this period, see Kathryn M. Olesko, *Physics as a Calling: Discipline and Practice in the Königsberg Seminar for Physics* (Ithaca, 1991), esp. chaps. 6–9.

2. For the various aspects of Weber's crucial work consult the indices to K. H. Wiederkehr, *Wilhelm Eduard Weber: Erforscher der Wellenbewegung und der Elektrizität, 1804–91* (Stuttgart, 1967); Christa Jungnickel and Russell McCormmach, *Intellectual Mastery of Nature: Theoretical Physics from Ohm to Einstein*, 2 vols. (Chicago, 1986), vol. 1, *The Torch of Mathematics 1800–1870*; and Olesko, *Physics as a Calling.*

3. Alfred Dove, "Heinrich Wilhelm Dove," *Allgemeine Deutsche Biographie*, vol. 48 (1904), 51–69. Hans Neumann, *Heinrich Wilhelm Dove. Eine Naturforscherbiographie* (Liegnitz, 1925). On multiple teaching positions held by Berlin physicists, see Jungnickel and McCormmach, *Intellectual Mastery*, 246–55.

4. Wilhelm Treue, *Wirtschafts- und Technik-Geschichte Preussens* (Berlin, 1984), 321–29. Jungnickel and McCormmach, *Intellectual Mastery*, 255, also describe the *Gewerbeinstitut* but strangely remark, perhaps without knowing of Dove's long tenure there, that academic physicists were not attracted to it.

5. Conrad Matschoss, *Die Entwicklung der Dampfmaschine: Eine Geschichte*

der ortsfessten Dampfmaschine und der Lokomobile, der Schiffsmaschine und Lokomotive, 2 vols. (Berlin, 1908), 1:185–86.

6. Ulla Galm provides a heroic portrait in *August Borsig* (Berlin, 1987), no. 18 in the series *Preussische Köpfe.*

7. Jungnickel and McCormmach, *Intellectual Mastery*, 189–92, 254.

8. Ibid., 107–10, 254.

9. Timothy Lenoir, "Helmholtz and the Materialities of Communication," in Thomas P. Hankins and Albert van Helden, eds., *Instruments and the Production of Scientific Knowledge*, special volume of *Osiris* (1994, in press).

10. David Cahan, "Werner Siemens and the origin of the Physikalisch-Technische Reichsanstalt, 1872–1887," *Historical Studies in the Physical Sciences* 12 (1982), 253–83; id., *An Institute for an Empire: The Physikalisch-Technische Reichsanstalt, 1871–1918* (Cambridge, Eng., 1989), chap. 2.

11. Weber followed a similar practice in Göttingen. Olesko, *Physics as a Calling*, 409–10. Graeme Gooday, "Precision Measurement and the Genesis of Physics Teaching Laboratories," *British Journal for the History of Science* 23 (1990), 25–51. On Thomson in Glasgow and Paris, see Crosbie Smith and M. Norton Wise, *Energy and Empire: A Biographical Study of Lord Kelvin* (Cambridge, Eng., 1989), chaps. 4, 5, 8; and M. Norton Wise, with the collaboration of Crosbie Smith, "Work and Waste: Political Economy and Natural Philosophy in Nineteenth Century Britain (III)," *History of Science* 28 (1990), 221–61.

12. Otto Sibum, "Reworking the Mechanical Value of Heat: Instruments of Precision and Gestures of Accuracy in Early Victorian England," *Studies in History and Philosophy of Science* (forthcoming), locates Joule's measuring practices in British brewing culture. Smith and Wise, *Energy and Empire*, chaps. 9, 10, 13, 19.

13. Robert W. Smith, "The Biggest Kind of Big Science: Astronomers and the Space Telescope," in *Big Science: The Growth of Large-Scale Research*, ed. Peter Galison and Bruce Hevly (Stanford, 1992), 184–211, on 187; and in the same volume, C.W.F. Everitt, "Background to History: The Transition from Little Physics to Big Physics in the Gravity Probe B Relativity Gyroscope Program," 212–35.

14. Simon Schaffer, "Late Victorian Metrology and its Instrumentation: A Manufactory of Ohms," in Robert Bud and Susan E. Cozzens, eds., *Invisible Connections: Instruments, Institutions, and Science* (Bellingham, Wash., 1992), 23–56. Lord Rayleigh did transform Terling into a large-scale, commercially successful, centrally organized dairy supplying milk to the city of London, a dairy of even larger proportions today, but that was farming.

15. Olesko, *Physics as a Calling.* "Ethos of Exactitude" is chap. 10.

16. In addition to Schaffer (this volume), see Kathryn Olesko, "Precision and Practice in German Resistance Measures: Some Comparative Considerations" (unpublished).

17. Fritz K. Ringer, *The Decline of the German Mandarins: The German Academic Community, 1890–1933* (Cambridge, Mass., 1969), 40–41; Martin J. Wiener, *English Culture and the Decline of the Industrial Spirit, 1850–1980* (Cambridge, Eng., 1981), 12–24, gives a useful, if polemical, summary and references.

18. Theodore M. Porter, *The Rise of Statistical Thinking, 1820–1900* (Princeton, 1986), 240–55. M. Norton Wise, "How Do Sums Count? On the Cultural Origins of Statistical Causality," in *The Probabilistic Revolution*, 2 vols. (Cambridge, Mass., 1987), I, 395–425.

19. Graeme Gooday, "Teaching Telegraphy and electrotechnics in the Physics Laboratory: William Ayrton and the Creation of an Academic space for Electrical Engineering in Britain 1873–1884," *History of Technology* 13 (1991), 73–111.

20. Joseph O'Connell, "Metrology: The Creation of Universality by the Circulation of Particulars," *Social Studies of Science* 23 (1993), 129–73.

21. From R. L. Hills and A. J. Pacey, "The Measurement of Power in Early Steam-driven Textile Mills," *Technology and Culture* 13 (1972), 25–43, on 41.

22. C. Walter, "Clapeyron," in *Männer der Technik*, ed. Conrad Matschoss (1925; reprint Düsseldorf, 1985).

23. Émile Clapeyron, "Sur la puissance motrice de la chaleur," *Journal de l'école polytechnique* 14 (1834); *Scientific Memoirs* 1 (1837), 347–76, diagram from p. 350; *Annalen der Physik und Chemie* 59 (1843), 446–51, 566–86. Rudolph Clausius, "Ueber die bewegende Kraft der Wärme und die Gesetze, welche sich daraus für die Wärmelehre selbst ableiten lassen," *Annalen der Physik und Chemie* 79 (1850), 368–97, 500–24, on 379. W.J.M. Rankine, "On the Geometrical Representation of the Expansive Action of Heat, and the Theory of Thermo-dynamic Engines," *Philosophical Transactions* 144 (1854), 115–75, on 116; see also Rankine's earlier Carnot diagram in "On the Economy of Heat in Expansive Machines" (1851), *Transactions of the Royal Society of Edinburgh* 20 (1853), 205–10, on 206f. and plate VIII, fig. 2 following p. 218. Balfour Stewart, *An Elementary Treatise on Heat* (Oxford, 1866), 312–13. P. G. Tait, *Sketch of Thermodynamics* (Edinburgh, 1868), 87–90, for "*Watt's Diagram of Energy*."

24. M. Norton Wise, "Work and Waste: Political Economy and Natural Philosophy in Nineteenth Century Britain (IV)," *History of Science* (in preparation), argues that the telegraph engineer Fleeming Jenkin adapted indicator diagrams for supply and demand curves in political economy.

25. Robert Brain and M. Norton Wise, "Muscles and Engines: Indicator Diagrams and Helmholtz's Graphical Methods," in Lorenz Krüger, ed., *Universalgenie Helmholtz: Rückblick nach 100 Jahren* (Berlin, 1994). H. E. Hoff and L. A. Geddes, "Graphic Registration before Ludwig: The Antecedents of the Kymograph," *Isis* 50 (1959), 5–21. Soraya Chadaravian, "Graphical Method and Discipline: Self-recording Instruments in Nineteenth-century Physiology," *Studies in History and Philosophy of Science* 24 (1993), 267–91.

PART THREE

MASS DISTRIBUTION

TEN

THE MORALS OF ENERGY METERING: CONSTRUCTING AND DECONSTRUCTING THE PRECISION OF THE VICTORIAN ELECTRICAL ENGINEER'S AMMETER AND VOLTMETER

Graeme J. N. Gooday

As the electrical industry develops, the methods employed in all its branches depart more and more from those in use in scientific laboratories. The vast strides made during the last ten years in the commercial use of electrical energy are marked by departures from the practice of the experimental physicist, and approach to those of the mechanical engineer . . . and this has brought about a complete change in the methods of measurement. There are now workshop, as opposed to laboratory, instruments. It must not for a moment be supposed, however, that they are therefore necessarily less accurate.
(James Swinburne, "Electrical Measuring Instruments," *1892)*[1]

. . . although the electrical is the latest developed branch of engineering, it is the most exact in its measurements, currents as small as 1/10,000,000 ampere and less, up to 2,000 or 3,000 amperes, being easily and accurately measured . . . in electrical work powers large enough to operate a train of some hundreds of tons weight, or powers so small that no ordinary mechanical device could measure them, are effected without special precautions or elaborate device.
(Tyson Sewell, The Elements of Electrical Engineering, *1903)*[2]

MEASUREMENT instruments have often had a more colorful career in the laboratory, workshop, and beyond than historians have hitherto suspected. Recent studies show that when scientists and technologists cannot agree about how to quantify the world, the credibility of even the highest-precision device can become problematic. In particular we see that the merits of competing mensurative schemes are not adjudicated solely by comparison of the precise numerical agreement between materialities and expectations that protagonists of each scheme can produce with the aid of their instruments. Schaffer and Mackenzie, for example, have respectively described how feats of precision in late eighteenth-century Italian eudiometry and mid-twentieth- century navigational gyroscopy held different "self-evident" meanings for differently aligned contemporary observers;[3] elsewhere in this volume Jan Golinski addresses a similar theme in eighteenth-century chemistry. If we follow these authors in suspending judgment about the epistemic persuasiveness of precision, two important questions arise: how do measurement instruments come to be regarded as productive of meaningful results if not through their precision alone? And, why is it that their precision is so readily mythologized as an incontestable guarantor of metrological credibility?

I shall address these questions in a comparison of the measurement activities of laboratory physicists and workshop electrical engineers in the late nineteenth century. Within their distinctive domains of practice both of these "metrological communities" were broadly recognized as holding a privileged expertise in the high accuracy[4] measurement of electrical energy movement (see Swinburne epigraph above). Yet their methods of achieving precision were not readily transferable from one disciplinary site to the other, and exploration of this nontransitivity reveals important generic differences in the material, managerial, and above all *moral* priorities of their measurement cultures. Earlier sections of this paper will document the implementation of these priorities in the everyday pedagogy and practice of measurement in the late Victorian physics laboratory and electrical engineering workshop. The latter discussion will include a brief account of the hitherto undocumented genesis of the ammeter and voltmeter in relation to the work of William Edward Ayrton and John Perry at Finsbury Technical College, London, between 1879 and 1884.[5]

From this discussion I shall explain why physicists' attempts to extend the applicability of the tangent galvanometer to the world of heavy electrical engineering were as unsuccessful as electrical engineers' attempts to establish the direct-reading ammeter and voltmeter as legitimate instruments of the academic physics laboratory. In the last part of the paper I will highlight the "moral" incommensurability of physicists and electrical engineers' cultures of precision by reference to their efforts to effect accurate electrical determinations of Joule's thermodynamic constant, known also as the "mechanical equivalent of heat." I shall focus on physicists' objections to Ayrton's use of the commercial electrical engineering practices—as embodied in the ammeter and voltmeter—in a student's apparatus for the ten-minute precision "determination" of this constant. From this discussion I will show the historical importance of understanding pedagogy as the means by which scientists and technologists seek to perpetuate their distinctive practices and interpretations of what can legitimately count as a precision measurement.

Prologue: The Transition from Telegraphic Instruments to Electrical Meters

> The number of electrical measuring instruments recently devised is very great. The practical man is not satisfied with the delicate instruments of the physicists, whilst the latter, of course, cannot be satisfied with the results of the measuring instruments arranged by engineers and technical electricians, however satisfactory for industrial purposes.
> *(The Telegraphic Journal, 1884)*[6]

I have argued elsewhere that British physicists and telegraphists shared a preoccupation in the practices of precision measurement after the rise of telegraphy in the mid-nineteenth century.[7] Both laboratory natural philosophers and telegraphic engineers had developed a common repertoire of delicate instruments for precisely measuring the tiny electrical currents that communicated messages through transcontinental and transoceanic cables and their pedagogical analogues the bench-top artificial testlines.[8] However, as a result of physicists' and electrician's differential responses to the development of heavy-duty electrical technologies of *energy transference* during the 1880s and 1890s, a deep fracturing occurred within this common culture.

Upon the public arrival of commercial arc lighting in 1878 and the arrays of filament lamps and the dynamo generators that powered them in the early 1880s, it was widely perceived that the management of nascent systems of electric light and power required radically new instrumentation. Knowing the precise size of the hundreds or even thousands of ampere currents, and the increasingly high-tension "pressures" involved in power distribution, was of critical importance in making such undertakings sufficiently safe and economical to be commercially successful. The development of accurate heavy-duty current- and voltage-measuring instruments rapidly became a critical issue for all involved in the neophyte electro-technology.[9] Thus the *Electrician* declared to its readership in 1883 that while in "the application of electricity to telegraphy the quantities to be measured have [hitherto] been wondrously small," and that "the majority of existing instruments" were "designed with a view to the measurement of these small quantities," there was no mistaking that "the present is a kind of transition period as far as the measurement of electrical quantities is concerned."[10]

Rhetorically posing the question "the general public ask, Can such measurements be made?" the editorial continued, "the scientific world answer they can, and already several inventors have devised apparatus for such work. . . ."[11] This judgment was a direct response to a widely hailed survey lecture by James Shoolbred on the extremely topical subject of "the measurement of electricity for commercial purposes," which detailed a burgeoning array of candidates for standard post-telegraphic electrical instruments. Commercial current measurers included Siemens's electrodynamometer, Obach's tangent galvanometer, Cardew's low resistance galvanometer, Ayrton and Perry's "am-meter," and Sir William Thomson's current galvanometer. Voltage measurers included William Thomson's potential galvanometers; Siemens's torsion galvanometer, and Ayrton and Perry's "volt-meter."[12]

In this new repertoire we see two different species of electrical instrumentation. Thomson, Siemens, and Cardew attempted to adapt such familiar and trusted *laboratory* devices as the (nondirect reading) tangent galvanometer and electrodynamometer to the harsher working conditions of commercial electrical engineering. By contrast, Professors Ayrton and Perry of Finsbury Technical College, London, elected to meet the requirements of day-to-day practice in the electrical workshop and factory by abandoning laboratory traditions of instrumentation in favor of radically new robust, portable, and *direct-reading* measuring devices. I shall discuss below how their ammeter and voltmeter rapidly eclipsed virtually all laboratory-gestated rivals to become electrician's (early) favored technology for "easily and accurately measuring" both small and vast electrical quantities, allegedly without "special precautions or elaborate device."[13]

To understand the distinctive precision practices implemented through electrical engineers' adoption of the ammeter and voltmeter—as "measuring" rather than mere "indicating" devices—we must first appreciate that physicists accepted electrical accuracy only as a skillful achievement with highly "elaborate" devices that required such "special precautions" that they were far from "easy" to use. This will serve as a background to my explanation of why physicists could not accept early forms of the ammeter and voltmeter as legitimate laboratory devices.

"Self-Reliance" and Absolute Measurement in the Utopian Physics Laboratory

> In most physical investigations the result aimed at is one in which practical absolute accuracy is attainable, although attainable only if infinite pains be taken to get it. It is the business of the physicist to control and modify his conditions, and to use only those which permit of the desired degree of accuracy being reached. In such investigations it sometimes becomes almost immoral to think of one condition as less important than another. Every disturbing condition must either be eliminated or completely allowed for. That method of making the experiment is the best which ensures the greatest accuracy in every part of the result.
> *(A.B.W. Kennedy, Address to Section G of the British Association, 1894)*[14]

Since the work of Pouillet in 1837, laboratory physicists had used the tangent galvanometer as their canonical device for the *absolute* measurement of "voltaic" currents. Its operation involved balancing the Oerstedian deflection of a compass needle in the vicinity of a current-carrying wire against the controlling force of the earth's magnetic field. Skilled use of this device could determine the strength of a current with considerable accuracy by observing the deflection of a needle placed at the center of the coiled wire carrying the current in question (fig. 1). To be effective this setup necessitated that (*i*) the plane of the current-carrying coil be exactly aligned with that of the geomagnetic meridian; (*ii*) the strength of the local horizontal component of the geomagnetic field "*H*" at the time and place of measurement be known with considerable precision; (*iii*) no "stray" disturbing magnetic fields vitiated this value of "*H*"; and (*iv*) sufficient leisure was available for the calculational reduction of the manifestly nonlinear measurement to give a value of absolute current.[15]

The lattermost point is important because tangent galvanometers were *never* directly calibrated; this was due only in part to the impossibility of constructing a trustworthy permanently engraved current scale while the "controlling" geomagnetic field was constantly varying. That this was the case is clear when we observe that all late Victorian precision laboratory instruments for absolute current measurement were *non*direct reading, most notably those standardizing devices constructed on the "square-law" principle of mutual current attraction. Sir William Siemens's elec-

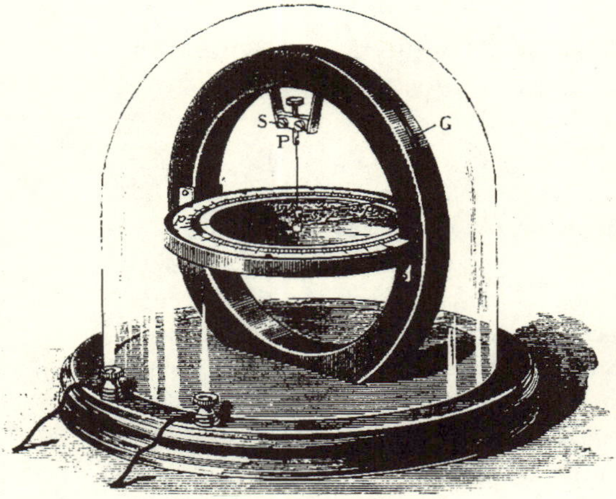

FIG. 1. Tangent galvanometer, ca. 1911. W. E. Ayrton and T. Mather, *Practical Electricity*, 3d ed., London, 1911, 37.

trodynamometer (fig. 2) and the current balance (fig. 3) were prominent among these during the 1880s, the latter especially after one such was used by Lord Rayleigh in 1884 to "weigh" currents to an accuracy of 0.01 percent.[16] Although differing in material construction, these instruments and the tangent galvanometer were similar in being productive of meaningfully precise results only in the idealized conditions of laboratory life. Indeed the use of these instruments exemplified three conventions specific to late Victorian laboratory culture.

First, wherever possible the measurements made were "absolute," that is, involved direct determinations of mass, length, and time alone. Since the work of Weber and Thomson in the 1840s and 1850s these parameters had been accorded a privileged status in physicists' articulations of the universality of their discipline, commonly through the auxiliary unifying discourse of energy.[17] Thus we see in the curricula of many physics laboratories and in the agenda of many physics textbooks that the first subject matter to be addressed was the grounding of all physics upon measurements of mass, length and time. For example, in Glazebrook and Shaw's widely consulted *Textbook of physics*[18] we find the whole gamut of all energetic and electromagnetic quantities categorized as calculational products of these three metrological fundamentals (table 1).

Clearly then an orthodox physicists' interpretation of current measurement was that it was "actually" a calculational *synthesis* of mass, length, and time. A significant interpretive "distance" thus existed between this "real" measure of a current and either the angular deflection of the tangent galvanometer's needle or the torsional force required to equilibrate a current balance/ electrodynamometer reading. For a physicist committed to such a view of the ampere as a "constructed" quantity it would

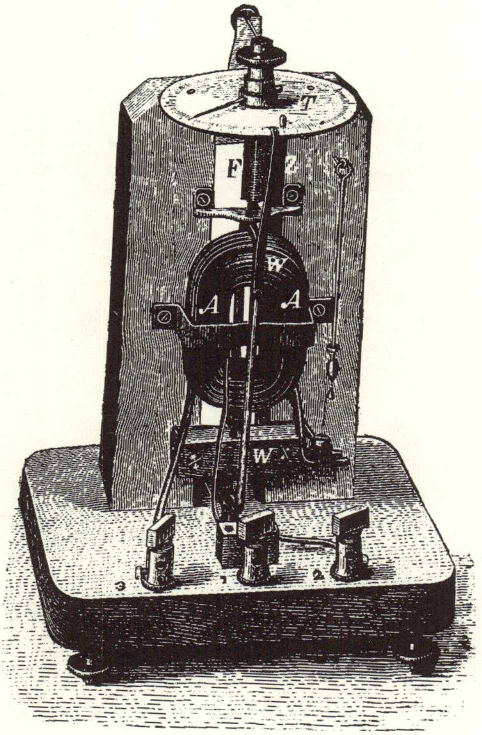

FIG. 2. Siemens Electrodynamometer, ca. 1897. R. Wormell (revised R. M. Walmsley), *Electricity in the Service of Man*, London, 1897, 167.

FIG. 3 (*below*). Thomson current balance, ca. 1888. W. Thomson, "On his new standard and inspectional electric measuring instruments," *Journal of the Society of Telegraph Engineers and Electricians* 17 (1885), 547.

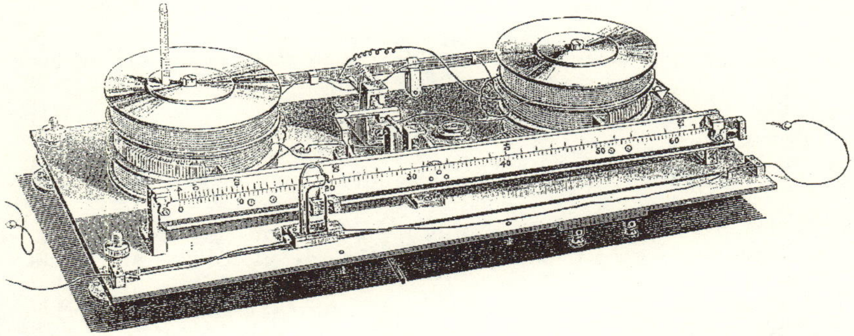

therefore have been quite meaningless to calibrate any of these instruments to give direct readings of a current. A meaningful measurement of current could stem only from an *indirect* procedure of absolute determination and ratiocination.

The second issue of such orthodox modes of current measurement

Name of quantity measured	Measurement actually made
MECHANICS.	
Area	Length (§ 1–6).
Volume	Length.
Velocity . . .	Length and time.
Acceleration . . .	Velocity and time.
Force	Mass and acceleration, or extension of spring.
Work	Force and length.
Energy	Work, or mass and velocity.
Fluid pressure (in absolute units) . . .	Force and area (§ 24–26).
Coefficients of elasticity .	Stress and strain, *i.e.* force, and length or angle (§§ 22, 23).
SOUND.	
Velocity . . .	Length and time (§ 29).
Pitch	Time (§ 28).
HEAT.	
Temperature . . .	Length (§ 32).
Quantity of heat . .	Temperature and mass (§ 39).
Conductivity . . .	Temperature, heat, length, and time.
LIGHT.	
Index of refraction . .	Angles (§ 62).
Intensity . . .	Length (§ 45).
MAGNETISM.	
Quantity of magnetism .	Force and length (§ 69).
Intensity of field . .	Force and quantity of magnetism (§ 69).
Magnetic moment . .	Quantity of magnetism and length (§ 69).
ELECTRICITY.	
Electric current . .	Quantity of magnetism, force, and length (§ 71).
Quantity of Electricity .	Current and time (§ 72).
Electromotive force	Quantity of electricity and work (§ 74).
Resistance . . .	Electric current and E. M. F. (§ 75).
Electro-chemical equivalent.	Mass and quantity of electricity (§ 72).

TABLE 1. Dimensional analysis of measurements, ca. 1884. R. T. Glazebrook and W. N. Shaw, *Practical Physics*, London, 1884, 6–7.

closely germane to the culture of laboratory physics related to the *unassisted* basis of proper experimental practice. A popular Smilesian discourse of practical self-help[19] was often deployed to establish that the integrity of a measurement derived from the expertise and exertions of the measurer *alone*. For example, of his recently built laboratory at King's College, London, in 1871, W. G. Adams stressed that his experimental curriculum involved "each student making in his turn and so checking every part of the measurement or determination" such that "the advance which may be made by him is dependent only on his own exertions."[20] Authenticity in acts of absolute measurement was thus premised on the deeply mythologized notion of an unmediated and asocial encounter with the natural world.[21]

However, as in all contemporary physics laboratories, Adams set his arriving students to make measurements of mass, length, and time with scales, rulers, and clocks that had been precalibrated in the hands of instrument makers.[22] The apparently unproblematic status of such "ready-to-use" direct-reading equipment reveals an ambivalence in the licensing of metrological integrity for the laboratory measurer. It was only by *convention* that instruments pregraduated to read mass, length, or time were invulnerable to demands for the laboratory worker to check their calibration for himself. Moreover, it was a convention with renegotiable boundaries since the otherwise self-helpful tyro in the later Victorian laboratory received an ever-broadening license to accept uncritically a number of metrological aids in electrical and thermal physics. Not only was he permitted to accept the authority of Kew Observatory's geomagnetic determinations and the veracity of a commercial prefabricated graduated thermometer, he was also (ironically) allowed the free use of "black-boxed" versions of notoriously corrigible B.A.A.S. standards of electrical resistance.[23] It is not clear, *prima facie,* therefore that physics students should necessarily have been denied access to the electrical engineers' standard direct measures of current and voltage: the ammeter and voltmeter.

The third generic feature of galvanometric practice that epitomizes contemporary laboratory measurement culture relates to the *nonindustrial* character of the physicist's experimental space. The physics laboratory was quite unlike the cable factory where all machines and workers were closely juxtaposed into a strict and immobilizing regime of efficient spatial and temporal organization. Equally much it was unlike the testing workshop or installation site where the haphazard mutual interactions of men and machines precluded any idealized delicacy in the execution of measurements. Working in a less constrained condition than either of these commercial regimes the physicist and his students characteristically endeavored—as expressed in the above epigraph from Kennedy[24]—to maintain a working environment that conveniently maximized absolutist *rigor* in their praxis of precision measurement. Such were the temporal and spatial priorities of the laboratory management of precision metrology for example, that a current measurement experiment could conveniently be structured around the meridianal orientation of the tangent galvanometer.[25]

A complementary aspect of the idealized laboratory usage of these instruments was the elimination of what James Clerk Maxwell famously dubbed "disturbing agents" that materially or socially inhibited the physicist from attaining full control over his environment.[26] Thus replications of standard results using the highly sensitive current balance, for example, had to be sited away from physical vibrations and movements of large masses of iron. Indeed as I have discussed elsewhere, much of the activity of physicists in late nineteenth-century metropolitan laboratories lay in their ongoing attempts to idealize their experimental conditions by wielding control over potential subversive human activities and technologies both within and without the walls of their institution. Clearly the effectiveness of such fragile metrological devices as tangent galvanometers and electrodynamometers was ineluctably tied to the sort of orderly regime sustainable only within a sympathetically organized and advantageously sited scientific institution.[27]

We can thus see that the deployment of the tangent galvanometer and current balance was constrained by their dependence upon three cultural features of precision metrology *local* to the physicists' domain of the institutional laboratory. Unsurprisingly then, attempts made by Sir William Thomson (fig. 4),[28] Dr. Eugen Obach, Henry Trowbridge, and others to deploy adapted forms of the tangent galvanometer to measure large electrical currents in the conditions of industrial practice were acknowledged failures, "owing," as the instrument maker Robert Paul later reported, "to the errors arising from stray magnetic fields."[29] As William Ayrton revealed about his own researches on the use of such instruments in 1881: "The actual [geomagnetic] field in a workshop is so modified by the iron machinery in the neighbourhood that we cannot, from the published tables of magnetic force, even approximate the strength of the magnetic field in which our experiments are made."[30]

Elsewhere Paul related that the fragility of electrodynamic devices "such as the absolute current balance of Rayleigh and the secondary standards balances of Thomson" during the 1880s was such that they received "comparatively little attention . . . except as laboratory instruments."[31] Nevertheless, according to Paul it was specifically in response to the failure of the extramural tangent galvanometer that electricians' attention was "turned first" to a completely new form of instruments "having a needle of soft iron polarized by a powerful magnet, and provided with a deflecting coil adjacent." It was in 1881 that his erstwhile

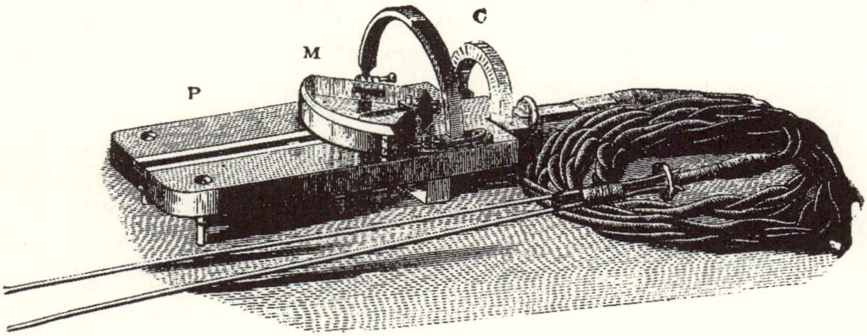

FIG. 4. Thomson's "portable" tangent galvanometer, ca. 1883. W. E. Ayrton, *Practical Electricity*, 1st ed., London, 1887, 54.

teachers Professors Ayrton and Perry, following the example of Marcel Deprez, designed "such instruments, the scales being graduated in degrees, which could be converted to amperes or volts by the application of a constant" (see below). To them, Paul added, "is due the abbreviation "ammeter," now in general use for [the older term] amperemeter."[32] I shall explain how Ayrton and Perry's development of these devices raised new problematic issues for what counted as "measurement" for the precision requirements of the electrical industry.

Making Meters: Constructing the Metrological Authority of the Direct-Reading Electrical Instrument

> Prof. W. E. Ayrton . . . was one of the first to realize that an [electrical] instrument should be direct-reading. To-day we are so accustomed to seeing the scale on an ammeter marked definitely in amperes that it is difficult to put ourselves in the position of those who in the early days had current meters with degree scales which were supplied with a curve to interpolate the readings, or had a series of constants to apply—involving skill in obtaining the result and delay in applying it.
> *(F. C. Knowles,* "The Growth of Electrical Measurement from its Commencement," *1931)*[33]

The public debut of Ayrton and Perry's so-called "direct-reading" ammeters and voltmeters in 1884 followed four years of intensive trials in their professorial engineering laboratories at Finsbury Technical College, and in the commercial context of electrical workshop and factory practice. From their experience as practitioners in this period,[34] they learned that the idealistically fastidious methods and values of physicists' laboratory metrology were impossible to replicate in the environments of industrial practice, namely, the installation sites and controlling stations of power and lighting networks.[35] Unlike physicists, electrical engineers needed instruments that could survive the hazardously peripatetic *modus vivendi* of the contracting electrician and tolerate the inescapable mechanical and electromagnetic disturbances from massive machinery plants.

Above all else, engineers' devices for measuring voltages and currents were required to be precise enough for commercial economy and safety while being easy and *quick* to use, yielding readings with the greatest possible immediacy. In what follows I shall explicate the sense in which Ayrton and Perry's ammeters and voltmeters embodied values of precision generic to electrical engineering in creating a new praxis of accurate electrical measurement. I shall give a brief account[36] of the material, marketing, and pedagogical strategies employed by them to construct the metrological authority of these instruments within the professional culture of electrical engineering.

Prior to Ayrton and Perry's prototype device of 1880–81, a number of instruments were developed for the industrial measurement of direct electrical currents. The majority of these deployed permanent magnets to replace the geomagnetic field as the controlling force over the dial indicator, these being generally of sufficient strength to dampen the indicator's oscillations to the ideal condition of "deadbeatness." However, in deploying controlling permanent magnets such instruments inconveniently lost the unique advantage of the tangent galvanometer's eponymously analytical relationship between current and deflectional readings. An empirical calibration chart or graph was thus typically drawn up in the laboratory with the aid of a standard cell and tangent galvanometer. However, a daily recalibration was usually essential to maintain any degree of accuracy in the measurements made with this device owing to the infelicitously variable qualities of permanent magnets, the unreliable behavior of which was, moreover, compounded by the prevalent idiosyncracies of manufacturing workmanship. As C. F. Knowles pointed out fifty years later, such measuring devices were far from easy or quick to use since they involved "skill in obtaining the result and delay in applying it."[37]

The first widely successful form of a "portable galvanometer for very strong currents" was the so-called "fishbone galvanometer" marketed in 1880 by the French *electricien* Marcel Deprez.[38] A permanent magnet provided the controlling force over a pointer which, at equilibrium, was balanced against the deflecting force from coils of current-carrying wire wound around a swiveling "fishbone" arrangement of soft iron-needles (fig. 5). Although described by *Nature* as a "handy instrument" for its portability, deadbeatness, and immunity to stray electromagnetic fields,

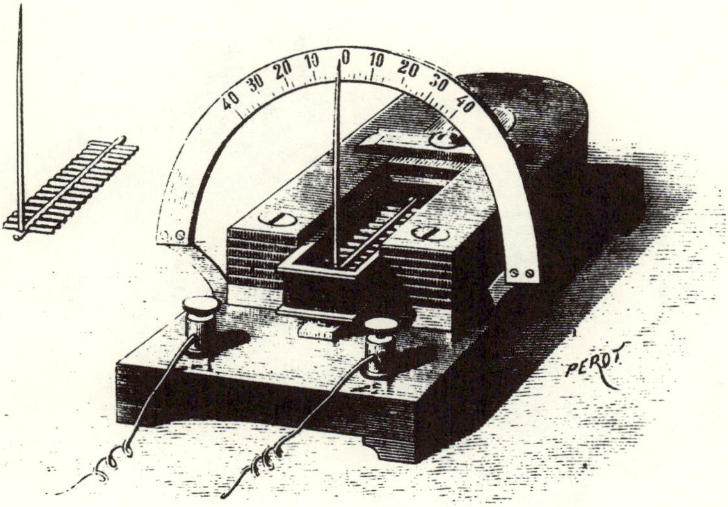

FIG. 5. Deprez fishbone galvanometer, ca. 1880. "Marcel Deprez's Galvanometer for Strong Currents," *Nature* 22 (1880), 246–47.

Ayrton and Perry criticized its reliance upon a calibration chart in the first paper they delivered on the subject at the Society of Telegraph Engineers and Electricians in April 1881. Their central contention was that electrical engineers really needed "a portable absolute galvanometer" that could be used to measure the "strongest electric light current" directly "without calculations or reference to any table," and which could be "calibrated absolutely with a single Daniell cell." As they "did not know of the existence of such an instrument," Ayrton and Perry had designed one to meet the demands of hundreds of trainee electrical engineers enrolling in Ayrton's day and evening classes in Electric Lighting at Finsbury."[39]

By late 1880 they had developed a permanent magnet device, modeled closely on an improved version of Deprez's device but with the crucial difference (so they claimed) that it gave dial readings directly *proportional* to the current flowing through it (fig. 6). Instead of a calibration chart, all that was needed was a single constant which could be used to multiply up its dial readings to yield absolute values of current—albeit a constant that as much needed daily adjustment as the calibration chart of the nonproportional Deprez instrument. Describing the coil of current-carrying wire somewhat evasively as being in a "proper arrangement,"[40] Ayrton and Perry claimed on behalf of their 1880/81 mechanism that "the deflections are about the same for any position of the instrument" on either of its nine or ninety weber scales up to a maximum of about forty-

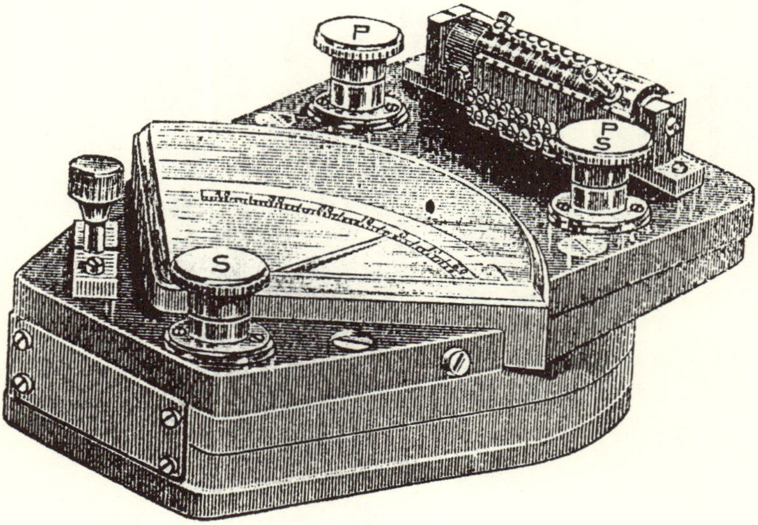

FIG. 6. 1880–81 Ayrton and Perry portable galvanometer for strong currents. W. E. Ayrton and J. Perry, "A Portable Absolute Galvanometer for Strong Currents, and a transmission Dynamometer," *Journal of the Society of Telegraph Engineers* 10 (1881), 157.

five degrees (fig. 7). Despite the lack of explicit guarantees that their "absolute portable galvanometer for strong currents" could indeed yield the alleged "proportional" readings with any accuracy, the inventors elicited a positive response. In fact, one practitioner declared that their device would be a "great help" to electric light engineers who were "greatly in want of a trustworthy galvanometer of sufficient portability to be carried down to the various installations as we fix them."[41]

Such was the confidence of the Finsbury professors in the utility and veracity of their instrument, and such was their determination to introduce it to as large a market audience as possible that over the following year they included it in their public appearances at the London Institution, the Society of Arts, and the Physical Society.[42] However, it was at a gathering of Section A of the British Association for the Advancement of Science in summer 1881, that Ayrton and Perry first publicly used a distinctively new taxonomy for their "complete set of instruments . . . for measuring purposes in electric lighting, and in the transmission of power." The former deadbeat Galvanometer was christened as the "Ammeter . . . a device which measures accurately a strong electric current in amperes"; when "wound somewhat differently"[43] this became their "Volt-meter, which measures instantaneously an electromotive force in volts."[44]

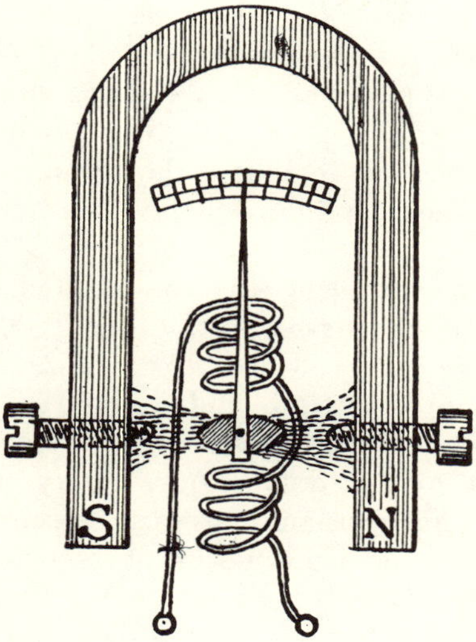

FIG. 7. Operational principle of moving-iron, permanent-magnet galvanometer. R. Kennedy, *A Handbook of Electrical Installations*, vol. 1, London, 1902, 102.

Despite the context of these devices' debut at a meeting of physicists (rather than of the engineers in Section G), Ayrton and Perry emphasized that the prerogative of the ammeter and voltmeter was unequivocally the "accurate" measurement of industrial energy transfer rather than of research. This message was reiterated at a meeting of the Physical Society of London in February 1882: having been commissioned by the Faure Accumulator Company to carry out some professional consultancy work on the working parameters of their secondary cells (i.e., efficiency, energy storage capacity, and longevity), the professors revealed that they had deployed their ammeter and voltmeter to "ascertain," for example, that at a power output of 44.25 foot-pounds per minute, the total energy extractable from this cell was 1,440,000 foot-pounds. Such was the precision they implicitly declared to be obtainable with these instruments in their Finsbury engineering laboratories.[45]

In May 1882 Ayrton and Perry repeated their message yet again at the Society of Telegraph Engineers & Electricians when displaying their rapidly expanding set of metrological inventions "for making accurately and at the same time quickly, the measurements that are necessary to be made, in order that an estimate may be made of the commercial success of any special system of electric lighting or transmission of power."[46] No less than thirteen instruments were presented in their spectacular display, including their new proportional ohm-meter and a now completely overhauled set of ammeters and voltmeters for, as Ayrton revealed, the proportionality he and Perry had claimed in the previous year had not been entirely fulfilled in practice. A reviewer for the *Electrician* remarked of this earlier arrangement that "to some extent" it had been "successful" yet "still the deflection was not proportionate along all of the scale."[47] The Finsbury professors had thus abandoned their first method "in favour of giving to the coil, needle and pole-pieces exactly the shape that theory shows to be necessary to obtain deflections exactly proportional to the current" (fig. 8). Unsurprisingly we find again that the pertinent theoretical analysis underpinning this alleged proportionality was not revealed.[48]

Ayrton and Perry instead used two other strategies for winning credibility for the precision achievable through the putative proportionality of their permanent magnet instruments. First they used a source of continuous current furnished by the Society to give practical validation of their claim that these "portable, and moderately cheap instruments" were indeed "exact" in their operation. Second, Ayrton took the opportunity to make a very favorable contrast between the precision of his own measuring instruments and those employed in the light works of Messrs. Siemens, examples of which had been rather briefly presented in the same evening by one of Siemens's technical staff, Dr. Eugen Obach. With regard to Obach's account of his dynamometer and two current meters,[49] Ayrton archly commented that although he would "not for a moment" be considered as "depreciating" these "admirable" instruments, he saw some "very serious objections" to claims for their reliability.

Unlike his own ammeter and voltmeters they were not "deadbeat"; they could not be used in any orientation; they were easily "disturbed" by powerful magnets and could not even be placed near dynamos; their movable parts were "exposed to the air" and liable to be "blown about."[50] Faced with such fatal criticisms of his instruments, Obach's subsequent career as an electrical engineer was not a lengthy one;[51] moreover, this was not an isolated example of Ayrton and Perry's domination of electrical engineering metrology. Just one week after this episode, another engineer bemoaned his "misfortune" at being summoned to show his construction of a current measurer to the Society so soon after the Finsbury professors. He confessed apologetically that had he "known that the ground would be so formidably occupied" he would have followed "such leaders" at a "more respectful distance."[52]

Not satisfied with such acknowledgments of their predominance in proportional instrumentation, throughout 1883 Ayrton and Perry continued to develop techniques for higher precision and also greater immediacy in dial readings—desiderata which tended to make conflicting demands upon the skills of both instrument builder and user. Most seriously, it transpired that in typical factory usage, the management of a large number of proportional instruments with different multiplying constants had proven to be intolerably problematic. Since many ammeters or voltmeters "looked alike" there was, Ayrton and Perry explained in the

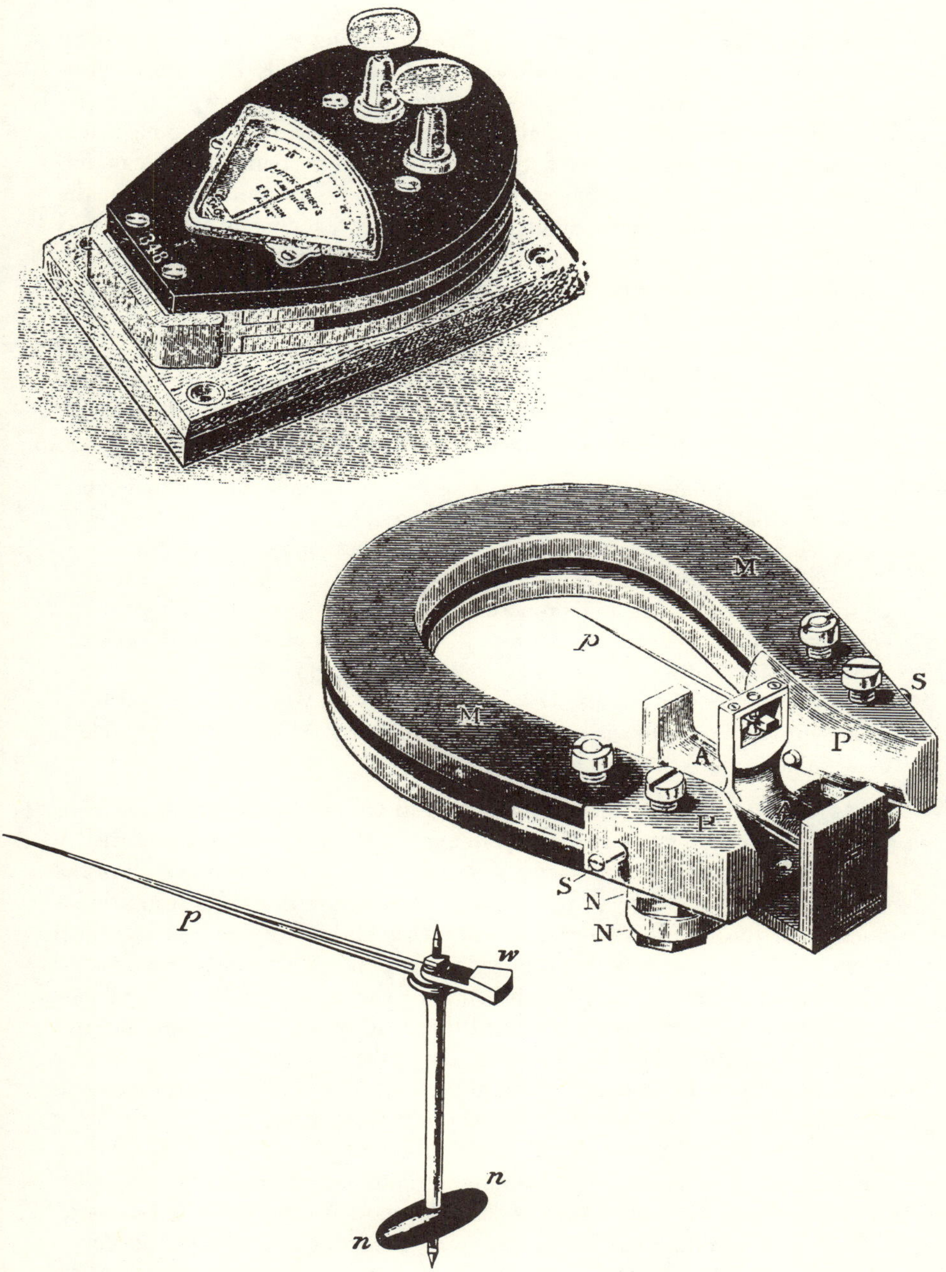

FIG. 8. 1882 Ayrton and Perry proportional-reading ammeter. *Top*, exterior view (W. E. Ayrton and J. Perry, "Measuring instruments used in electric lighting and transmission of power," *Journal of the Society of Telegraph Engineers and Electricians* 11 [1882], 254–78); *middle*, interior mechanism displaying shape of pole pieces and adjustment (Ayrton and Mather, *Practical Electricity*, 1911, 110); *bottom*, moving iron piece attachment to pointer (ibid.).

following year, a "great danger" of confounding the constants of two different instruments, with awkward and occasionally calamitous consequences. Even when the correct calibration constant was used, accurately multiplying it by a dial reading to obtain an absolute electrical measure was "troublesome except for those expert in mental calculation"—a quality, they dryly added, "not always predominating in the men in charge of electric light circuits." However such expertise was no longer necessary by early 1884, for the technology of *instantaneous* absolute precision measurement was now available from their Finsbury workshops.

The *direct-reading* ammeter and voltmeter were revealed to the world at a meeting of the Physical Society of London in January of that year. Indeed they took this opportunity to *redefine* the character of measurement achievable with their instruments, for as they declared,

> The names "Ammeter," "Voltmeter," by which we ventured to christen two of our children, have found favour in the eyes of Electricians, and are now current in electrical literature as the typical names, or surnames, of instruments used to measure strong currents and large electromotive forces, respectively. But . . . we now propose to take advantage of certain improvements that we have recently effected . . . for the future we propose to confine the definition of an "Ammeter," a "Voltmeter," and an "Ohmmeter" to instruments on which respectively amperes, volts, and ohms can be immediately read off without any calculation or any reference to a constant or table of values.[53]

Ayrton and Perry achieved their bold move into devices that were directly and permanently graduated in electrical units by displacing the responsibility of instrumental reliability from the constructional skill of the manufacturer's workforce to a new user expertise in calibration adjustments. In the permanent magnet ammeter and voltmeter the technical basis of this user-interactive quality was a pair of "specially-constructed charcoal-iron cores" fitted to the ends of the current-carrying coils. By conjoint or differential alteration of the position of these cores, the user could adjust the deflecting power of the coils, however idiosyncratically these had been wound in the manufactory. This deflecting force could thus be balanced against that of the controlling force of the permanent magnet until the preinscribed scale matched the external calibration device; this adjustment could be made regularly thereafter to allow for secular changes in the strength of the permanent magnet. A ten-ampere, direct-reading ammeter of this type was, according to Ayrton and Perry, "correct to one-tenth of an ampere" its maximum obtainable precision thus being 1 percent[54] (fig. 9).

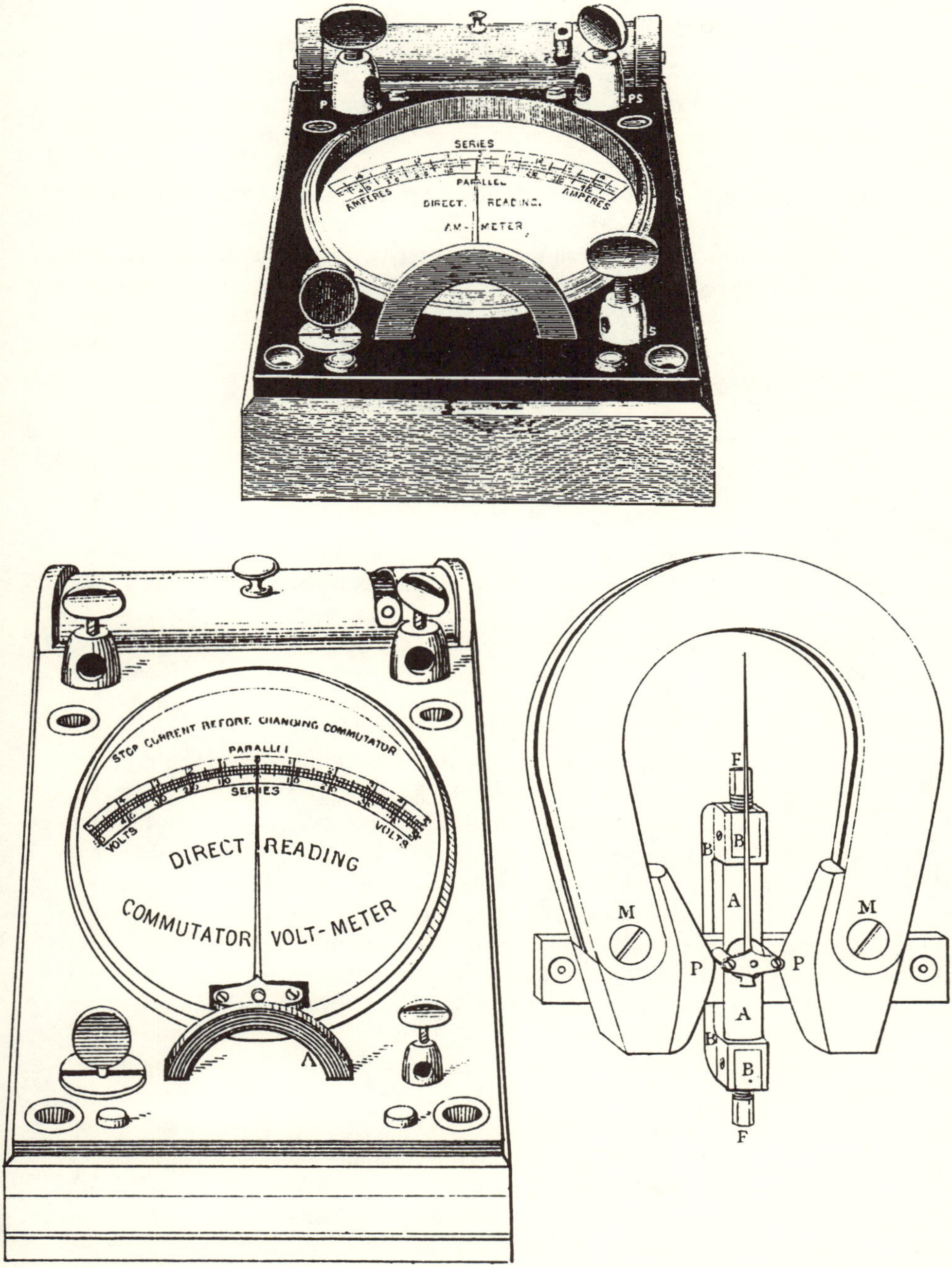

FIG. 9. 1883–84 Ayrton and Perry permanent-magnet, direct-reading devices. *Top*, ammeter exterior; *left*, voltmeter exterior; *right*, interior mechanism displaying movable charcoal-iron cores. W. E. Ayrton and J. Perry, "Direct-Reading Instruments, and a Non-Sparking Key," *Proceedings of the Physical Society* 6 (1884), 59; reproduced in *The Electrician* 12 (1884), 347, 370–71.

A similar technique was applied by Ayrton and Perry later in the same year to a new form of ammeter/voltmeter based upon their patent magnifying spring. As they pointed out in a paper on the theory of this device presented at the Royal Society in June 1884, the "pressure indicators" long used by steam and gas engineers had traditionally produced dial readings by a mechanism of a combined piston, spiral spring, and a magnifying lever. Ayrton and Perry amalgamated the latter two aspects of this mechanism in a spiral spring constructed with a mathematically optimized cross section and 45° pitch so as to convert, with rigorous proportionality, a small longitudinal movement into a large axial rotation between its ends. This convenient principle was deployed in their new ammeters and voltmeters: by mounting the current coils as a solenoid, an iron core drawn into it caused an attached spiral spring to move a pointer over a calibrated radial dial commensurately with the size of the current. Such direct-reading devices, like their earlier permanent magnet counterparts, were fitted with a user-adjustable calibration for secular changes in their electromechanical properties, for example, post-manufacture "setting" of the spring[55] (fig. 10).

No explicit claims were made by Ayrton and Perry for the degree of precision obtainable with these spring instruments. All we can discern is an implicit confidence on the order of 1 percent in, for example, the 1888 researches in which these devices were used to determine the efficiency of incandescent lamps.[56] However in the same year we know that the American Professor William A. Anthony of Cornell University had found that imported British spring ammeters—almost certainly Ayrton and Perry's patented devices—were reliable to no more than about 5 percent. Anthony, nevertheless, candidly remarked to his audience at the Franklin Institute, "Do not understand from this that I consider these instruments of little value; they are the best we have."[57] It was only by 1888 that the predominance of Ayrton and Perry's 1884 models in both British and American markets was eclipsed by devices generally held to be superior in precision: Edward Weston's moving-coil ammeter and voltmeters patented and produced in substantial numbers by the British emigré's company in New Jersey.[58]

Whatever could be said of the accuracy achievable with spring ammeters and voltmeters in the hands of those who had not acquired their skills under the direct tutelage of Ayrton and Perry, their own usage of their 1884 models had effectively set and mastered the agenda of 1880s electrical engineering metrology. The majority of subsequent current and voltage measurers, most notably Weston's, emulated their direct-reading technology and their taxonomies of "ammeter" and "voltmeter," respectively. Their enormous early impact upon the field and its lingering significance can be judged from the *Electrician*'s celebration of Ayrton's

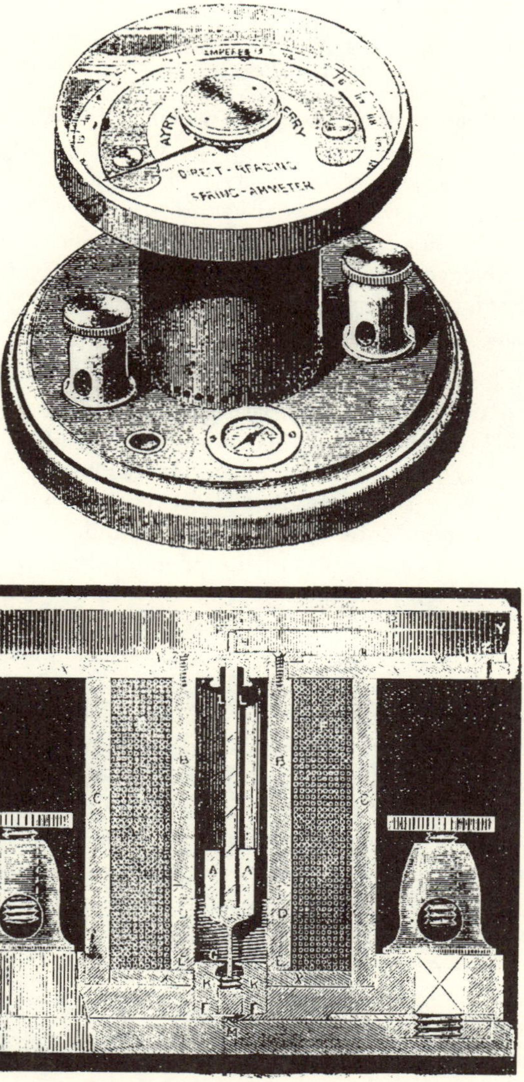

FIG. 10. 1883–84 Ayrton and Perry spring ammeter. *Top*, exterior view; *bottom*, cross section of interior mechanism. W. E. Ayrton and J. Perry, "A new form of spring for electric and other measuring instruments," *Proceedings of the Royal Society* 36 (1884), 279–319. Also in *Nature* 30 (1884), 205–9.

preeminence as president of both the Institution of Electrical Engineers (formerly the Society of Telegraph Engineers and Electricians) and the Physical Society of London in 1892:

> As an inventor his name, in association with that of Prof. Perry, is known all over the world in connection with the direct-reading electrical measuring instruments. Their pioneering labours in instrument making . . . at a time when electrical engineering was in its earliest infancy are apt to be overlooked, because they do not occupy a prominent position in modern work; but the Ayrton and Perry ammeters and voltmeters were invaluable, even in their crudest form, and hundreds of the oldest types are in daily use at the present time."[59]

This widespread usage resulted from Ayrton's role in the systematic training of a very large number of electrical engineers in the use of the ammeter and voltmeter during his enormously prolific years as Professor of Technical Physics/Electrical Engineering at Finsbury and South Kensington. Of the year 1886 John Perry later recalled that, sitting at a meeting of the I.E.E. and "looking over the large audience," he was "able to say that nearly three-quarters of the people present were Ayrton's old students."[60] Whatever the rhetorical exaggeration of his old ally's claim, Ayrton had indeed been obliged to meet a demand for "scientific" training from *thousands* of apprentices in the electrical industry.[61] The scheme of technical education that he developed was—unlike the rigorous and painstaking syllabus of the physicists—consequently one of industrial efficiency and expediency: the factory-style mass production of identically competent practitioners able rapidly to replicate quasi-mechanical routines of electrical determination. As he declared in his presidential address to the I.E.E. in 1892 of his nineteen-year "experiment" in "how best to train the young electrical engineer" ". . .[it] involves the preparation of the machine that is daily used alike by the dynamo constructor, the cable manufacturer, the central station engineer, and the lamp maker—viz., the human machine—the problem of fashioning this tool, so that it may possess sharpness, an even temper, moral strength, and a mental grain capable of taking a high polish."[62]

The organization of Ayrton's electrical engineering laboratories indeed bore a very close resemblance to London's mechanized dynamo, electromotor, and lamp factories, the shop floors of which many of his students were destined to populate. His system for maximizing the throughput of this skill-manufacturing environment had been developed at Finsbury Technical College contemporaneously with the ammeter and voltmeter. It facilitated large numbers of students working simultaneously in groups around a circus of preprepared and precalibrated experiments on rigidly configured apparatus boards.[63] Ayrton extended this scheme directly to his laboratories at the Central Institution at South Kensington in 1885.[64]

Preempting unfavorable contrasts between his laboratory "manufactory" and the morally rigorous self-helpful regime of the conventional physics laboratory discussed above, Ayrton defended his scheme in the introduction to his 1887 text *Practical Electricity: A Laboratory and Lecture Course*:

> At first sight it might appear that the student finding each set of apparatus joined up quite complete, with current laid on all ready for the carrying out of the experiment, would prevent his learning to adopt expedients for overcoming experimental difficulties, and would retard his acquiring habits of originality. For first year's students, however, I have found it a good plan to have each set of apparatus complete in position . . . because it is only with some such arrangement that fifty or more students can commence work almost simultaneously, and in the course of two or three hours have all performed some quantitative experiment.[65]

As important to Ayrton as students' rapid acquisition of basic metrological expertise was their ability to make precision replications of fundamental electrical experiments early in their training.[66] Importantly, he argued that this process should begin by inculcating secure notions not of mass, length, and time but of the electrical quantities encountered by the electrical engineer in his regular labors. Thus without the physicist's constant recourse to the balance, ruler, and clock, the electrical novice in Ayrton's electrical engineering laboratory learned to use the tangent galvanometer from experiments with the voltameter, and thence to calibrate and use ammeters as direct-reading indicators of current.[67] Having been quantitatively inculcated with the primary status and meaning of the ampere, and having learned the integrity of the ammeter as a trustworthy device for the measurement of current, the apprentice then learned all else of the electricians' "science" in the execution of Ayrton's well-supervised curriculum:

> Experience has shown me that after a student has gone intelligently through this course, under proper direction, he has obtained clear notions of the meaning of the ampere, the volt, the ohm, the coulomb, the farad, and the watt, and feels himself familiar with their connection with one another, and with the modes of employing them in actual practice. He has, in fact, mastered the basis of the exact commercial measurement of electrical quantities.[68]

It was in the context of training future workers in the skills of exact commercial measurement for the British energy industry that Ayrton first conceived his extension of the ammeter and voltmeter into the milieu of precision research physics. I will now discuss the controversy that surrounded his attempt to use these engineering instruments in a pedagogical "determination" of the "mechanical equivalent of heat."

Precision, Propriety, and Pedagogy in "Determining" Joule's Constant

> We can imagine a scientifically minded engineer who has sent his son to the Central Technical College for a course of physics, saying, "Why home so early, my son? . . . I thought you had laboratory work this afternoon." "Yes, father, the lecture was over at three; but as I had only to make a determination of Joule's equivalent, I did the trick and caught the 3:15 at South Kensington easily. I worked out the result on the train, and it comes out within point nought, nought, nought two of the mean between Griffiths (who is the nailer at Joule's equivalent) and the chap they call Michael S. Q. [Micelescu]" . . . Well may the father fear that he is not up to date, but at the same time he may comfort himself with the reflection that these technical colleges must be wonderful institutions.
> ("A Student's experiment and its teaching," *The Electrician, 28 December 1894*)[69]

The interest of late Victorian experimental physicists in the precise measurement of electrical energy focused upon the "mechanical equivalent of heat," known also as Joule's constant in deference to the claims of the Manchester natural philosopher who first established its existence.[70] After a turbulent phase in the early 1840s,[71] the significance of this constant was established for the nascent science of thermodynamics by Thomson, Rankine, Clausius, Helmholtz, and others. The exact determination of the "mechanical equivalent" was a goal zealously pursued by experimental physicists, and among a variety of techniques developed for the task, one of the most popular was the electro-calorimetric method. Their concern to increase the precision of this constant was notably heightened when a consensus emerged in the late 1870s that Joule's original figure of 772 foot-pounds per degree F had been a considerable underestimate. This was due in part to the instability of the value of the British Association ohm that had featured so centrally in Joule's own 1867 electrical redetermination of his "constant"[72] (table 2).

Such problems of unreliable auxiliary standards were recurrent throughout later (none too convergent) attempts at stabilizing a value of the Joule constant to a replicable degree of precision. A notable example of this in 1893 was the announcement by E. H. Griffiths, a Fellow of Sidney Sussex College, Cambridge, after five years of continuous labor on one hundred separate evaluations, that he had used the electro-calorimetric method to

Date.	Observer.	Method.	Ergs per gramme-degree C.	Foot-lbs. per lb.-degree F.
1867.	Joule.	Electrical.	$4·155 \times 10^7$	772
1870.	Violle.	Heating of a disk between the poles of a magnet.	$4·269 \times 10^7$	793
1875.	Puluj.	Friction of metals.	$4·179 \times 10^7$	776
1878.	Joule.	Friction of water.	$4·159 \times 10^7$	772
1878.	Weber.	Electrical.	$4·145 \times 10^7$	770
1879.	Rowland.	Friction of water 15°.	$4·189 \times 10^7$	778
1888.	Perot.	By the relation $L = r(v_2 - v_1)\frac{dp}{dt}$.	$4·167 \times 10^7$	774
1889.	Dieterici.	Electrical.	$4·232 \times 10^7$	786
1891.	D'Arsonval	Heating of a cylinder in a magnetic field.	$4·161 \times 10^7$	773
1892.	Miculescu.	Friction of water.	$4·186 \times 10^7$	778
1893.	Griffiths.	Electrical.	$4·198 \times 10^7$	780
1894.	Schuster and Gannon.	Electrical.	$4·192 \times 10^7$	779

The results obtained since 1879 by stirring water and by electrical methods may be tabulated thus :—

Rowland (friction)	$4·189 \times 10^7$ or 778,
Dieterici (electrical)	$4·232 \times 10^7$ or 786,
Miculescu (friction)	$4·186 \times 10^7$ or 778,
Griffiths (electrical)	$4·198 \times 10^7$ or 780,
Schuster and Gannon. (electrical)	$4·192 \times 10^7$ or 779.

The result obtained by our students is greater than the mean of these results by just one half per cent. It is probably not generally known that such accuracy can be obtained by the use of such simple apparatus as that here described, and the fact is we think sufficient justification for bringing the result under the notice of the Society.

TABLE 2. Table of determinations of the mechanical equivalent of heat up to 1894, with comparative comment from Ayrton and Haycraft. W. E. Ayrton and H. C. Haycraft, "A student's simple apparatus for determining the mechanical equivalent of heat," *Proceedings of the Physical Society of London* 13 (1894–95), 306.

determine the mechanical equivalent as 778.99 foot-pounds per thermal unit F in the latitude of Greenwich (*g* = 32.195). Read at the Royal Society by his well-placed ally R. T. Glazebrook, Griffith's paper explicitly revealed the reliance of his figure upon other contemporary feats of standardization: (*i*) the ohm as defined by the B.A and legalized by the Board of Trade in 1892 as that of a column of mercury of mass 14.4521 grams, of length 106.3 centimeters, and of uniform cross section at 0°C; and (*ii*) the electromotive force of a standard Clark cell at 15°C = 1.4342 volts.[73]

The young Rollo Appleyard also reported to the *Electrical Review* that the highest order of contemporary mechanical engineering had been utilized by Griffiths in placing the calorimeter within a steel chamber that had been "constructed with extreme precision by Messrs. Whitworth & Co."[74] The unfortunate Griffiths nevertheless had in the following year to recant a reading as mundanely incompetent as an "incautious acceptance of the indications of a thermometer" and thence issue a revised value for the mechanical equivalent of 779.77 foot-pounds per thermal unit F.[75] The use of such an instrument as a direct-reading thermometer in these precision determinations was clearly not without its problems. Indeed broader doubts were often entertained by late nineteenth-century physicists about the suitability of the direct-reading instrument for experimental research and *a fortiori* for the training of young laboratory researchers.

Such reservations were made graphically apparent in William Ayrton's 1894 development of a pedagogical apparatus by which student novices could make an electrical determination of Joule's constant using an ammeter and voltmeter. By using heavy-duty industrial measuring apparatus with a commercial power supply Ayrton reckoned that he and his junior partner H. C. Haycraft could radically reduce the time and labor involved in the laboratory process of determining Joule's constant. In effect Ayrton was emulating the ventures of his mentor Sir William Thomson during the 1850s and 1860s in investigations that freely interchanged electrical measurement instruments, especially his mirror galvanometer, between natural philosophy classes and practical engineering projects.[76] Ayrton was clearly reckoning upon a similar degree of success in attempting to introduce the electrical engineer's workshop ammeter and voltmeter into the physicist's repertoire of precision laboratory instruments when he and Haycraft presented their paper "A Student's Simple Apparatus for Determining the Mechanical Equivalent of Heat" at a meeting of the Physical Society of London on 23 November 1894.

Their principal goal was to ensure the experiment was completed with such rapidity—*in under ten minutes*—that "troublesome corrections" for heat losses were preempted. This was achieved by using an industrial intensity of current (30 amperes) measured by a Weston moving-coil ammeter accurate to 1/3 percent, and a potential difference of 8.65 volts measured by a Weston moving-coil voltmeter accurate to 1/5 percent. Such a degree of precision was obtainable, Ayrton argued, since "now the commercial values of the electrical units are known with considerable accuracy in the C.G.S. system, it is possible to measure energy in foot-pounds by means of a good commercial ammeter [and] voltmeter with greater ease and certainty than by any mechanical dynamometer." With an ordinary laboratory stopwatch and thermometer and a preprepared calorimeter, Ayrton's students were able, with "no special manipulative skill," to achieve the result of 779 foot-pounds (Greenwich) degrees F—only 1/2 percent greater than the mean of the results of Rowland, Griffiths, and others. The authors considered a "sufficient" reason to bring their result under the notice of the Society that it was "probably not generally known that such accuracy can be obtained by the use of such simple apparatus" (fig. 11).[77]

The response to this "ten-minute determination" that they received from physicists present at the meeting was not, however, as hospitable as they had obviously expected. While the fortnightly gatherings of the Physical Society were known for their friendly exchanges of "unusual animation,"[78] this particular meeting revealed differences rather of unusual sharpness between physicists' and electricians' criteria for what counted as a meaningfully achieved degree of precision in laboratory measurement.[79] The more traditional members of the physics community were appalled by the incursion of manufacturer's values, instruments, and practices into their domain of teaching and research. In their view, by transgressing the divide between commercial industrial practices and rigorous etiquettes of experimental science, Ayrton and Haycraft rendered nugatory any claim to precision made by either student or qualified scientist on behalf of their meter-based "determination" of the mechanical equivalent of heat.

Among the indignant physicists present was the author of one major recent determination of the Joule constant that had taken five years to effect, E. H. Griffiths himself. While admiring the design of the apparatus, Griffiths attacked its pedagogically "harmful" failure to address the "intelligent application of corrections" since this would lead students "to underrate the difficulty of an experiment and the care required to obtain reasonable accuracy. . . ." Targeting the moral laxity of Ayrton's "smoothing out" the "difficulties" of an experiment, Griffiths specifically argued that "if a student need, as it were, only turn a handle, and the result comes out, he is tempted to think that he is doing good work and would become conceited." Another like-minded physicist, Professor George Carey Foster of University College, London, pointedly objected to students' use of direct-reading instruments alleging that these devices yielded only "unscientific information." He had "never" given his students direct-reading instruments to use, preferring that they should use

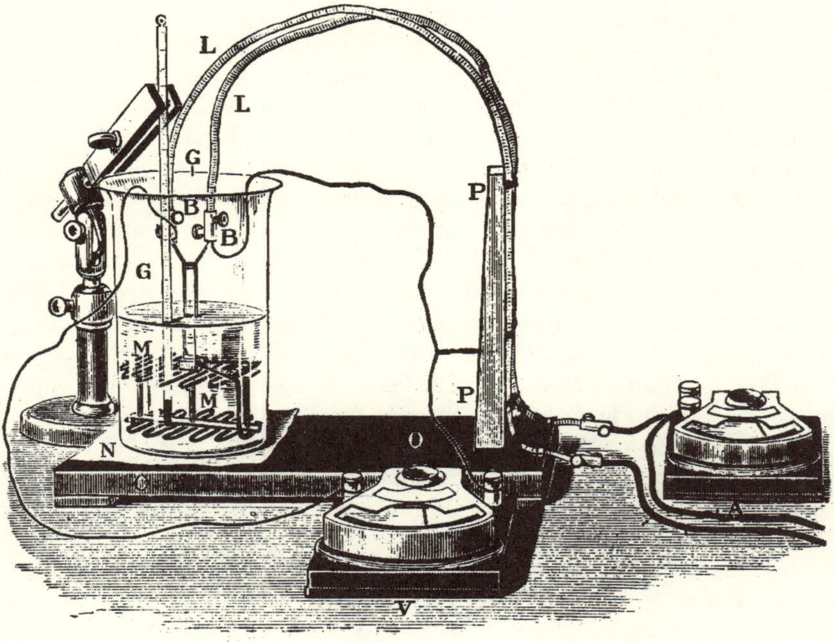

FIG. 11. Ayrton and Haycraft's "Student's simple apparatus for determining the mechanical equivalent of heat." Ayrton and Mather, *Practical Electricity*, 1911, 276.

"old-fashioned instruments" such as a tangent galvanometer, or a potentiometer and Clark Cell to measure current, insisting always that each of them "reduce their observations to absolute measure for themselves."[80]

However, Professor Silvanus P. Thompson, inheritor of Ayrton's methods of training in Technical Physics at the Finsbury Technical College, came to Ayrton's defense. He challenged the arbitrariness of Foster's injunction against the laboratory use of direct-reading ammeters and voltmeters while placing no similar prohibition on the deployment of such nonelectrical direct-reading instruments as stopwatches and thermometers. Most specifically he "differed" with Foster over the use of direct-reading electrical instruments. Flatly denying that they were "harmful" he demanded to know why it was that Foster did not instead insist upon laboratories being equipped with such "old-fashioned instruments" as water clocks for the absolute measurement of time or why the student making absolute determinations of mass should be allowed to luxuriate in a precalibrated set of weights? "Where," Thompson enquired, "should the line be drawn?"[81]

Ayrton's indignant response to the calumnies of Griffiths and Foster was to reply first as a strict manager of an educational factory that the time of his busy students was "not wholly spent in determining the mechanical equivalent of heat." As trainee industrial workers their business was also to learn the professional *trust* of the coworkers who had constructed and graduated their electrical instruments—an essential feature of the values and practices of precision measurement in electrical engineering: "When a student was using—not calibrating—an instrument, he did not assume that it was wrong." Indeed the trust of the electrical engineers began from the realities of precision electrical measurement established by the researchers of the B.A.A.S. committees; as the exasperated Ayrton put it: "If a student was not allowed to take an ohm as an ohm, how was he to begin at all? He must start somewhere."

Second, Ayrton opined that the student's goal in executing the experiment was not, like their painstaking predecessors in the field, to "learn how to make corrections," but to "learn what was meant by the mechanical equivalent of heat . . . by measuring it for himself." He should have no more qualms about using an ammeter and voltmeter without corrections for this purpose than a child should have in using a "foot rule" instead of refined trigonometrical methods to learn the size of a room.[82] Ayrton's defense consisted of the claim that the direct-reading ammeter and voltmeter were direct mensurators of real electrical quantities just as the scales, ruler, and clock were (respectively) so for the physicists' privileged parameters of mass, length, and time. Moreover, the force of Ayrton's argument was that the greater *precision* available from his electrical methods—even when carried out by a lowly inexperienced student—imbued his "determination" of the Joule constant with greater certainty than any that preceded it.[83]

The debate at the Physical Society on "Student's Simple Apparatus for Determining the Mechanical Equivalent of Heat" provoked a lengthy editorial response from the *Electrician*[84] commencing with the satirically fictionalized anecdote cited in the epigraph to this section. The journal fully supported Ayrton's goal of "impressing" on the student's mind the "nature of the important quantity which is called the mechanical equivalent of heat." For a future industrial worker an experiment "conscientiously" executed with a voltmeter and ammeter in under a few minutes to achieve a "sufficiently accurate result" of 1 percent error was a laudable endeavor, so long as he later learned the principles upon which these instruments were constructed. For Ayrton's experiment, however, it skeptically queried the 1/3 percent and 1/5 percent accuracies claimed, respectively, for measurements made with the ammeter and voltmeter. The editorial alleged that "good commercial" versions of these devices had only come to be "relied upon" to produce accuracies of 1 percent "within the last year or so—if, indeed th[at] time had already come."

Indeed it was difficult to forget that the integrity of these meters as significantly accurate direct-reading instruments had been hard-won over a considerable time:

> Direct-reading galvanometers . . . developed and barbarously christened as "ammeters" by Prof's Ayrton & Perry, were for many years unworthy of the appellation "direct reading." A calibration which had to be torn up and recalculated from time to time would have been much better than a scale which pretended to read in amperes, but which was always wrong, and which sometimes required a variable correction throughout all its length. Unworthy, too, were such instruments of a place in a laboratory except as shocking examples; and voltmeters which began to warm up almost before the needle had stopped fidgeting were perhaps worse. No wonder that they failed to gain an entrance into many a laboratory, and are now even excluded from some.[85]

Cognizant, however, that the best ammeters and voltmeters were increasingly held to be reliable to within half a percent, it pondered whether their continued absolute exclusion from the physical laboratory was not "undue conservatism." Surely, it explained, "the time has come when they may be accepted as ordinary laboratory instruments for work in which a one per cent error is allowed to pass." After all, the editorial pungently asserted in sympathy with S. P. Thompson, "in no physical laboratory is the attempt made to refer all measurements to length, mass and time; and if the thermometer is accepted, if a resistance box is used, if a Clark cell is recognised, where shall the line be drawn?"

Yet while giving license to use of the ammeter, voltmeter, and other direct-reading instruments for limited application in laboratory endeavors of teaching and research, the *Electrician* hesitated to acknowledge that they could ever legitimately be used in physical "determinations" such as that of the Joule constant. Certainly a purist investigator like E. H. Griffiths would turn to the "much derided tangent galvanometer" for this purpose and "would never think of using a ready-made ammeter or voltmeter." More fatally still the editorial dismissed the hubris of Ayrton's claim that students' experiments with electrical engineering metrology could amount to physical research: the half percent "approximation" of his "educational experiment" to "high-class determinations" of Joule's constant did not in itself "justify a comparison." Clearly no degree of precision agreement—*qua* numerical similarity between the results of Griffiths et al. and the work of his students with an ammeter and voltmeter—was ever going to qualify the latter as a "replication" of the former.[86]

Such themes were immediately taken up in a colorful bout of correspondence solicited by the *Electrician* from leading physicists and electrical engineers engaged in educational matters. The vehemence with which the majority attacked the propriety of Ayrton and Haycraft's "labour-saving"[87] apparatus reveals that much was at stake for physicists in the matter of establishing what counted as a properly achieved precision laboratory determination. G. C. Foster wrote on 1 March 1895 to argue that the epistemic status of their pedagogical experiment indeed had nothing to do with its accuracy. Had the title of their paper referred not to "determinations" but rather to the "Verification of a direct-reading ammeter and voltmeter" then "one should not have had a word of fault to find." Reiterating the physicists' protocol that the "determination" of a physical constant could be based only upon "absolute" measurements of mass, length, and time, Foster's argument rested upon his perception of the necessarily inferior status of the engineer's ammeter and voltmeter. Precision agreement between Ayrton and Haycraft's results and the received value(s) of the "mechanical equivalent" in his view served only to confirm the *calibration* of their direct-reading instruments.[88]

Foster's position supported that presented earlier by Charles Vernon Boys, assistant professor at the Royal College of Science, a near neighbor of the Central Institution on Exhibition Road. On 25 January 1895 Boys wrote to "protest" against the utter meaninglessness of the title of "Student's Simple Apparatus for Determining the Mechanical Equivalent of Heat." Boys considered that in relying upon the mere "hearsay" of direct-reading instruments, the student did not "determine" anything in the sense that he understood the word. Deconstructively historicizing the integrity of the ammeter and voltmeter, he asserted that

> The measurement attributed to the student is really made by the original investigators who made the absolute determinations, e.g., of the electrochemical equivalent of silver, employed later for standardising a Kelvin balance, employed later for calibrating an ammeter—who made the absolute determination of the electromotive force of a Clark cell, or who found the value of the resistance of a piece of wire called an ohm. These people did the work of making the determination. Various middlemen passed this work along, and copied or compared, and after a time the actual ammeter and voltmeter worshipped by the student became calibrated, and finally a dummy in the form of a student, who, according to the provision of the Paper, is carefully guarded from doing anything which involves either [*sic*] knowledge, pain, or skill, shakes out the magical figure, possibly under the impression that he has created it.[89]

Many other correspondents echoed the message that a precision measurement could only be claimed as a "determination" by the experimenter who had "created" the "magical" figure literally by doing all of the work himself. In so doing they attacked Ayrton's notion that hard laboratory labor could in any way be replaced by the electrical engineers' trust in the expeditious exactitude of prefabricated direct-reading industrial instruments. Specifically targeting the pedagogical imperatives implied in Ayrton's metrological scheme, many criticized the idea of training the next generation of both physics researchers in the skills of accurate measurement without also instilling in them the *moral* qualities of the "self-helpful" man in order to become a true laboratory investigator. As G. H. Robertson wrote on 28 December 1894, "It is extremely doubtful whether the beginner does not gain more by performing (unaided by such appliances) experiments which yield less scientifically accurate results, but require more mental effort on his part. A little manly enthusiasm for his work, and courage gained from difficulties overcome, are worth more as a start in life than much acquaintance with refined apparatus."[90]

The anonymous but evidently eminent "J" wrote on 11 January 1895 to complain that it was of no avail for a student "on entering the physical laboratory" to attempt to make "determinations which occupied masters of research during years of patient work . . . with still greater accuracy in a few minutes." He would crucially have failed to be instilled with Smilesian moral robustness germane to authentic acts of measurement:

> When a boy who has gone through a course of "automatic labour-saving experiments finds himself face to face with a practical problem, which he has to think out for himself and solve practically with scarcely any of the resources of the physical laboratory at his command, and no one to make his connections for him, will the labour-saving apparatus avail him much? Will not a few old-fashioned experiments with jam jars and sawdust, and bits of copper and zinc, and tangent galvanometers and so forth have done more to develop self-reliance and resource?[91]

An identical use of Smilesian moral discourse can be seen again in C. V. Boys's brisk dispatch of the integrity of the work carried out by his South Kensington neighbors. If, he contended, the authorities at the Central made the "staring at direct-reading instruments their chief aim and object, the result would be the production of a kind of glorified plumber without the skill of a real plumber. . . ." So far as Boys was concerned Ayrton had illicitly broken the important *moral* as well as *technical* demarcation between the academic laboratory and the manufactory: a man who had learned to use conveniently direct-reading ammeters and voltmeters would probably "be to his instruments" no more than "what the machine minder is to some marvellous automatic machine-tool." Such an individual, Boys argued, had not learned to make "determinations" any more than a man he had met in a steelworks in Sheffield, who had "ridiculed the fuss that is made about squaring the circle, because he could do it with a slide-rule." The factorylike metrology embodied in Ayrton's South Kensington electrical engineering laboratory was thus for Boys not a legitimate mode of producing new physical knowledge.[92]

For E. H. Griffiths, the saving of time and effort effected by using ammeters and voltmeters told against Ayrton's claims that—their accuracy notwithstanding—such instruments could communicate the "reality" of the mechanical equivalent. Only arduous character-building exercises could communicate such a reality: "Let the junior student slip down the rope a little too quickly; let him hammer a piece of lead until it becomes so hot that it ignites some phosphorous placed on it; then let him raise the temperature of water by means of a frictional machine, and I think he will acquire a greater 'grip' of the real meaning of the phrase 'mechanical equivalent.'"[93] Griffiths's point that precision of measurement was of subordinate importance to "self-helpfulness" in a proper laboratory education is starkly plain in G. C. Foster's account of the metrological activity at University College:

> In my laboratory students continually make, as laboratory exercises, what we call "absolute determinations" of *H*, of electrochemical equivalents, of differences of potential, of magnetic susceptibility, sometimes of resistance, in which the only measurements made are measurements of length, mass and time. For this purpose we use commercially accessible standards [*sic*]. Our results have not always been very good—indeed, they are often very bad—but . . . even a rough result obtained in this way is, I believe, of greater educational value than a far more accurate one got by help of a direct-reading contrivance, for the accuracy of whose indications the experimenter is not personally responsible. My contention is very much that we ought to try to teach students of physics to rely as much as possible on themselves and to depend as little as they can upon the accuracy of their instruments.[94]

For Foster, then, self-reliance in experimentation was more important as a *moral* guarantor of metrological authenticity than the degree of *precision* achieved through inappropriately assisted measurement techniques.

Ayrton's technological allies wrote in to the *Electrician* to defend the "mechanical equivalent" experiment against the "educationalists" moral critique. They did so by broad reference to the *efficiency* of the "factory" scheme of mass education which Ayrton and Haycraft's setup so manifestly epitomized. In a reply solicited from Ayrton's early collaborator John Perry, this scheme was portrayed as being so obviously "effective" that it had been "enthusiastically copied" in the "great majority of teaching institutions all over the world." Giving every one of a large number of students "effective" laboratory instruction in electrical physics on "any other system" would have an appalling effect on the "mental fatigue of the demonstrators" if it were "not to be a farce." More truly farcical, argued Perry, was the unhistorical pretence of the "self-help" school of physics that mass, length, and time were the only quantities that could legitimately be measured directly with laboratory instruments for the purposes of precision research. While he was aware that the direct-reading ammeter was a "complex," that is, indirect, mensurator of electrical reality "because I know the steps by which it has been evolved, and it was not common in my youth," the only reason Perry did not perceive a two-foot rule to be a problematic measuring device was simply that he had "always had one."[95]

This crucial difference between what physicists and electrical engineers counted as legitimate means of effecting precision electrical measurements in both pedagogy and practice was also addressed on February 1 by John Ambrose Fleming, Professor of Electrical Engineering at University College, London. Grounded as the electrician's precision measurements were in the order of everyday industrial life so remote from that of the academic laboratory, Fleming contended that it should not be necessary for a novice to make his point of "departure" from the contrived abstractions of "absolute" physics. Thus for a prospective electrical practitioner seeking a training in either a physics or engineering laboratory, little was to be gained by a Smilesian self-denial of the instrumental assistance that the engineering community could offer; on the contrary there was "every possible advantage to be gained" by introducing a student to quantitative physical work with "good instruments" at a very early stage. Moreover it was "sheer nonsense" to say that the work was done "for him" by the instrument maker for the boy had to "learn to read an instrument accurately first before he can get any results at all."[96]

Fleming was therefore acerbically critical of the irrelevance of skills cultivated in the moral order of the conventional physics laboratory for a novice about to be launched into industry. Not being skilled in the use of direct-reading ammeters and voltmeters,

> he is not trained in becoming an exact observer, and he is or may be as ignorant as a babe of actual and practical instruments as used in the workshop or testing room, although he may be familiar with rough laboratory equivalents of them. Place such a student in the electrical testing rooms of a cable factory or dynamo building, and he has to begin his acquaintance with the actual tools of his trade. "Yes, but," say the advocates of the jam-pot battery and home-made ohm, "he has had his mind so much strengthened by his previous training that he very readily assimilates the new knowledge."

Fleming's dessicated conclusion upon the matter was that "we may overdo this cult of mind-strengthening education."[97]

The most pungent counterattack on the position of Ayrton, Perry, and now Fleming in this conflict between physicists and electrical engineers came in a letter solicited by the *Electrician* from the *physicien* Alexandre Cornu. Cornu's tirade of 15 March 1895 was directed against the vulgar presumptuousness of Ayrton as a professorial and practicing electrical engineer. Such a man, wrote Cornu, considered he had "no need to trouble his head" over the nature of the ampere or volt since, for example, "the idea of measuring the electric pressure in volts is as familiar to him as is the measurement of the length of his cables in feet or metres." Far removed by his everyday labors from such "scientific meditations" as the "scaffolding" of C.G.S. Units, he would give his students direct-reading instruments and allow them to end up in "persuading themselves that the volt is an irreducible unit of electric pressure in the same way as the pound or the kilogramme is an irreducible unit of force, and the foot or the meter length." These students, Cornu complained, would naturally have no difficulty in seeing Ayrton's experiment as a "veritable determination" of the mechanical equivalent of heat.[98]

The more "cautious" observer would, however, see that the meaning of the experiment was not so "simple" for he would discern that, in relying only upon Ayrton's authority for the premise that mechanical work was directly measurable as the product of readings from an ammeter and voltmeter, "words are offered him in the place of things." To recognize the physicist's true "idea of mechanical work," the student had instead to "separately [*sic*] see *force* in the shape of a weight, and *displacement* in the shape of a path along a divided rule." The prospect of such rigorous practice being polluted by the spreading of electrical engineering pedagogy into the physics laboratory was deeply disturbing to Cornu. He envisioned that "the technical professor, instead of taking as his basis our knowledge of the laws of physics and mechanics, will lay down as primordial definitions the indications of commercial instruments. The danger is a grave one. The movement has already begun." From the point of view of the "facilitating of commercial transactions" Cornu accepted that this would "prove to be a grand piece of progress"; he had severe "doubts" however, as to whether the "spirit of discovery and practical applications" would be a "gainer" in the same degree.[99]

On 29 March the *Electrician* concluded the debate by drawing a strong distinction between the kinds of pedagogy in electrical measurements that were relevant to the production of meaningful precision for the communities of physicists and electrical engineers. Inevitably sympathizing with the mass of artisans undergoing Ayrton's metrological training scheme for the high-efficiency production of fresh electrical engineers, it pronounced that for "students in such classes the volt must be merely a unit of pressure seven-tenths that of a Clark Cell; the ohm a resistance of 10ft of 10mil copper wire." Yet while having "no sweeping objection" to this particular pragmatic deployment of "apparatus screwed down to a board, with connections already fixed," the editorial felt obliged to "object to call[ing] it science teaching." As far as it was concerned, Ayrton had broken the protocols of physical science in trying to pass off the ammeter and voltmeter as legitimate instruments of "scientific" activities. Whatever degree of instantaneous precision could be wrought from their direct-reading scale, these instruments were irredeemably incompatible with the moral order of the physics laboratory. As such their readings were excluded from the exalted catalogue of determinations of the "mechanical equivalent of heat."[100]

Epilogue: The Changing Meanings, Methods, and Sources of Precision

This paper would not be complete without a discussion of the historical instability of the practices and interpretive conventions that constrain the meaning(s) of precision and the means for achieving it. For the late twentieth-century historian of metrology two remaining issues throw the above account into graphic relief. Both points arise from events conducted under the aegis of the Royal Society and publicly communicated through the general science journal *Nature* between November 1894 and May 1895, contemporaneously with the controversy over Ayrton's apparatus.

During this period an acute crisis developed among physicists about the *identity* of the determinations undertaken in thermal metrology over the preceding fifty years. This was first manifested in the work of Arthur Schuster and William Gannon, cited by Ayrton and Haycraft as the most recent determination of the "mechanical equivalent of heat," and read at the Royal Society on 22 November—the day immediately before the latter authors' demonstration of their "Simple Apparatus" at the Physical Society. Ayrton's citation notwithstanding, Schuster and Gannon's paper had actually been entitled "A Determination of the Specific Heat of Water in terms of the International Electric Units."[101] As hinted in the title, their proposal was to systematize energy measurement with the commercial units of electricians by replacing the calorie, non-absolute in its dependence upon the thermal properties of water, by the Joule, the absolute "dynamical" unit of heat.

The need for a conversion "equivalent" being thereby eliminated, all previous evaluations of Joule's constant were henceforth to be redesignated as determinations of the "Specific Heat of Water," the palpable temperature-variation of which accounted for the notorious non-convergence of previous high-precision researches on the subject! This suggestion, in a modified form, was vigorously pursued by none other than E. H. Griffiths in the pages of *Nature*,[102] and through his relentless agitation, rendered statutory by the B.A.A.S. Electrical Standards Committee.[103] From this episode we learn not only of the indeterminate relationship between the identity and the precision of a Victorian physicists' measurement, but also of the mediated role played by commercial interests in such transformations of metrological identity.

Second, in the same issue of *Nature* in which Griffiths's letter was published (9 May 1895), we see a report that W. E. Ayrton had recently presented his controversial apparatus for "Determining the Mechanical Equivalent of Heat" at a Royal Society Conversazione—manifestly undeterred by the opposition encountered at the Physical Society and in the columns of the *Electrician*. Indeed, so resistant was Ayrton to criticisms of ammeters and voltmeters to effect precision measurements within the disciplinary territory of physics that this experiment became a permanent feature in both his South Kensington teaching laboratories and in successive editions of *Practical Electricity*. Notably, however, we find that by the posthumous 1911 edition, the experiment had been retitled "Measuring the heat equivalent of electric energy," and the "considerable accuracy" obtainable in a beginner's results was more modestly claimed as "not differing by as much as one per cent from the truth."[104]

Following this example, within only a few years of Ayrton's death we find the ammeter and voltmeter being assimilated into the instrumental repertoire of teaching and research in laboratory physics, albeit with a considerable ambivalence. In their 1916 *Textbook of Practical Physics*, two physics lecturers at Kings College, London, H. S. Allen and H. Moore, argued first that "although measuring instruments of precision are essential, the use of elaborate apparatus, sometimes almost automatic in its actions, is to be deprecated in a laboratory for students."[105] Nevertheless, accepting that a physics laboratory in a metropolitan world densely populated by electrical machinery could no longer rely upon the "inconvenient" tangent galvanometer for attaining the requisite precision, the direct-reading ammeter had now become a "very important instrument."[106]

Conclusion

In this paper I have shown how, for late Victorian electrical engineers, the credibility of the ammeter and voltmeter as the generic industrial measurers of energy depended upon material constructions and usages that were commensurate with the moral order of commercial electrical metrology. Contemporary physicists, however, decried the credulous indolence of those who followed William Ayrton's suggestion of employing such "labour-saving apparatus" for both teaching and research in the morally incommensurable milieu of the physics laboratory. The expeditious exactitude which these instruments could be made to yield in the engineering context was for physicists utterly valueless in the academic laboratory because their alleged precision had not been achieved by practitioners, practices and instrumentation that had legitimacy in that domain. For late Victorian physicists, the precision with which an instrumental measurement had been executed was not a *sufficient* guarantee that its numerical value could count as an evidential claim about the world.

As far as they were concerned, the precision achieved by using a metrological instrument could be discredited as a source of authoritative research if (*i*) it were claimed as a direct measure of a non-fundamental quantity, such as an electrical current, that was a synthesis of mass, length, and time; (*ii*) the alleged precision were achieved only by an unskilled student in the irrelevant context of a scheme of technical education; (*iii*) unsubstantiated claims made in the past on behalf of its reliability were at a high profile in the disciplinary memory of practitioners; and (*iv*) the *moral* qualities required of its user fell short of disciplinary standards of self-reliance. Any or all of these critical points could be invoked to resist the incursion of a measurement instrument of whatever precision from an alien engineering domain into the laboratory apparatus of the physicist. Without acknowledged legitimacy in this domain, the numerical values it could be made to produce were the subject only of calibration against standard devices and could not be harnessed to the production of new physical knowledge.

We are now in a position to propose an answer to the first question raised in my introduction. Measurement instruments become credible as meaningful and trustworthy sources of precision values by fulfilling the material, managerial, and moral priorities local to a metrological community's domain of practice. Their credibility can only be universalized to other metrological communities when sympathetic shifts in these cultural priorities occur within their domains of practice. If we are to explain the eventual acceptance of the ammeter and voltmeter as instruments fit for laboratory teaching and research in physics, at least three factors are likely to be relevant: (*i*) the elevation of electrical quantities to a new status of fundamentality after the "arrival" of the electron ca. 1900 in the ontology of laboratory reality; (*ii*) the onset of disciplinary amnesia about deficiencies in the precision of early ammeters and voltmeters; and (*iii*) a transition in the moral order of the physics laboratory that loosened restrictions upon the range of standardized metrological aids in order to expedite rapidly growing demands for research and teaching in the subject.

Historians of electrical instrumentation could fruitfully consider how these transitions in the moral order of physics laboratories come to pass, for such episodes can tell us much about the cultural processes by which metrological precision comes to have a stable meaning. Then we may begin to understand how an instrument's precision has so often been mythologized as the universal arbiter of measurement-mediated claims about the world.

I would like to thank Jeff Hughes, Norton Wise, Joseph O'Connell, Andrew Warwick, Simon Schaffer, Jerry Ravetz, Bruce Hunt, Sophie Forgan, Crosbie Smith, Ben Marsden, William J. Ashworth, Tom Wright, Otto Sibum, Kathy Olesko, and others present at the Princeton Workshop for their valuable comments and criticisms of earlier drafts of this paper.

I am grateful for the financial sponsorship provided by the British Academy and the Institution of Electrical and Electronic Engineers, New York.

Notes to Chapter Ten

1. J. Swinburne, "Electrical Measuring Instruments," *Proceedings of the Institution of Civil Engineers* 110 (1892), 1.

2. T. Sewell, *The Elements of Electrical Engineering* (London [2d ed.], 1903), xiv–xv.

3. S. Schaffer, "Measuring virtue: eudiometry, enlightenment and pneumatic medicine" in Cunningham and French, eds., *The Medical Enlightenment of the Eighteenth Century*, Cambridge, 1990, 281–318; D. MacKenzie, "Missile accuracy: a case study in the social processes of technological change," in W. Bijker, T. Hughes, and T. Pinch, eds., *The Social Construction of Technological Systems* (Cambridge, Mass., and London, 1987), 195–222.

4. Following typical nineteenth-century conventions I shall use "accuracy" and "exactitude" as synonyms for "precision." Note, for example, that the earliest citation given by the *Oxford English Dictionary* (2d ed., 1989), for the modern usage of "precision" as "number of digits given" vis-à-vis "accuracy" as "nearness to the truth value" is 1948. No such differentiation is made in the 1st edition of 1933.

5. For a more detailed analysis see G. Gooday, *The Morals of Measurement* (forthcoming).

6. "Uppenborn's Electrical Measuring Instruments," *The Telegraphic Journal and Electrical Review* 5 (1884), 103.

7. G. Gooday, "Precision Measurement and the Genesis of Physics Teaching Laboratories in Victorian Britain," *British Journal for the History of Science* 23 23 (1990), 25–51.

8. G. Gooday, "Teaching Telegraphy and Electrotechnics in the Physics Laboratory: William Ayrton and the Creation of an Academic Space for Electrical Engineering 1873–84, *History of Technology* 13 (1991), 73–111; C. W. Smith

and M. N. Wise, *Energy and Empire: A Biography of Lord Kelvin*, Cambridge, 1989, 669–78; B. Hunt, "The Ohm is where the Art is," *Osiris* (forthcoming, 1994). Gooday, *The Morals of Measurement*, 33–35.

9. P. Dunsheath, *A History of Electrical Engineering*, London, 1962, 305; this point is not acknowledged in T. P. Hughes's seminal monograph, *Networks of Power* (Baltimore and London, 1983).

10. "The Measurement of Electricity for Commercial Purposes," *Electrician* 10 (1883), 372.

11. Ibid.

12. J. Shoolbred, "On the measurement of electricity for commercial purposes," *Journal of the Society of Telegraph Engineers* 12 (1883), 84–107, discussion on 107–22.

13. Cf. Sewell's remarks, *The Elements of Electrical Engineering*, 2.

14. A.B.W. Kennedy, "The Critical Side of Engineering Training," *B.A.A.S. Report*, 1894 (pt. 2), 743–44.

15. This method employed the formula (in C.G.S. electromagnetic units) for absolute current "i," π Hrtan(A)/2nc, where "A" was the angular displacement, "r" the coil radius, "n" the turnage, and "H" the local horizontal component of the geomagnetic field. See Maxwell's succinct analysis of the galvanometer family in *A Treatise on Electricity and Magnetism* (II) (3d ed., 1891 [reprinted 1954]), 708–29.

16. Rayleigh and E. Sidgwick, "On the electro-chemical equivalent of silver . . . ," *Philosophical Transactions of the Royal Society* 175 (1884), 411–60. *A History of the Cavendish Laboratory* (London, 1910), 67–69.

17. Gooday, "Precision Measurement," 33, 34; Cf. James Clerk Maxwell's energy-based analysis of scientific apparatus: "General considerations concerning scientific apparatus," *Handbook to the Special Loan Collection of Scientific Apparatus, [South Kensington Museum]* (London, 1876), 15–21.

18. R. T. Glazebrook and W. N. Shaw, *Textbook of Physics*, 1884. Table is from the 4th edition of 1918.

19. S. Smiles, *Self-Help* (London, 1859).

20. W. G. Adams, "Physical Laboratories," *Nature* 3 (1871), 323.

21. See S. Shapin, "The Mind is its Own Place: Science and Solitude in Seventeenth Century England," *Science in Context* 4i (1991), 191–218.

22. W. G. Adams, evidence to *Royal Commission on Technical Instruction*, 1884, 257 (Q.2653), reprinted I.U.P., Shannon, 1970.

23. See discussion elsewhere in this volume by Simon Schaffer of how, after the scandalous aberration of the 1866 B.A.A.S. determinations, Lord Rayleigh's 1881 Cavendish ohm was consensually adopted as the "real" B.A.A.S. standard.

24. A.B.W. Kennedy, "The Critical Side of Engineering Training," 743–44.

25. Some authorities cite comparable restrictions upon the meridianal orientation of the electrodynamometer, M. Maclean, ed., *Modern Electric Practice*, vol.1 (2d ed., London, 1909), 37.

26. "In designing an Experiment the agents and phenomena to be studied are marked off from all others and regarded as the Field of investigation. All agents

and phenomena not included within this field are called Disturbing Agents, and their effects Disturbances; and the experiment must be so arranged that the effects of these disturbing agents on the phenomena to be investigated shall be as small as possible," Maxwell, *Handbook*, 1. Discussed in P. Galison, *How Experiments End* (Chicago, 1987), 24–25.

27. See G. Gooday "Edifice, Artifice and Artefact: the social architectonics of physics laboratories in Victorian Britain," unpublished paper from conference "Writing the History of Physics," St. Johns College Cambridge, April 1991.

28. S. P. Thompson, *Life of Lord Kelvin* (London, 1910), 755, 1275–76.

29. R. W. Paul, "Electrical Measurements before 1886," *Journal of Scientific Instruments* 13 (1936), 51.

30. W. E. Ayrton and J. Perry, "A Portable Absolute Galvanometer for Strong Currents, and a Transmission Dynamometer," *Journal of the Society of Telegraph Engineers* 10 (1881), 157.

31. R. W. Paul "Some Early Electrical Instruments at the Faraday Centenary Exhibition," *Journal of Scientific Instruments* 8 (1931), 342–43.

32. Ibid.

33. F. C. Knowles, "The Growth of Electrical Measurement from its Commencement," *J.I.E.E.* 78 (1931), 36.

34. J. Perry, "William Edward Ayrton, 1847–1908," *Electrician* 62 (1908), 187–88.

35. For a broader view of the British context of systems installation, see T. Hughes, *Networks of Power: Electrification in Western Society, 1880–1930* (Baltimore, 1983).

36. A fuller account of the genesis of heavy-duty electrical engineering metrology will appear in Gooday, *The Morals of Measurement.*

37. Knowles, "The Growth of Electrical Measurement."

38. "Marcel Deprez's Galvanometer for Strong Currents," *Nature* 22 (1880), 246–47. Discussed in J. T. Stock and D. Vaughan, *The Development of Instruments to Measure Electric Currents*, Science Museum, London, 1983, 18.

39. Ayrton and Perry, "A Portable Absolute Galvanometer."

40. From later sources we know this to mean that the coil was mounted at an acute angle to the moving iron piece, rather than the conventional orthogonality shown in fig. 7; W. E. Ayrton, *Practical Electricity* (London, 1887), 71–72.

41. *Journal of the Society of Telegraph Engineers* 10 (1881), 169.

42. J. Perry, "The future development of electrical appliances," *Nature* 24 (1881), 20; *Nature* 25 (1882), 426–27; W. E. Ayrton, "The storage of energy," *Nature* 25 (1882), 498.

43. Their voltmeter had the same construction as the ammeter except that it was wound with high-resistance wire and was used in parallel with a circuit element.

44. Two compounded forms of the ammeter and voltmeter were also announced: a "photometer" and an "ergometer." W. E. Ayrton and J. Perry, "On a dynamometer coupling," *British Association Report* (pt. II, 1881), 553.

45. W. E. Ayrton and J. Perry, "Experiments on the Faure Accumulator," *Proceedings of the Physical Society of London* 5 (1884), 104–9.

46. W. E. Ayrton and J.Perry "Measuring instruments used in electric lighting and transmission of power," *Journal of the Society of Telegraph Engineers and Electricians* 11 (1882), 254–78.

47. "The Society of Telegraph Engineers and of Electricians," *Electrician* 9 (1882), 17.

48. W. E. Ayrton and J. Perry, "Measuring Instruments Used in Electric Lighting and Transmission of Power," *Journal of the Society of Telegraph Engineers and Electricians* 11 (1882), 254–55.

49. E. Obach, "Description of apparatus and diagrams," *Journal of the Society of Telegraph Engineers* 11 (1882), 279–85.

50. Discussion, *Journal of the Society of Telegraph Engineers and Electricians* 11 (1882), 290.

51. One obituarist remarked, "after spending a few years in electrical engineering, Dr. Obach devoted himself mainly to the study of the chemistry of gutta-percha and india-rubber," *Nature* 59 (1899), 254. However, see E. Obach, "Improved construction of the moving-coil galvanometer for determining current strength and electromotive force in absolute measure," *Proceedings of the Physical Society* 5 (1884), 289–303.

52. Lieut. P. Cardew "A New Method of Determining Large Electric currents and Very Low Resistances, *Journal of the Society of Telegraph Engineers* 11 (1882), 301–41.

53. W. E. Ayrton and J. Perry, "Direct-Reading Instruments, and a Non-Sparking Key," *Proceedings of the Physical Society* 6 (1884), 59.

54. Ibid., 61.

55. W. E. Ayrton and J. Perry, "A new form of spring for electric and other measuring instruments," *Proceedings of the Royal Society* 36 (1884), 297–319. Reproduced in *Nature* 30 (1884), 205–9.

56. W. E. Ayrton and J. Perry, "The efficiency of incandescent lamps with direct and alternating currents," *Proceedings of the Physical Society of London* 9 (1888), 211.

57. Cited by R. W. Paul, "Some Early Electrical Instruments," (31), 343. For details of Anthony's career, see Hughes, *Networks of Power* 9, 146–47.

58. Dunsheath, *A History of Electrical Engineering* 9, 308; Stock and Vaughan, *The Development of Instruments*, 27.

59. "William Edward Ayrton, F.R.S." *Electrician* 28 (1892), 347.

60. J. Perry, "Prof. William Edward Ayrton, F.R.S.," *Nature* 79 (1908), 75.

61. See Student Record Books for Finsbury and the Central in the City and Guilds of London Institute Archive, Guildhall Library, London.

62. W. E. Ayrton, "Electro-Technics," *Journal of the Institution of Electrical Engineers* 21 (1892), 5.

63. See Gooday, "Teaching Telegraphy," 104.

64. G. W. von Tunzellman, "The Central Institution No. III: The Physical Department," *Engineering* 46 (1888), 497–98, 523, 559–61. In 1893 the Central Institution was renamed the Central Technical College.

65. Ayrton, *Practical Electricity* 40, viii.

66. Ayrton knew all too well from his experiences in 1881–82 that fallible

experiments undermined the credibility of his whole enterprise of technical education, Gooday, "Teaching Telegraphy," 100–103.

67. Ayrton, *Practical Electricity* 40, 1–79.

68. Ibid., iii. N. B. Tunzellmann, "The Central Institution," 498, significantly uses an almost identical wording.

69. "A Student's experiment and its teaching," *Electrician* 34 (1894), 218.

70. See Heinz Otto Sibum, "Reworking the Mechanical Value of Heat. Instruments of precision and gestures of accuracy in early Victorian England" (forthcoming).

71. Smith and Wise, *Energy and Empire*, 302–16.

72. See Schaffer elsewhere in this volume.

73. E. H. Griffiths (with G. M. Clark) "The Value of the Mechanical Equivalent of Heat, deduced from some experiments performed with the view of establishing the Relations between the Electrical and Mechanical Units, together with an Investigation into the Capacity for Heat of Water at different Temperatures," *Philosophical Transactions of the Royal Society* 184 (1893), 361–504; Reported in *Nature* 47 (1893), 476–78.

74. R. Appleyard, "A New Determination of the Mechanical Equivalent of Heat," *Electrical Review* 33 (1893), 146–47.

75. E. H. Griffiths, "Appendix to a communication entitled 'The Mechanical Equivalent of Heat,'" *Proceedings of the Royal Society* 55 (1894), 23–26. The comment is cited from an editorial in the *Electrician*, "A Student's Experiment," 69, 218.

76. See Gooday, "Precision Measurement"; Smith and Wise, *Energy and Empire*, 8.

77. W. E. Ayrton and H. C. Haycraft, "Student's Simple Apparatus for Determining the Mechanical Equivalent of Heat," *Proceedings of the Physical Society* 8 (1894–95), 295–309.

78. See, for example, "The Physical Society," *Electrician* 34 (1894), 136–37.

79. *Electrician* 34 (1894), 220–21.

80. "The Physical Society," *Electrician* 34 (1894–95), 220–21.

81. Ibid.

82. Ibid.

83. Ibid.

84. "A Student's Experiment," 69, 216–18.

85. Ibid.

86. Ibid. As a rare exception however to physicists' disdain for the ammeter and voltmeter, note that Ayrton and Perry's friend and colleague, Professor Oliver Lodge had used their 1882-type models in the whirling machine experiments of 1891. See P. Rowlands, *Oliver Lodge and the Liverpool Physical Society*, Liverpool, 1990 for illustrations, front cover, and p. 67; B. Hunt, "Experimenting on the ether: Oliver Lodge and the great whirling machine," *Historical Studies in the Physical Sciences* 16 (1986), 111–34.

87. For an economic historian's perspective on this under-researched subject, see H. J. Habakkuk, *American and British Technology in the Nineteenth Century: the search for labour-saving inventions* (Cambridge, 1962).

88. G. C. Foster, "Direct-reading instruments and the teaching of physics," *Electrician* 34 (1894–95), 548–49.

89. C. V. Boys, "'Labour-saving apparatus' as an instrument of education," *Electrician* 34 (1894–95), 376.

90. G. H. Robertson, "Refined apparatus as an instrument of scientific education," *Electrician* 34 (1894–95), 257.

91. "J," "'Labour-saving apparatus' as an instrument of education," *Electrician* 34 (1894–95), 309–10. Editorial comment revealed that this pseudonymous correspondent had "good and sufficient reasons for not wishing to write over his own signature."

92. Boys, "Labour-saving apparatus," 89.

93. E. H. Griffiths, "'Labour-saving apparatus' as an instrument of education," *Electrician* 34 (1894–95), 283.

94. Foster, "Labour-saving apparatus."

95. J. Perry "'Labour-saving apparatus' as an an instrument of education," *Electrician* 34 (1894–95), 342–43.

96. J. A. Fleming, "'Labour-saving apparatus' as an instrument of education," *Electrician* 34 (1894–95), 407–8.

97. Ibid.

98. A. Cornu, "'Labour-saving apparatus' as an instrument of education," *Electrician* 34 (1894–95), 618–19.

99. Ibid.

100. "Science teaching vs. technical education," *Electrician* 34 (1894–95), 674–76.

101. A. Schuster and W. Gannon, "A Determination of the Specific Heat of Water in terms of the International Electric Units," *Philosophical Transactions of the Royal Society* 186 (1896), 415–67; abstracted in "Societies and Academies," *Nature* 51 (1894), 214.

102. "The Unit of Heat," *Nature* 52 (1895), 30 (correspondence from E. H. Griffiths and Oliver Lodge).

103. *B.A.A.S. Report* (pt. I), 1896, 150–61; *B.A.A.S. Report* (pt. I), 1897, 206–9. Note that in Griffiths's subsequent publications, "Recent investigations into the numerical value of the 'mechanical equivalent,'" *Nature* 56 (1897), 258–59, and "Note in recent investigations on the mechanical equivalent of heat," *Proceedings of the Royal Society*, the terms "mechanical equivalent of heat" and "capacity for heat of water in the C.G.S. system" are used *synonymously.*

104. W. E. Ayrton (3d ed., revised T. Mather), *Practical Electricity* (London, 1911), 275–76.

105. H. S. Allen and H. Moore, *Textbook of Practical Physics* (London, 1916), vii.

106. Ibid., 462–63.

ELEVEN

PRECISION IMPLEMENTED:

HENRY ROWLAND, THE CONCAVE DIFFRACTION GRATING, AND THE ANALYSIS OF LIGHT

George Sweetnam

IN 1882, Henry Augustus Rowland (1848–1901) established a dynasty in instrumentation. With the production of the first concave diffraction gratings, Rowland created a standard that would endure to the Second World War. After 1901, others would refine his methods and improve on his spectroscopic innovation. Yet the most successful pursued the path cleared by Rowland, and until the 1940s, many of the best diffraction gratings in the world continued to come from the Johns Hopkins University,[1] indeed from the very engines Rowland constructed, in the basement workshop where his own ashes were interred. It is remarkable that, over an ensuing period marked by great insights and innovations in physics, a standard of instrumentation should remain so influential, and that Rowland's technique for producing gratings should fail for so long to be superseded. Until 1949, at least, all successful ruling engine designs were derived from that used by Rowland.[2]

The present paper has three foci. I will first discuss the value of the Rowland grating to physical science. Although Rowland's own principal achievement with his invention concerned the sun, most of the major initial ramifications arising from grating use came from studies of the elements in a laboratory environment; some became important to the new quantum theory. Spectral lines, which Rowland gratings made more distinct, were fundamental manifestations of chemical elements, but did not arise independent of the agency of physicists and their instruments. Rowland's products became part of the common culture of spectroscopy, when that field was still a major specialty of the physics community in the U.S. Consideration of the demand for Rowland gratings points to a second major theme, that of supply, concerning the production of the instrument. The dominance in grating supply held by Johns Hopkins calls for an examination of technical, practical, and other obstacles to production that may have acted as constraints. Mechanical engineering, especially the control of fine motion, was essential to grating production, and led to developments significant to the physics of the new century. Production and distribution, which never turned a profit under Rowland, nevertheless had to be managed with some regard to revenue. The nonprofit nature of Rowland's instrument-making raises the question of his motivation, and so leads to the issue of his values, both as an individual and as a scientist, which form the third focal point of the paper. Rowland characteristically placed an emphasis on action, as opposed to pure contemplation, and on the moral value of precise observation. Both pursuits contributed to the creation of a better instrument.

Setting Standards

The initial reaction to Rowland's invention in England and on the Continent was captured in letters from Harvard professor John Trowbridge, who accompanied Rowland there in the 1880s, to Johns Hopkins president Daniel Coit Gilman. In Paris, E. Mascart, a photographic spectroscopist, "kept repeating with bated breath, '*Il faut qu'on commenserat* [*sic*] *encore*' or words to that effect."[3] "The Germans spread their palms, looked as if they wished they had ventral fins and tails to express their sentiments."[4] "The English men of science were actually dumbfounded."[5] One anecdote concerns professor James Dewar of Cambridge, a leading spectroscopist who heard Rowland at a meeting of the London Physical Society: "Professor Dewar arose and said, 'We have heard from Professor Rowland that he can do as much in an hour as has been done hitherto in three years. I struggle with a very mixed feeling of elation and depression. Elation for the wonderful gain to science and depression for myself, for I have been at work for three years in mapping the ultra violet.' "[6] Rowland gratings were especially effective in the ultraviolet.

The European reception of the gratings is also illustrated by the warm welcome that was extended to John A. Brashear, who had worked in the Pittsburgh iron mills for twenty years before opening an optical shop and subsequently supplying the blanks from which Rowland would make gratings. As compensation for making concave blanks and acting as distributor of the finished product, he retained one-half of gross receipts, to be applied against his expenses.[7] Because he sold the gratings, both concave and plane, that Rowland ruled, the gratings were sometimes taken to be Brashear's products. It was Brashear who dispersed the instruments of dispersion. "The spectroscope he sent to Dr. Hans Hausewalde [*sic*] of Magdeburg, Germany, with a focus of twenty-one feet, made Brashear famous all over Europe, for Hausewalde was outfitting his private observatory at great expense and his wants were exacting. He had searched Germany before sending his order to America."[8] Hauswaldt was a businessman and an amateur spectroscopist. (Gerhard Herzberg, who worked on atomic spectroscopy in Germany in the 1930s, later recalled that there was no accessible source of comparable gratings in Europe before World War Two—"They were all Rowland gratings then."[9])

Brashear, a man of humble origins, reaped the gratitude of the European scientific community, being received graciously by such notables as London instrument maker Adam Hilger, Sir James and Lady Dewar, Alexander Herschel, and astronomers in Dublin, Leipzig, and Potsdam.[10] Victor Schumann of Leipzig, an amateur who pioneered in photography of ultraviolet spectra, was among the most appreciative. Brashear reported to Rowland, "Previously he worked with a battery of 18 quartz prisms . . . when your grating was received he was the most enthusiastic German I ever heard from. Got up out of a sick bed to see the spectra & said 'his eyes were *glued* to it for two hours' "[*sic*].[11] The acclaim may not have surprised Rowland himself. In 1876 he had written to James Clerk Maxwell from Graz: "I am surprised to find the instruments of research used here often quite poor. The observatories I have visited are very far inferior to many in America, and yet the work done in them is superior in quantity and *perhaps* in quality to most of it in our country."[12]

The best appreciation of Rowland's product, however, came from an American, George Ellery Hale, in 1909: "During the last quarter of a century the study of spectroscopic phenomena in the laboratory has been completely transformed. It may well be said that this transformation, which has involved such discoveries as spectral series, the effect of pressure on wave-length, and the Zeeman effect, has been directly due to the use of Rowland's concave gratings."[13] In 1890, after graduating from MIT, Hale had upgraded his backyard observatory and spectroscopic laboratory into the Kenwood Physical Observatory, which possessed a large concave grating.[14] At Kenwood, Hale in many ways created a prototype of the Mount Wilson Observatory, which he would help to found, with its distinctive combination of astronomical and laboratory spectroscopy. This combination was characteristic of the newly emerging discipline of astrophysics, which sought to explain celestial light by comparison with light produced on earth—"It was because of this great interest in astrophysics that abundant spectroscopic materials were readily available for a Balmer or Rydberg, whose tools were pencil and paper rather than spectroscope and photographic plate."[15]

For those who used spectroscopes, Rowland instruments became an essential part of their material culture, but one which received, for the most part, only passing mention in their published works. Awareness of

Rowland devices was largely tacit, for it was on the highly reflecting surface of speculum metal that Rowland's reputation shone most brightly, in both the observatory and the laboratory. The reach of Rowland's physics depended as much on the distribution of his gratings as on the readership of his published papers, which described Rowland's investigations of the ohm, the mechanical equivalent of heat, and the magnetism of a rotating, electrified disk. In spectroscopy, Rowland reached his colleagues in a more material way: by January of 1901, 250–300 gratings had been sold. As historian Deborah Jean Warner observes, "There was probably no physics laboratory of note which did not have at least one Rowland grating. All the major studies of diffraction spectra around the world were done with them."[16]

In the two decades after the introduction of the gratings to Europe, theory evolved from speculation about harmonic ratios among spectral lines to J. J. Balmer's formula for one series of hydrogen, and on to the more generalized spectral series formulas of J. R. Rydberg and others. Balmer did not, in deriving his formula, consider any data that had been obtained with a Rowland grating. And although Rowland had produced a highly accurate chart of the lines in the solar spectrum by the time of Rydberg's work, Rydberg used as a measuring scale an earlier chart, created with a plane grating by Anders Ångström, which Rowland deemed to have been replaced by his own. But between 1884, the date of Balmer's publication, and 1890, when Rydberg published, the Rowland gratings made their mark, both in Europe and the United States. The chief rival to Rydberg's formula was put forward by Heinrich Kayser and Carl Runge, who worked with standard values that Rowland published.[17] During a period of collaboration at Hannover Technical University, from 1887 to 1894, Kayser and Runge obtained their own wavelength data; they "determined the spectra anew, using a Rowland concave grating, and found the results to be much more reliable from this method." During this period they found that "for many elements a regular structure [of spectral lines] could indeed be demonstrated."[18]

A number of observations that would become fundamental explananda of quantum theory arose at least in part from use of Rowland gratings. Advanced observations of fine structure in atomic spectra resulted from the collaboration of Kayser and Runge, as well as the work of Rydberg.[19] Later, the lines constituting fine structure were explained by the concept of electron spin. Pieter Zeeman indeed used a Rowland grating when he discovered the splitting of spectral lines in a magnetic field,[20] and Johannes Stark later used one to observe the analogous splitting of lines in an electric field.[21] Miguel Catalán, working in London, used a Rowland grating as well as a spectrograph with a glass prism and one with a quartz prism when he discovered the spectral structures which he dubbed "multiplets."[22] Hyperfine spectral structure (line-splitting on the order of 0.5 angstroms or less) was observed by means of concave gratings, as well as interference spectroscopes. This phenomenon was to be explained, eventually, with reference to properties of the nucleus.[23] In at least one instance, in 1915, the grating was used explicitly to test Niels Bohr's theory of the atom.[24] Spectroscopy in general placed stringent constraints on atomic theory. Under Bohr's theory, the Rydberg constant, *R*, essential to series calculations, could be derived either from studies of spectra or from the charge and mass of the electron, and Planck's constant. The former method, Bohr acknowledged, gave a figure accurate to six places, while the latter justified only three places.[25]

Rowland's invention was also used simply to catalog spectral lines. (A "Rowland mounting" for the grating, light source, and photographic plate is shown in fig. 1.) Theodore Lyman evaluated the grating for his 1900 doctoral thesis at Harvard, differentiating between valid spectral lines and those he held to result from flaws in the grating. Lyman also later used a Rowland grating to extend the known ultraviolet to 500 angstroms and record the spectral series that bears his name.[26] This work corroborated the relatively new formulas for the spectral series of hydrogen. The gratings, which mostly eliminated the need for lenses, were especially useful in studies of ultraviolet light, for which absorption by glass is strong. After the discovery of terrestrial helium in 1895 (it was detected in the solar spectrum in 1868), Friedrich Paschen, working with Runge, used a six-inch Rowland grating to study the helium spectrum in the laboratory.[27] "Overnight Paschen acquired an international reputation."[28]

While spectroscopy received impetus from the newly emerging discipline of astrophysics, the usefulness of Rowland gratings to that field was at first limited chiefly to solar studies, because they were in a sense too good. Their dispersion was so great that any stellar spectra produced were extremely faint, and difficult to photograph. In 1905 Hale and Walter Adams hoped to use a Rowland concave grating with a horizontal telescope at Mount Wilson to compare spectra of the sun with that of Arcturus. However, at the time they only had access to a Rowland plane grating, which they used to take a cumulative twenty-three-hour exposure of Arcturus over five nights, while the grating temperature was kept constant within less than a tenth of a degree Centigrade. They compared the result with sunspot spectra and also with laboratory spectra taken from elemental samples heated by electric arc or furnace, and so concluded that sunspots are relatively cool compared to the sun overall.[29] In 1908, when he detected the Zeeman effect, and thus a magnetic field, in sunspots, Hale also used a grating manufactured under Rowland.[30] But

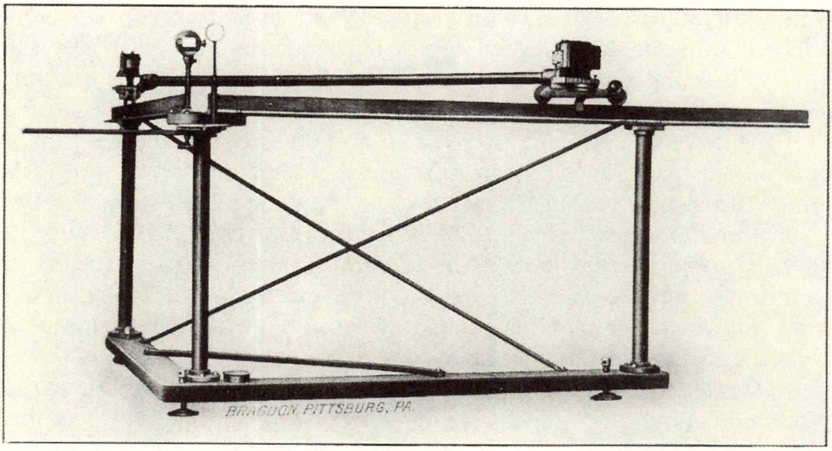

FIG. 1. A mounting for a concave grating. The crosspiece kept the grating and a photographic plateholder a fixed distance apart, keeping the spectrum in focus. Motion of the grating and holder on the perpendicular rails brought different portions of the spectrum onto the photographic plate. University of Pittsburgh Archives.

diffraction gratings (mostly plane) became truly useful for nonsolar astrophysics only when, in the 1930s, improvements brought about much brighter spectra.

As for Rowland himself, after setting a high standard in instrument production, he used the end product to elevate standards in solar observation. In the 1880s and 1890s, with the instrument, Rowland made detailed, high-dispersion photographic maps of the solar spectrum, superseding Ångström's smaller-scale engraved maps. At a meeting of the National Academy of Sciences in New Haven, Rowland "astounded and amazed his scientific audience with a photograph of the visible solar spectrum in ten sheets, making a total length of sixty feet."[31] From his maps, Rowland and assistants such as Lewis Jewell tabulated over the years some twenty thousand spectral lines. The tables were a landmark in spectroscopy, advertising his instruments. (Historian Klaus Hentschel states that Rowland's efforts unexpectedly led to the discovery of solar redshift, even though Rowland "did not accept the shift of spectral lines as a real effect."[32])

When, in the 1890s, Hale organized the *Astrophysical Journal*, he and Rowland agreed that Rowland's tables would appear, in installments of sixteen pages each, in the first issues. The tables gave wavelengths to the nearest thousandth of an angstrom unit. In a preface to the tables, Rowland stated his intent, eventually, to "publish a standard list of the lines of the solar spectrum with all the elements to which they belong."[33] The list could unify diverse areas of spectroscopic research and, if made truly standard, promote agreement. (This program of research was not wholly new—it dated back to the work of Gustav Kirchhoff of 1859—but it long remained a desideratum of astrophysics. By 1965, the number of observed solar lines had increased to twenty-four thousand, but only 73 percent had been identified.[34]) His map of the solar spectrum took the place of Ångström's map of 1868, which had marked spectroscopy's "birth as an exact physical science."[35] By "common consent," Rowland's "scale was universally adopted as the standard of reference."[36]

The origins of Rowland's standard-setting pursuits can be located in space and time to Baltimore in 1876, the year of inauguration of his university and department. Effects, such as instruments, spectrum maps, and wavelength tables, spread out like waves from there. After establishing the department, Rowland organized a "sub department of standards," where outsiders' instruments could be calibrated "in absolute measure by comparison with our own instruments."[37] Like the meter-bar in Paris, Rowland's instruments would, he envisioned, provide an absolute reference.

Maintenance of absolute standards would also act locally to uplift the physics students to be trained at Johns Hopkins. The research university was new to the U.S., and Rowland was an idealist who wished to elevate minds by means of a more precise understanding of nature, in a new discipline which needed standard values of professionalism as well as standard values of wavelength. Rowland, like his Johns Hopkins colleague Ira Remsen in chemistry, saw the importance of a professional ethos. Both men felt that Johns Hopkins had a mission to create challenging and uplifting graduate training.[38] This emphasis offers a partial explanation of Rowland's achievement in spectroscopy, for his diffraction grating was above all a matter of standards: a standard of instrumentation in itself and a means of establishing standards of analysis.

The Johns Hopkins laboratory, led by Rowland, became one standard U.S. source on precise determination of wavelengths of light from both the sun and chemical elements in the laboratory. Leaders in spectroscopy, such as George Ellery Hale, long deferred to the Hopkins laboratory on questions of exact wavelength determination. Eventually, in the 1920s and 1930s, the National Bureau of Standards, with which Johns Hopkins cooperated, assumed standard-setting functions with respect to light from chemical elements. Physics graduate students of Johns Hopkins in that later time frequently worked at the Bureau; the geographical proximity of the two laboratories smoothed the transition. In mapping the solar spectrum, Mount Wilson Observatory for a while assumed the lead. But both Mount Wilson and the Bureau had been "colonized" by one or more Hopkins-trained physicists, a sign that Rowland's school of physics acted across space and time. And Rowland's very spectrum map long remained a milestone. The Mount Wilson work was considered a revision of his map, and when an updated solar spectrum catalog appeared as late as 1965, it was called a "Second revision of Rowland's preliminary table."[39]

Spectroscopy of the sun and of the chemical elements already constituted two well-established empirical sciences toward the end of the nineteenth century, but with his invention Rowland brought into the field thousands of previously undifferentiated spectral lines. For every solar line cataloged by Ångström, Rowland found twenty. He created and disseminated an improved technology of knowledge that dispersed solar and other spectra far more strongly than previous instruments. With the grating he revealed that some spectral lines previously thought to be simple had, in fact, multiple components. Rowland set out to achieve more precise wavelength determinations of spectral lines, but refinement brought new discoveries. The extension of quantitative precision had brought into physical science new phenomena in the form of more spectral lines.

The Grating Business

In an age in which American manufactures of many kinds reached distant parts of the world, Rowland, through Brashear, exported one means by which others could acquire precision measurements similar to his own. His map of the solar spectrum served as a catalog of lines that might also be obtained in the laboratory. While Rowland for the most part disdained applications of science to practical, commercial ends,[40] he was more than willing to dedicate his own talent in engineering to the cause of experiment and measurement in physical science. In spectroscopy, Rowland did not seek monetary gain, but if, instead, account is made of renown and standing in the world scientific community, he appears as very much an American entrepreneur of the late nineteenth century. Perhaps the currency he sought was a level of scientific achievement high enough to be noticed in Europe. Rowland lamented that "we have taken the science of the old world, and applied it to all our uses, accepting it like the rain of heaven, without asking whence it came."[41] In this respect, he felt that America was a nation in intellectual poverty: "Shall our country do its share, or shall it still live in the almshouse of the world?"[42]

Rowland posed the question in 1883, one year after he had entered production with his concave grating. The diffraction grating was already a well-established tool before Rowland—in the 1870s the New York amateur Lewis M. Rutherfurd distributed fifty or more plane gratings around the world, making diffraction spectroscopy more widely accessible.[43] Rowland's improvements on Rutherfurd, however, were threefold:

his gratings were larger (intensifying the light and improving the resolving power) and more finely ruled (giving greater dispersion), and they focused the spectra they produced (eliminating the need for collimating and focusing lenses, which absorb light, although a condensing lens might still be used). Not for many decades afterward could any commercial firms profitably enter the field; to manufacture the gratings, Rowland required subsidies from his university, Johns Hopkins. But the benefits that accrued to Rowland and to Hopkins were renown, prestige, and a sense of usefulness, and seen in this way the products of Rowland's shop were indeed part of a much larger trend in which, by "the end of the nineteenth century, American products were flowing to many points on the globe."[44]

While the gratings made possible a new standard of precision in wavelength measurement, production of the gratings required a new standard of precision in the control of manufacturing, comparable to some of the most precise machine-shop work of the 1880s. In order to divide the solar spectrum more completely, Rowland created gratings ruled with as many as forty-three thousand lines to the inch, on a slightly concave speculum metal surface. To make these lines, Rowland's ruling engine had to move the plate a uniform distance, on the order of one micron, after each pass of the diamond ruling head along the reflecting surface. Any deviation from parallelism or uniformity in spacing could render a grating unusable, and the engine had to maintain this precision for five to six days of nonstop ruling to complete one grating four to six inches across, with a total of about ten miles of lines. A machinist tended the engine at night. To find wavelengths accurate to hundredths of an angstrom unit, Rowland had to inscribe lines spaced less than 0.0001 inch apart.

Rowland's point of departure, in production, was the work of William Rogers, an astronomer at Harvard, whose interest in fine machining and measurement arose from observational needs in astronomy. Rogers had planned to improve on the ruling engines of Rutherfurd and others, and he established a new and precise way to cut the screw which moved the ruling head the minute distance between one line and the next. When Rowland turned to the creation of his own engine, he had the benefit of Rogers's work. Rowland examined the latter's engines and acquired ruling implements from him. Very soon, Rowland gratings supplanted those of Rogers in the academic marketplace.[45]

Displaced from the manufacture of spectroscopic instruments, Rogers then pursued precise measures of length, and the development of precise measuring bars. Together with George Bond, a machinist from Hartford, Rogers worked near the highest levels of mechanical precision measurement (excluding interferometry). The two created a master measuring device, the Rogers-Bond comparator, which was finished in 1882, the same year Rowland began to make gratings. The comparator was accurate to one millionth of an inch (0.025 microns). It was used to develop marketable measuring machines that were accurate to one hundred-thousandth of an inch, which were exported to many parts of the world.[46]

Rowland and Rogers remained in contact, Rogers once supplying Rowland with a measuring bar at the price of $100. Like Rowland, Rogers carried out work in an underground room of constant temperature; the Rogers workroom was "in the underground passage of the Observatory and is surrounded by 1000 tons of granite." At Harvard, Rogers measured Rowland gratings as a check for accuracy. The process was, he said, "a real pleasure." He wrote to Baltimore: "It is surprising how nearly you keep your temperature constant." There was apparently no rancor over Rowland's ascendancy in grating manufacture.[47]

The work of Rogers and Bond was unique, and that of their nearest rivals rare. A later shop expert wrote: "The importance of precision workmanship is out of all proportion to its volume."[48] Those who worked at the extremes of precision worked apart from conventional commercial machine shops. Except in the case of gauges, work of the highest precision "is not, to any large extent, offered for sale. It is usually done at home to meet the problems in hand. So thoroughly apart from ordinary commercial workmanship is it, that one may live a life time in a mechanical atmosphere and scarcely see it."[49]

Rowland constructed the ruling engine in his own department, with the assistance of a machinist, Theodore Schneider, whom he hired away from a commercial firm. For the reflecting metal blanks, Rowland relied on an instrument maker in Pittsburgh whose salary was paid by a wealthy sponsor, William Thaw.[50] When this maker, Brashear, was once asked to build a micrometer for a Swedish physicist, he wrote to Rowland: "Personally I wish the money question could be eliminated. I never have and never will make any money at this work—and don't care so I can come out square. I would like to do work of the highest precision—and am using every effort to bring my work up to the highest class."[51] Thus Thaw's subsidy was essential.

The fineness of motion in ruling engines can be compared with fineness of motion in contemporary nonspectroscopic astronomical instruments. Until the full development of astronomical photography in the late nineteenth century, many astronomical measurements had to be taken through a telescope eyepiece, often with the aid of micrometers featuring cross hairs or engraved measuring scales. In the taking of optical measurements with these devices, as in the operation of Rowland's ruling engine, precision screws were essential for controlling fine motion. In the best telescope micrometers of the late nineteenth century, the level of accuracy was approximately 2–3 microns,[52] the same order of magnitude as spacing by the Baltimore ruling engines.

Rowland did not realize mechanical precision by himself; to build and operate the engines, he employed a staff of one, the machinist Schneider, formerly of W. & L. E. Gurley, a firm in Troy, New York, where he had performed work for Rowland, when the latter was teaching at Rensselaer Polytechnic Institute. Schneider worked at Johns Hopkins for twenty-five years—very nearly the same twenty-five years as Rowland. Described by Rowland as "the instrument maker of the university,"[53] Schneider made most of the working parts of the ruling engine, including the critical feed screw, and oversaw the production of all gratings that left Johns Hopkins.[54] Rowland originally planned that Schneider would also sell the instruments,[55] but in fact Brashear in Pittsburgh handled most of the marketing.

When the engine was finished, someone was required to tend it, and Rowland had to convince university president Gilman that the operation was worthwhile: "It must be remembered that the machine is *unique* and so gratings cannot be obtained except here. Besides, it will not look badly to see in every important paper on the spectrum, the name of this University."[56] Schneider worked on many projects for the physics department, and tended the ruling engine as an extra duty. For the latter he earned about $1000 a year, which was often more than the university received in revenue from sales of the gratings.[57]

Evidently Gilman began to doubt Schneider's value, because in 1888, writing to Gilman, Rowland came to Schneider's defense: "Mr. Schneider is an absolute necessity to the Physical laboratory. . . . The engine cannot be run without a person near it all the time to oil it & keep it in repair, to put the belts on & off the dynamos & to see that no accident happens. . . . If Schneider is to go, all spectrum work must cease." Rowland had to persuade Gilman that the grating manufacturing operation, running a deficit, was still justified. He wrote: "This is the only place in the world where they can be made well & every spectroscopist in the world is dependent on them. It would be a considerable blow to science if this was interfered with. There are people all over the world waiting for them." And Schneider was needed to support other research initiatives as well: "Every original experiment in physics requires new apparatus or changes in the old: also the apparatus constantly needs repairs."[58]

Rowland prevailed, and Schneider remained. But despite Rowland's hopes, grating production ran a deficit in nine of the ensuing ten years.[59] Rowland did not have the instincts of a businessman. In 1884 he suggested that a number of finished gratings should simply be given away.[60] In 1888, with Schneider's position in question, Rowland finally proposed to raise the prices of his gratings.[61] Schneider remained, even looking after Rowland's house, and continued to monitor the engines, but by 1897 Schneider's doctor warned him not to work in the ruling engine vault when the weather was hot.[62] By 1900 the demands of precision manufacturing extracted a toll. Schneider wrote to Rowland's summer address in Maine: "I have had a very bad cold, and am not feeling well at all. I have had a hard time finding points, as the weather has been so intensely hot, and it is almost impossible to work in the vault any length of time, and I attribute my sickness to this, and casting the plates."[63]

Rowland also depended on Brashear to perform the precision shaping of the concave blanks and to act as distributor. Brashear stated that the production of blanks was so demanding at first that it nearly put him out of business; in fact, early on, he had to convince Rowland that the blanks being produced were true to form.[64] As distributor, Brashear mediated between the demands of the scientific community and the often reserved Henry Rowland. Brashear sold plane gratings produced in Rowland's workshop as well as concave. Runge wrote to Brashear regarding a concave grating that would be on display at a Berlin photographic exhibition. He asked, "Would you be so kind as to give the necessary instructions in order that it may not be given to any body else."[65] After obtaining the grating, Runge arranged for payment of $300.[66]

Prospective buyers also contacted Rowland directly. In 1891, Hermann von Helmholtz expressed an interest in acquiring the kind of large grating, five inches across rather than two inches, that only the Hopkins shop could then provide, for experiments on radiation. Henri Becquerel inquired, too.[67]

Pricing varied according to the number of lines to the inch, proportional to resolving power, as well as overall size, proportional to brightness. Occasionally there were complaints. Rydberg wrote to Brashear to differ about a bill charging $250, when he thought the price for his grating should be $170.[68] Henri Deslandres, one of the foremost French spectroscopists, wrote directly to Rowland stating that his grating produced ghosts ("spectres supplementaires"). Yet after Rowland apparently offered to replace the grating, Deslandres, at the Observatory of Paris, declined, saying the grating was still quite useful.[69] (Some "ghosts," false lines caused by imperfections in the instrument, plagued production into the 1920s.) Arthur Gamgee of Lausanne returned a grating in 1895, complaining of tarnish, short focus, and high price.[70]

Complaints were the exception rather than the rule, however; requests for price quotes were common. A 1911 Brashear catalog shows that Rowland gratings came in a range of sizes, from 1.1 inches of ruled surface to 5.8 inches, and a range of grades. They were priced accordingly, the smallest and most coarse being offered at $15, and the largest and sharpest at $400. Rowland gratings with mountings ranged from $195 to $1,150.[71] All gratings were ruled at Hopkins. A price list appears in table 1.

In the marketplace, Rowland and Brashear had few competitors, per-

Price List of Gratings

Diameter of Surface	Character of Surface	Ruled Surface	Radius of Curvature	D	C	B	A
1.4-inch	Flat........	0.75x1.2 -in.	....	$ 15	$ 20	$ 25	$ 30
		0.9 x1.03-in.					
	Concave...	0.5 x1.25-in.	4-ft.	20	25	35	40
		0.75x1.12-in.	3-ft.	25	30	35	40
2.5-inch	Flat.........	1.25x1.9 -in.	....	30	40	50	60
		1.5 x1.75-in.					
	Concave....	1. x2.1 -in.	6-ft.	40	50	60	70
		1.4 x1.9 -in.	4-ft.	50	65	75	85
4 - inch	Flat.........	1.75x3.2 -in.	....	75	85	100	115
		2.25x3. -in.					
	Concave....	1.5 x3.5 -in.	10-ft.	90	110	130	150
		(2 x3.4)-in.	7-ft.	100	125	150	175
5 - inch	Flat.........	2.5 x3.9 -in.	......	125	150	175	200
		2.9 x3.6 -in.	......	135	165	190	215
	Concave....	1.75x4.3 -in.	15-ft.	150	175	200	225
		(2.5x3.9)-in.	10-ft.	175	200	225	250
6 - inch	Flat.........	3 x5.4 -in.	......	200	250	275	300
		3.5 x5.1 -in.	......	225	275	300	325
	Concave....	2 x5.8 -in.	21-ft.	250	300	325	350
		(3 x5.4)-in.	15-ft.	300	350	375	400

Explanation of Symbols

Gratings marked "B", "C" and "D" are all of the same general quality, as free from "Ghosts" as possible, but of varying degrees of excellence from the standpoint of brightness and definition.

Gratings marked "A" are distinguished as being particularly good in definition and in all other respects.

The width of the ruled space (length of ruled lines) is optional, except where a separate price is given in the case of a flat grating, and where the dimensions are given in parenthesis in the case of the concave grating. In this latter case a somewhat higher price may be charged on account of the greater difficulty of ruling perfectly a grating with long lines.

Gratings with length of lines or radius of curvature different from those specified in this list will be given a price proportional to these factors.

These gratings are ruled on Professor Rowland's engine, under the care of the department of physics of the Johns Hopkins University, Baltimore, Maryland. The plates are prepared by the John A. Brashear Co., Ltd., Pittsburgh, Pennsylvania, to whom all communications should be addressed.

TABLE 1. A page from the 1911 catalog of the John A. Brashear company. University of Pittsburgh Archives.

haps because the interests of the principal producer, Rowland, were noncommercial. Even in the mid-twentieth century, gratings were apparently not always traded entirely according to free-market principles. In the 1950s a British writer reported that to acquire a grating "one may have to wait years, and even then obtain it only because one is able to interest someone, who controls their disposal."[72] Thus the marketing of gratings brought a measure of power. Yet even where other people and institutions may have been ready to invest extensive resources for intangible returns, success was rare.

By 1915, A. A. Michelson, working at the University of Chicago, had manufactured gratings of a size and resolving power to rival or even better the productions of Johns Hopkins. However, Michelson's gratings did not eclipse those of the Rowland tradition. Certainly, Michelson achieved a comparable fame for his work in their time,[73] but he produced a smaller number of gratings than the Johns Hopkins operation, concerned as he was more with improving the performance limits of his ruling engine than with continual grating production. Of his two best gratings, 8 inches and 9.3 inches across, one was ruled on a blank that was too thin to hold its shape and so became deformed and useless. The other was accidentally broken soon after it was completed.[74] But users of sturdier Michelson gratings later included the demanding staff at Mount Wilson Observatory.

Michelson initially requested a grating from Rowland in 1883.[75] By the end of that decade, probably with ruling engines, among other things, in mind, Michelson informed Rowland that he was looking for "a private assistant—a man handy with his fingers and a good observer—who will do just what he is told—and is not too ambitious."[76] Apparently Michelson's interest in ruling engines grew over the years, even after he began to spend time at Pasadena in the 1920s: "Besides teaching, his main work for almost a decade had been to perfect ruling engines for the production of better diffraction gratings."[77] Michelson's daughter later explained how the challenge of grating production could take hold: "Ruling engines have had a curious way of dominating their makers. A man who was by nature gregarious and trusting of his fellows became secretive after spending much time in the company of a machine."[78] Michelson did not necessarily operate the machine himself, but had to contend with an attendant who was, at times, surly. In general, Michelson could not sustain production, for "during thirty years of trial and error . . . he was never completely satisfied with his results."[79] In 1930, Henry Gale took charge of the ruling operation at Chicago, but the engine fell into disuse again in 1940. Bausch & Lomb acquired it in 1947 and restored it,[80] and began ruling again about 1950. Michelson had considered interferometric control of his engine in the early 1900s, but did not enter full production with this method, either.[81]

In 1921, the University of Chicago lost Robert Millikan and his graduate student, Ira Bowen, to Caltech, where George Ellery Hale supplied them with a grating measuring 10 cm by 12 cm. (Bowen, once a student of Michelson, became director of Mount Wilson Observatory in 1946.) Hale had long deemed the gratings central to physical science as he practiced it, and in 1916 hired physicist John A. Anderson away from Hopkins, where he oversaw ruling operations after Rowland's day, to provide in-house production at Mount Wilson. Anderson attempted to construct an engine at Mount Wilson that would manufacture gratings twenty inches in width, but for him once again some six inches, the size of Rowland's largest, proved to be the practical limit.[82] Ruling engine operation hardly made Anderson a minor figure—he eventually went on to oversee the 200-inch Mount Palomar telescope project, after it was approved in 1928.

Johns Hopkins, the Mount Wilson Observatory, and the University of Chicago long remained the only sources in the world of large machine-ruled gratings.[83] At Hopkins the engines sat idle from 1916 until 1923, when Robert Wood assumed control of production. Like Rowland, Wood had an assistant monitor actual operation; like Rowland, too, Wood was a professor of physics at Hopkins and served as president of the American Physical Society. But it was Wood who made the process self-supporting for the first time.[84] In Europe, smaller gratings were produced at the National Physical Laboratory in England and at the Nobel Institute in Stockholm. A British writer has suggested that ruling engines remained scarce in part because they are "notoriously temperamental," and in part because of the reluctance of their operators to discuss them.[85] Others tried to produce large gratings, but to little avail. The efforts of one group in the 1890s were most notable. Brashear wrote: "Sir Archibald Campbell had Hilgers Brothers working for four or five years on a ruling engine at his place . . . and Hilger showed me some of his work. When I was there he had spent $10,000 on his grating machine, but had not ruled a single good plate on it."[86] Eventually, Otto Hilger did complete a ruling engine, the one taken over by the National Physical Laboratory. However, as late as 1951 British writers were still bemoaning the U.S. lead.[87]

Clearly, Rowland had little to gain from a proliferation of ruling engines beyond his own department; the kind of monopoly he envisioned in his letter to Gilman would best be served by guarding, not spreading, the technical know-how necessary to build and operate a ruling engine. Michelson surely had Rowland and Hopkins in mind in 1912 when he said that the difficulties of production "would doubtless have been diminished if my predecessors in the field had been more communicative."[88] Later, annual reports from Mount Wilson made public various improvements in production there, "but so briefly that they give no real guidance to others."[89] In his own lifetime, Rowland published neither a complete description of his engines nor a schematic diagram. When the gratings were shipped, packed in paraffin for safekeeping,[90] the details of their production did not travel with them.

Rowland made a gesture of openness when he made public the manufacturing steps necessary to produce the component which he stated was critical to the whole engine. Rowland here relied on the most basic sort of mechanical engineering, "the simplest instrument for converting a uniform motion of rotation into a uniform motion of translation." This component was the feed screw which advanced the ruling diamond the minuscule distance from one line to the next. Rowland initially cut the screw using a lathe procedure developed by Rogers.[91] This screw was then machined so finely that he could detect no error "so great as one one-hundred-thousandth part of an inch at any part."[92] The uniformity was achieved simply by running the screw repeatedly—for two weeks—through a split grinding nut containing oil and emery powder. Water kept the temperature constant. Rowland described the procedure in the article "Screw" for the ninth edition of the Encyclopedia Britannica.[93] His innovation, which provided underpinning to a most modern science, had a strikingly nineteenth-century character. At a time when electromagnetism and thermodynamics were advancing rapidly, Rowland's contribution (apart from the idea of concavity) belonged distinctively to the machine shop. It was engineering and workmanship that brought him scientific fame. A more complete sharing of details could have compromised his standing as a unique provider. Rowland did not wish to publicize the production process, but only the product.

"Eliminate All Drunkenness"

Much of the explanation for Rowland's success lies in technical obstacles surmounted, but part also lies in Rowland's temperament. In the days of "little science" in the U.S., what impelled this physicist? Why was Rowland so determined that he could overcome the tedious process of grinding perfect screws, inscribing countless miles of lines on small pieces of metal, and measuring and recording thousands of wavelengths to the second or third decimal place?

That Rowland "was by nature a mechanician of the highest type"[94] marks one important difference between Rowland and physicists of a later time. To Rowland there was little separation between knowing and doing; to graduate students he emphasized the importance of action over contemplation, and this reflected a personal bent:

> Rowland's interest in applied science cannot be passed over, for it was constantly showing itself, often, perhaps unbidden, an unconscious breaking forth of that strong engineering instinct which was born in him, to which he often referred in familiar discourse. . . . Although everywhere looked upon as one of the foremost exponents of pure science, his ability as an engineer received frequent recognition in his appointment as expert and counsel in some of the most important engineering operations in the latter part of the century.[95]

Johannes Rydberg described Rowland as an authority in "Practical Physics," especially on spectra.[96] Rowland applied his talent as an engineer to the service of pure science. The problem of refining and normalizing spectroscopic observations was a problem not of metaphysics but of optics and mechanical engineering. Light was a thing to be acted upon, not just passively received. Similarly, the scientists whom Rowland aimed to train in his department "must try experiment after experiment and work problem after problem until they become men of action and not of theory."[97]

This proclivity for action and application is most apparent in Rowland's work for the laboratory and the classroom rather than his public rhetoric, in which he always promoted "pure science."[98] For Rowland, nature was a repository of virtue, of right in opposition to wrong, as well as truth in opposition to falsehood. The only way to gain access to the truth inherent in nature was through interactive engagement, not passive reflection. The physicist knew nature through active contact, during which nature corrected the multitudinous errors of the human mind. (Mathematics offered another way to test the powers of reason.) To lose touch with nature, as it existed in the laboratory or workshop, was to lose touch with the reservoir of truth. A physicist had to know what would work, as well as what ought to work. An experiment was a practical attempt to come to terms with, and to grips with, nature. If there was disagreement, nature was necessarily right, although the question of which spectral lines were natural and which were artifacts of imperfect instruments could be perplexing.[99]

Even to arrive at disagreement, however, required active steps; in that regard, Rowland was not alone in the U.S. As Robert Kargon put it, "American physics seems to have cultivated a curious type of borderland physics-engineer hybrid, not unlike Rowland himself."[100] Evidence for this hybrid type includes the education of Michelson at Annapolis, where he favored optics and heat but had to learn seamanship and gunnery as well;[101] or the graduate thesis of Willard Gibbs at Yale, on the design of gear teeth; or Rogers's grasp of industrial mechanics as well as astronomy. In the United States, "the heritage of the past and the thrust of historical development did not neatly separate grand savant and practitioner; theoretician and earnest mechanic; the abstruse and the vernacular."[102] Ideas and machines developed alongside each other. (Rowland and his ruling engine appear in fig. 2.)

Rowland's ideal of action was in fact embodied in the instruments he produced. The gratings performed a number of operations at once: dispersing and focusing light, and normalizing the resultant spectra. Any observer or experimentalist who employed a grating automatically performed these operations, which are therefore important to the epistemology of physical science in the twentieth century. The grating design incorporated principles from both wave and geometrical optics: the ruled surface dispersed the spectrum, and the spherically concave shape of the surface brought that spectrum to focus. The design of the concave grating saved users time and money. Spectroscopists previously busied themselves with the complicated mathematics necessary to interpret nonnormalized spectra dispersed by prisms, and paid high prices for prism spectroscopes. Rowland, however, by shaping the reflecting surface, eliminated the need for lenses previously used to collimate and focus light respectively approaching and leaving flat gratings. He also found that, with a mounting of suitable proportions, the spectrum would be normal, that is, the space between spectral lines would be proportional to the difference in their wavelengths.[103] This was not true of prismatic spectra, from which linear measures of line position were converted to wavelength only through complex calculations.

W. D. Hackmann points out that instruments "can either be passive or active when exploring Nature." Often, apparatus devoted to the handling of light are placed in the passive category, which is where Hackmann puts the microscope and the telescope.[104] But with the active steps of dispersion and focusing inherent in their design, Rowland's gratings likely belonged to the second category. Active instruments, in Hackmann's view, "made it possible to isolate phenomena in a controlled laboratory environment."[105] Rowland's instruments isolated specific wavelengths of light. Hale's spectroheliograph built on Rowland's achievement by using single wavelengths of light to create a monochromatic image of the sun. Rowland used mathematics and engineering to produce a better instrument; others used mathematics to find unifying patterns in the spectra produced. To Rowland, mathematical operations alone did not constitute sufficient "action"; he sought to make phenomena orderly in the laboratory even more than on paper, and the laboratory order extended to six or seven significant figures, in measurements of wavelength.

The attainment of precision, first in grating manufacture and second in

FIG. 2. Rowland examining one of his ruling engines. The Johns Hopkins University Archives.

grating use, was the accomplishment of a man for whom precision in the study of nature possessed moral as well as technical value. Lorraine Daston has found, in the history of science more generally, a moral economy associated with precision measurement which "cultivated certain personal idiosyncrasies, namely those of skill and, especially, the character traits of diligence, fastidiousness, thoroughness, and caution."[106] Certainly these are consonant with the probity and bearing, the sobriety and uprightness, of Rowland. With his assistant, Rowland made equipment that would furnish better access to nature, and thus to moral uplift for toilers in physics laboratories throughout the world. The lathe-cut feed screw for the ruling engine was ground to practical perfection, through the narrow straits between two halves of a split nut, held together elastically. The lengthy perfecting was essential to Rowland's success, and evinced his diligence. Machining the screw to perfection meant, in principle, that spectral observations could be precise. Simultaneously, the grinding of the screw meant that certain human-sounding traits were removed from the screw and expelled from the laboratory workshop: "all errors of run, drunkenness, crookedness, and irregularity of size."[107] Only vigilant attention to continuously operating machines could remove these failings. The perfect screw held none of them. Nor, of course, would a perfect physicist.

In his terms for potential flaws, Rowland used the improvised vocabulary of the machine shop; for the most part his published experimental results did not use terms that were potentially as rich in connotations. Rather, his quantitative results appeared as testimony to the integrity of himself and the physical laboratory of his university, as opposed to other laboratories. Daston again: "The more precise the measurement, the more it stands as a solitary achievement of the measurer, rather than as the replicable common property of the group."[108] (In Rowland's case, he had assistants.)

The *actions*, including measurement, which Rowland emphasized as so important to the practice of science, were intended to improve human character and at the same time to bring human knowledge into closer agreement with nature. An emphasis on empirical knowledge provides an important insight into Rowland's own aims as the leader at Hopkins. For Rowland, knowledge of nature was not justified only in terms of amassing information to be exploited; his early disdain for practical applications of science is well known. Rather, the experience of nature served as a standard to guide the straying human mind. The unscientific outlook "is an irresponsible state of mind without clearness of conception," and "the unguided mind goes astray almost without fail."[109] Aspiring men of science must learn "that their own mind is most liable to error."[110] To remove error, Rowland asserted, "it is necessary that we have some standard of absolute truth: that we bring the mind into direct contact with it and let it be convinced of its errors again and again."[111] Where is this standard? The scientist "must enter the laboratory and stand face to face with nature."[112]

It was Rowland's brief to inculcate new habits of work and thought into his students. His approach to pedagogy was consonant with that of Ira Remsen, the first professor of the Hopkins chemistry department. Each man produced the greatest number of Ph.D.s in his field in the U.S. in the late nineteenth century,[113] but each also envisioned his role to be larger than simply the training of practitioners of science. Morality ruled: "Remsen's laboratory was as much a place for instilling the virtues of work and discipline as it was a nursery of new knowledge."[114] The situation in Rowland's department was much the same.[115] Accordingly, Rowland could be severely critical, but he was hardly a martinet. Asked early on what he planned to do with his graduate students, he replied, "Do with them?—*I shall neglect them*."[116] Yet the neglect was apparently benign: between 1879 and 1901, Rowland oversaw the completion of forty-five Ph.D.s.[117] "To be neglected by Rowland was often, indeed, more stimulating and inspiring than the closest personal supervision of men lacking his genius and magnetic fervor."[118]

There was at once a sense of pride and a sense of humble obligation in Rowland's view of laboratory physics. He believed that the scientific mind "is destined to govern the world in the future and to solve problems pertaining to politics and humanity as well as to inanimate nature."[119] Rowland's contempt for mediocrity made him uncomfortable with some aspects of democracy, for he felt that "the doctrine of the equal rights of man has been distorted to mean the equality of man in other respects."[120] Scientists, he proclaimed, were "a new variety of the human race."[121] Curiously, this same world-ruling human variety endlessly submitted its opinions to the test of the laboratory, where "the result is invariably humility." The laboratory taught that nature's laws "must be discovered by labor and toil and not by wild flights of the imagination and scintillations of so-called genius."[122]

The idea of the necessity of toil can be viewed partly as a work ethic, and partly as a way to reduce a debt Rowland felt had been incurred. In science, the New World owed the Old for all the learning that Americans used and profited from without returning much to the source. Part of the particular obligation of American scientists was to give something back to science.[123] In this connection, Rowland was in a special position to feel obliged, because of his training in Europe, in 1875–76. At that time he met Maxwell in person, and spent four months working in the laboratory of Helmholtz. He also, while in Europe, identified almost one hundred pieces of apparatus, costing more than $6000, as suitable for his new physics department, and received authorization to buy.[124] Even before his trip, Rowland would have felt a sense of community with European science, because it was Maxwell, rather than any American, who had first recognized the importance of Rowland's work on electromagnetism, and Rowland's reputation in Europe helped to secure the Johns Hopkins post.

With respect to balancing the ledger of achievement between America and Europe, the most symbolic moment came when William Thomson lectured at Johns Hopkins in 1884, and Rowland proudly presented him with a grating, as well as models of molecules made from wood and piano wire.[125] Rowland provided material embodiments that would articulate abstract ideas. The molecular models represented sources of light; the gratings handled actual light emissions, regardless of whether light was understood according to an elastic-solid ether theory or James Clerk Maxwell's electromagnetic theory.[126] The wood-and-wire models lent substance to an analogy current to the late nineteenth century, that molecules were comparable to mechanical resonators. Thomson's Baltimore lectures, which attracted many notables, brought Rowland and his department firmly into the world scientific community.

By 1900, American physical science no longer lived in "the almshouse of the world." Rowland, as well as Michelson, had attracted the interest of physicists abroad. He was a foreign member of scientific societies in London, Edinburgh, Rome, Berlin, Stockholm, and other European capitals. The U.S. by 1900 ranked with Germany and Britain in having the greatest numbers of academic physicists of various ranks;[127] Hopkins had led the way in the American expansion. Theoretical leadership still came from abroad. None of the major formulas or theories that guided the mathematical analysis of spectra in this period originated in the U.S. Yet the proliferation of new spectral phenomena that had to be ordered by theory depended in part on the ruling engines in a vault in the sub-basement of Rowland's department. In 1908 Hale recorded that Rowland gratings "have gone into observatories and laboratories in all parts of the world, where they have been the principal agents of spectroscopic research in the last quarter of a century."[128] The concave diffraction grating made manifest a multitude of new lines; the agency of Rowland in producing the lines was less manifest, but essential.

I am grateful to Johns Hopkins University for permission to quote from the Rowland Papers and the Gilman Papers, and to the University of Pittsburgh for permission to reproduce parts of a catalog of the John A. Brashear Co., Ltd., in the Archives of Industrial Society.

I would like to thank John Servos, for starting me on this study, and Gerald Geison, Charles Gillispie, and Norton Wise for encouragement and criticism. I am also grateful for comments from other members of the Princeton Program in History of Science, and for the remarks of Deborah Jean Warner and others at the May 1992 precision workshop.

Notes to Chapter Eleven

1. George W. Stroke, "Diffraction Gratings," *Handbuch der Physik* (New York: Springer-Verlag, 1967), 429.

2. George R. Harrison, "The Production of Diffraction Gratings," *Journal of the Optical Society of America* 39 (1949), 419.

3. John Trowbridge to Gilman, 12 November 1882, in *Science in Nineteenth-Century America: A Documentary History*, ed. Nathan Reingold (New York: Hill and Wang, 1964), 271–72.

4. Trowbridge to Gilman, 30 November 1882, ibid., 273.

5. Ibid.

6. Ibid.

7. Executive Committee agreement, 20 December 1884, in Correspondence—Rowland, Daniel Coit Gilman Papers, Ms. 1, Special Collections Department, Milton Eisenhower Library, Johns Hopkins University.

8. Harriet Gaul and Ruby Eiseman, *John Alfred Brashear: Scientist and Humanitarian, 1840–1920* (Philadelphia: University of Pennsylvania Press, 1940), 131.

9. Gerhard Herzberg, "Rowland Gratings, Molecular Hydrogen, and Space Astronomy," *Vistas in Astronomy* 29 (1986), 220. (Special issue of *Vistas* devoted to "Henry Rowland and Astronomical Spectroscopy.")

10. John A. Brashear, *John A. Brashear: The Autobiography of a Man Who Loved the Stars*, ed. W. Lucien Scaife (New York: The American Society of Mechanical Engineers, 1924), 99–100, 104, 111, 114.

11. John A. Brashear to Rowland, 8 February 1887, Henry A. Rowland Papers, Ms. 6, Special Collections Department, Milton Eisenhower Library, Johns Hopkins University.

12. Rowland to James Clerk Maxwell, March 1876, in *Science in Nineteenth-Century America*, 269.

13. George Ellery Hale, "Solar Vortices and Magnetic Fields" (1909), in Helen Wright, Joan N. Warnow, and Charles Weiner, eds., *The Legacy of George Ellery Hale* (Cambridge, Mass.: MIT Press, 1972), 164.

14. Hale, *The Study of Stellar Evolution* (Chicago: University of Chicago Press, 1908), 107–8.

15. William McGucken, *Nineteenth-Century Spectroscopy: Development of the Understanding of Spectra, 1802–1897* (Baltimore: The Johns Hopkins University Press, 1969), 133.

16. Deborah Jean Warner, "Rowland's Gratings: Contemporary Technology," *Vistas in Astronomy* 29 (1986), 129. See also Gerard L'E. Turner, *Nineteenth-Century Scientific Instruments* (Berkeley: University of California Press, 1983), 162–63. Brashear stated that he distributed "several thousand" ruled plates over the years, but his count probably included plane gratings. *John A. Brashear*, 76.

17. E.C.C. Baly, *Spectroscopy* (London: Longmans, Green, and Co., 1905), 311–12.

18. H. C. Freiesleben, "Kayser, Heinrich Johannes Gustav," *Dictionary of Scientific Biography*, ed. Charles C. Gillispie, vol. 7 (1973), 267–68.

19. H. E. White, *Introduction to Atomic Spectra* (New York: McGraw-Hill, 1934), 11. Michelson and Morley detected the simplest form of fine structure, the double components of one red hydrogen line, by interferometer prior to 1890. Albert A. Michelson and Edward W. Morley, "On a Method of Making the Wave-length of Sodium Light the Actual and Practical Standard of Length," *Philosophical Magazine* 24 (1887), 466.

20. P. Zeeman, "On the Influence of Magnetism on the Nature of the Light Emitted by a Substance," *Philosophical Magazine* 43 (1897), 227.

21. J. Stark and H. Kirschbaum, "4. Beobachtungen über den Effekt des elektrischen Feldes auf Spektrallinien. IV. Linienarten, Verbreiterung," *Annalen der Physik* 43 (1914), 1023.

22. Miguel A. Catalán, "Series and Other Regularities in the Spectrum of Manganese," *Philosophical Transactions of the Royal Society*, Series A, 223 (1922), 128–29. Catalán's discovery brought prestige to Blas Cabrera's research school in physics in Madrid upon Catalán's return there. José M. Sánchez-Ron and Antoni Roca-Rosell, "Spain's First School of Physics: Blas Cabrera's Laboratorio de Investigaciones Físicas," *Osiris* 8 (1993), 146–47.

23. Gerhard Herzberg, *Atomic Spectra and Atomic Structure* (New York: Dover Publications, 1944; first published in German 1936), 182.

24. E. J. Evans, "The Spectra of Helium and Hydrogen," *Philosophical Magazine* 29 (1915), 288–89.

25. Niels Bohr, "On the Spectrum of Hydrogen" (1913), in *The Theory of Spectra and Atomic Constitution* (Cambridge: Cambridge University Press, 1922), 13.

26. P. W. Bridgman, "Theodore Lyman, 1874–1954," *Biographical Memoirs, National Academy of Sciences* 30 (1957), 239, 243.

27. C. Runge and F. Paschen, "On the Spectrum of Cleveite Gas," *Astrophysical Journal* 3 (1896), p. 6 note. Cleveite is a mineral which yields helium.

28. Paul Forman, "Paschen, Louis Carl Friedrich," *Dictionary of Scientific Biography*, vol. 10 (1974), 346.

29. Donald E. Osterbrock, "Failure and Success: Two Early Experiments with Concave Gratings in Stellar Spectroscopy," *Journal for the History of Astronomy* 17 (1986), 125–26; Hale, *The Study of Stellar Evolution*, 168.

30. Hale to Joseph S. Ames, 16 November 1909, George Ellery Hale Papers, Microfilm Edition, California Institute of Technology, 1968.

31. *Baltimore American*, 18 November 1883.

32. Klaus Hentschel, "The Discovery of the Redshift of Solar Fraunhofer Lines by Rowland and Jewell in Baltimore around 1890," *Historical Studies in the Physical and Biological Sciences* 23, pt. 2 (1993), 277, 249.

33. Rowland, "Preliminary Table of Solar Spectrum Wavelengths. I," *Astrophysical Journal* 1 (1895), 29.

34. J. B. Hearnshaw, *The Analysis of Starlight: One Hundred and Fifty Years of Astronomical Spectroscopy* (Cambridge: Cambridge University Press, 1986), 422.

35. Baly, *Spectroscopy* (1905), 33.

36. Ibid. (London: Longmans, Green and Co., 1927), 25.

37. Rowland, "Guidelines for Laboratory Work" [undated], Box 39, Rowland Papers.

38. On Remsen, see Owen Hannaway, "The German Model of Chemical Education in America: Ira Remsen at Johns Hopkins (1876–1913)," *Ambix* 23 (1976), 152–53, 158.

39. Charlotte E. Moore, M.G.J. Minnaert, and J. Houtgast, *The Solar Spectrum 2935 Å to 8770 Å: Second Revision of Rowland's Preliminary Table of Solar Spectrum Wavelengths* (Washington: National Bureau of Standards, 1966). This second revision was prepared under the auspices of the International Astronomical Union. The first revision, in 1928, was published as part of the papers of Mount Wilson Observatory.

40. Two significant exceptions were consultation for a hydroelectric power project near Niagara Falls, and an effort to make patentable improvements in the telegraph.

41. Rowland, "A Plea for Pure Science" (1883), in *The Physical Papers of Henry Augustus Rowland* (Baltimore: The Johns Hopkins University Press, 1902), 594.

42. Ibid., 613.

43. Warner, "Lewis M. Rutherfurd: Pioneer Astronomical Photographer and Spectroscopist," *Technology and Culture* 12 (1971), 210.

44. Eugene S. Ferguson, "History and Historiography," in *Yankee Enterprise: The Rise of the American System of Manufactures*, ed. Otto Nathan and Robert C. Post (Washington, D.C.: Smithsonian Institution Press, 1984), 11.

45. Chris J. Evans and Deborah Jean Warner, "Precision Engineering and Experimental Physics: William A. Rogers, the First Academic Mechanician in the U.S.," in *The Michelson Era in American Science, 1870–1930*, ed. Stanley Goldberg and Roger H. Stuewer (New York: American Institute of Physics, 1988), 4–5; and Warner, "Rowland's Gratings," 127.

46. Paul Uselding, "Measuring Techniques and Manufacturing Practice," in *Yankee Enterprise*, 120–21.

47. William Rogers to Rowland, 27 March 1884, 8 February 1885, Rowland Papers.

48. Frederick A. Halsey, *Methods of Machine Shop Work* (New York: McGraw-Hill Book Co., 1914), 27. My thanks to Michael S. Mahoney for bringing this work to my attention.

49. Ibid.

50. Thaw paid Brashear $600 per year because he did not think Brashear could turn a profit. Gaul and Eiseman, *John Alfred Brashear*, 81.

51. Brashear to Rowland, 18 February 1889, Rowland Papers.

52. Randall C. Brooks, "The Development of Micrometers in the Seventeenth, Eighteenth and Nineteenth Centuries," *Journal for the History of Astronomy* 22 (1991), 168.

53. Rowland, "Preliminary Notice of the Results Accomplished in the Manufacture and Theory of Gratings for Optical Purposes" (1882), in *Physical Papers*, 487.

54. [J. S. Ames and committee], "A Description of the Dividing Engines Designed by Professor Rowland," in *Physical Papers*, 692.

55. Rowland to Gilman, 25 September 1882, Gilman Papers.

56. Rowland to Gilman, 1 May 1883, ibid.

57. "Receipts and Expenses of Gratings from November 1888–January 31, 1901," Box 15, Rowland Papers.

58. Rowland to Gilman, 16 June 1888, Gilman Papers.

59. "Receipts and Expenses," Rowland Papers.

60. Rowland to Gilman, 1 January 1884, Gilman Papers.

61. Rowland to Gilman, 16 June 1888, ibid.

62. Schneider to Rowland, 8 September 1893, 9 August 1897, Rowland Papers.

63. Schneider to Rowland, 18 July 1900, ibid.

64. *Brashear*, 75, 77.

65. Runge to Brashear, in Brashear-Rowland Correspondence, 10 July 1889, Rowland Papers.

66. Runge to Brashear, 9 October 1889, ibid.

67. Helmholtz to Rowland, 14 May 1891; Becquerel to Rowland, 25 March 1884, Rowland Papers.

68. J. R. Rydberg to Brashear, 5 August 1889, ibid.

69. Henri Deslandres to Rowland, 15 June 1889 and 4 February 1890, ibid.

70. Queens & Co., Inc. [agent] to Brashear, 28 August 1895, ibid.

71. *Catalogue—Optical, Physical, Astrophysical and Astronomical Instruments* (Pittsburgh: John A. Brashear Company, Ltd., 1911), 31, 40, at Archives of Industrial Society, Hillman Library, University of Pittsburgh.

72. C. Candler, *Modern Interferometers* (Glasgow: Hilger & Watts Ltd., 1951), 412.

73. Robert A. Millikan stated in a 1931 obituary that Michelson had been the best-known American physicist of his day in 1880. Former Rowland student Edwin H. Hall took exception to this, however, saying the distinction belonged to Rowland instead. Hall, "Michelson and Rowland" [letter], *Science* 73 (1931), 615.

74. Harrison, "The Production of Diffraction Gratings," 414.

75. Michelson to Rowland, 26 January 1883, Rowland Papers.

76. Michelson to Rowland, 5 September 1899, ibid.

77. Loyd S. Swenson, "Michelson, Albert Abraham," *Dictionary of Scientific Biography*, vol. 9 (1974), 373.

78. Dorothy Michelson Livingston, *The Master of Light: A Biography of Albert A. Michelson* (New York: Charles Scribner's Sons, 1973), 193.

79. Ibid., 94, 192.

80. Harrison, "The Production of Diffraction Gratings," 415.

81. Stroke, "Diffraction Gratings," 429–30.

82. Candler, *Modern Interferometers*, 415.

83. Harrison, "The Production of Diffraction Gratings," 415.

84. Ibid., 414.

85. Candler mentions a "security black-out" between 1923 and 1944 surrounding the engines at Mount Wilson. *Modern Interferometers*, 349–50.

86. Brashear to Rowland, 25 March 1901, Rowland Papers.

87. "Experiment is essential if the lead established by the United States is to be challenged." Candler, *Modern Interferometers*, 349.

88. A. A. Michelson, "Recent Progress in Spectroscopic Methods," *Nature* 88 (1912), 364.

89. Candler, *Modern Interferometers*, 350.

90. Brashear to Rowland, 27 June 1889, Rowland Papers.

91. Rowland, "Screw," in *Physical Papers*, 506–7.

92. Rowland, "Preliminary Notice of the Results Accomplished in the Manufacture and Theory of Gratings for Optical Purposes" (1882), in *Physical Papers*, 488.

93. Reprinted in *Physical Papers*, 506–11.

94. Thomas C. Mendenhall, "Henry Rowland: Commemorative Address," in *Physical Papers*, 8.

95. Ibid., 10.

96. Johannes Rydberg to Rowland, 11 December 1899, Rowland Papers.

97. Rowland, "The Physical Laboratory in Modern Education" (1886), in *Physical Papers*, 617.

98. The public stance is highlighted in Daniel J. Kevles, *The Physicists: The History of a Scientific Community in Modern America* (Cambridge, Mass.: Harvard University Press, 1987), in which Kevles ascribes to Rowland a "best-science program," which serves as a recurrent theme.

99. The ideas in this paragraph occur most strongly in Rowland, "The Physical Laboratory in Modern Education," 614–18. For Rowland's ideas about the place of mathematics, see "The Triumphs of Mathematics" [undated], 40, Box 42, Rowland Papers.

100. Robert H. Kargon, "Henry Rowland and the Physics Discipline in America," *Vistas in Astronomy* 29 (1986), 136.

101. Livingston, *Master of Light*, 35. Kathryn Olesko has written that reform of physics instruction at Annapolis, which began when Michelson was still a student there, was expected to enhance "the application of science to military ends," although Olesko also makes it clear that there was still a place for basic science. Kathryn M. Olesko, "Michelson and the Reform of Physics Instruction at the Naval Academy in the 1870s," in *The Michelson Era*, 111, 118.

102. Nathan Reingold, *Science, American Style* (New Brunswick: Rutgers University Press, 1991), 22.

103. Rowland, "Preliminary Notice of Results Accomplished in the Manufacture and Theory of Gratings for Optical Purposes," in *Physical Papers*, 488–89.

104. W. D. Hackmann, "Scientific Instruments: Models of Brass and Aids to Discovery," in *The Uses of Experiment: Studies in the Natural Sciences*, ed. David Gooding, Trevor Pinch, and Simon Schaffer (Cambridge: Cambridge University Press, 1989), 39–40.

105. Ibid., 40.

106. Lorraine Daston, "The Moral Economy of Science," in *Critical Problems and Research Frontiers* (Madison, Wisc., 1991), 433.

107. Rowland, "Screw," 507.

108. Daston, "The Moral Economy," 432.

109. Rowland, "The Physical Laboratory," 615.

110. Ibid., 617.

111. Ibid., 615.

112. Ibid., 618.

113. Hannaway, "The German Model," 155; Kargon, "Henry Rowland," 131.

114. Hannaway, "The German Model," 153.

115. Rowland and Hopkins president Daniel Coit Gilman "saw advanced education and research not merely as *training*, but as *mental and moral discipline*. It is impossible to overstress the *moral* component of their approach to science; it is but one of the firmly-held viewpoints that separates their age from ours." Kargon, "Henry Rowland," 132.

116. Mendenhall, "Commemorative Address," 16.

117. Kargon, "Henry Rowland," 131.

118. Mendenhall, "Commemorative Address," 16.

119. Rowland, "The Physical Laboratory," 618.

120. Rowland, "The Highest Aim of the Physicist" (1899), in *Physical Papers*, 668.

121. "[W]e form an aristocracy, not of wealth, not of pedigree, but of intellect and of ideals." Ibid.

122. Rowland, "The Physical Laboratory," 616.

123. On this point, see Rowland, "A Plea for Pure Science."

124. John D. Miller, "Henry Augustus Rowland and His Electromagnetic Researches," Ph.D. dissertation, Oregon State University, 1970, 185, 188.

125. Miller, dissertation, 297.

126. In his teaching Rowland "developed the electromagnetic and elastic-solid theories side by side, pointing out their similarities and differences and noting the difficulties inherent in each." Henry Fielding Reid, "Henry Augustus Rowland," *American Journal of Science* 11 (1901), 461.

127. Paul Forman, John L. Heilbron, and Spencer Weart, "Physics *circa* 1900," *Historical Studies in the Physical Sciences* 5 (1975), 12.

128. Hale, *The Study of Stellar Evolution*, 58.

TWELVE

THE LABORATORY OF THEORY

OR

WHAT'S EXACT ABOUT THE EXACT SCIENCES?

Andrew Warwick

And not only has an incredible amount of labour been saved,
but a vast number of calculations and researches have been
rendered practicable which otherwise would have
been beyond human reach.
(J.W.L. Glaisher, "Mathematical Tables," *1910)*[1]

1. Introduction: Between Theory and Experiment

WHEN we consider the role of precision in the history of the physical sciences, mathematical theory is not generally the first topic to spring to mind. The difficulty of assigning precise numerical values to physical quantities is more commonly associated with the practical problems of exact measurement than with the predictive power of mathematical theories. Traditionally, such theories *generate*, rather than *limit*, the exactitude of the so-called exact sciences. According to this view, once the fundamental laws of a physical theory are stated in bare mathematical form, testable predictions can, at least in principle, be calculated to any level of accuracy. The only limits placed upon such calculations are those imposed by the patience and competence of the human calculator. Mathematical theory, then, is definitive of the exactitude to which precision measurement can aspire.

In practice, however, the process of generating testable predictions from mathematical theories is much less straightforward than is generally supposed. Cartwright has pointed out, for example, that most testable predictions are arrived at via a series of approximations whose precise form is not dictated by the theory itself.[2] Indeed, such approximations are designed precisely to ensure that the theory *does* generate the correct predictions. Cartwright's point is, primarily, a philosophical one; she seeks to show that the chain of reasoning from fundamental to phenomenological laws is, as it were, inferentially irreversible.[3] By the time we have deduced a testable prediction from a fundamental theory, we have made so many approximations and auxiliary assumptions that it is unclear what is being supported if the prediction turns out to be correct. Cartwright concludes that any correspondence between theoretical predictions and empirically generated data might just as well be taken as evidence for the truth of the approximations involved as for the truth of the fundamental laws themselves. The exactitude of the exact sciences derives, it seems, more from the fact that the chains of reasoning (from physical laws to precise predictions) are expressed in mathematical form, than from the epistemic security of the laws themselves.

Cartwright's work is of considerable interest to historians of science because it draws attention to the role of professional skill and judgment in the work of the physicist. It reminds us that most mathematical physicists are not concerned primarily with the invention of grand new theories, but with the much more workaday problems of adapting extant theoretical tools to new applications. It is, moreover, the general invisibility of this important work that gives rise to the sense that fundamental theories make precise and unequivocal predictions across a wide range of physical phenomena. As Hacking has argued in a similar vein, the seemingly tight hypothetico-deductive structure of arguments (from theory to prediction) commonly found in physics textbooks, derives from the playing down of the epistemological significance of the approximations and other auxiliary assumptions being made along the way.[4] In practice, the range of skills and artifices required to develop useful mathematical models of physical phenomena are as important to the physicist as are the fundamental laws themselves.[5]

In this essay I shall discuss a form of approximation in theoretical work which, although rather different from that discussed by Cartwright, can usefully be approached via her work. Our faith in the exactitude of the exact sciences relies on the existence of a correspondence between either a theoretically generated numerical value and its experimentally measured counterpart, or a theoretically generated algebraic expression and the phenomenological law that is its counterpart. Cartwright claims that these correspondences cannot be said to provide direct evidence for the truth of the fundamental laws themselves because the process of prediction requires the adoption of approximations that are not dictated by the laws. What Cartwright does not explicitly discuss is: first, how theoretical predictions in the form of algebraic expressions (however obtained) can be made to generate actual numbers that can be compared with observational data; and, second, how so-called phenomenological laws (in the form of algebraic expressions) are actually generated from empirical data. I shall argue that these computational procedures form a crucial part of the interface between mathematical theory and experiment, and must be included in any discussion of the role of precision in the historical development of the exact sciences.

My argument is developed in two stages. In the section that follows I discuss the problem of practical computation and suggest, following Wittgenstein, that mathematical calculation has much more in common with precision measurement than is generally supposed. I argue that the activity of calculating is neither more certain in outcome, nor more independent of the physical world, than is the activity of measuring in the laboratory. Having developed the similarities between these activities I then show, in the balance of the paper, how a symmetrical understanding of calculating and measuring opens up the history of exactitude in the exact sciences to much broader analysis. The historical discussion is focused primarily on Britain in the second half of the nineteenth century, the period when both mathematical physics and precision measurement became professional activities. I shall suggest that precision measurement in the physical laboratory developed coextensively with new techniques of calculation in mathematical physics, and that it was the development of these techniques that kept the interface between theory and experiment manageable. I shall also argue that the new technologies of calculation employed in physics, especially the mathematical table and the mechanical calculator, were not developed primarily for scientific use, but for much wider commercial and bureaucratic purposes. What I offer is by no means a complete history of scientific calculation in the nineteenth century—that would be a mammoth project. I offer, rather, a speculative framework within which the exactitude of the exact sciences can be understood in broader historical context.

2. Calculation and Measurement

> Four applicants for a vacant clerkship came before [Sir William Curtis] in succession. He asked each of them, "What is the sum of 4 and 3?" Three of the applicants answered at once "Seven"; but the fourth (who had probably got a hint) said, "I will tell you in a minute, Sir!" and taking out his pocket-book, put down the numbers and their sum in figures, "You are the man for me," said Sir William, "I like no mental calculation."
>
> *(The Spectator, 1878)*[6]

At first glance, the problem of generating numbers from algebraic expressions appears somewhat prosaic. We can, surely, calculate such numbers to any desired level of accuracy simply by following basic arithmetic operations. Unlike our interest in precision measurement, which derives from the belief that such measurements can never be error-free, the study of calculation appears otiose because it ought always to be error-free. But, as Wittgenstein has argued, the problem of error arises in the practice of calculation just as it does in the practice of measurement (albeit in a different way). The following practical example is perhaps the easiest way of introducing Wittgenstein's line of argument.[7] Imagine that I have 25 empty boxes and I decide to place 25 objects, say beads, into each box. How many beads will I have used altogether when I have finished the task? The most obvious way of finding out—I am tempted to say the most fundamental—is to count them. Having counted 25 into each box, I then begin with the first box and count the beads consecutively until I have counted all of the beads in all of the boxes. Let us assume that I arrive at the answer 625. How do I know I have reached the right answer? Perhaps I have miscounted. Perhaps some of the beads vanished spontaneously before I had a chance to count them. In order to allay these fears I might recount the number of beads in each box and then recount the total. I might ask another person to repeat the exercise for me. If the same answer, 625, emerged each time (or at least most times) I might eventually decide that it was the correct answer after all.

There is, of course, another way of working out the total number of beads in the 25 boxes; I can calculate the product of 25 x 25. I might not be able to do this calculation by mental arithmetic, but I can use the computational aid of pencil and paper. I write down the numbers in the way I was taught at school and go through the steps of the calculation: 5 x 5 makes 25 (I have learned that off by heart); I write down the five and carry the two; 5 x 2 makes 10, plus the 2 I carried makes 12; and so on. Eventually I arrive at an answer, say 625. But how do I know I have arrived at the right answer? Perhaps the rules I followed are not reliable. Perhaps I have misremembered the multiplication table. Perhaps I made a mistake in my calculation. Philosophers of mathematics have long wrestled with this general problem, but we need only consider the solution offered by Wittgenstein.[8] Let us assume, in the spirit of common sense, that I get around the problem as follows. I first claim that counting the beads is the most reliable method of deciding how many there are. I therefore conclude that the correct number is 625. I then justify my calculating technique by noting that it too gives me the figure 625. Furthermore, provided that my two answers routinely agree in such trials I can claim that my calculating technique is generally justified because it fulfills my practical needs: it allows me to calculate how many beads I would have for any given distribution without actually undertaking the laborious task of counting them.

But suppose that in some of the trials the two answers did not agree (as in practice they probably would not). Then experience (actually my training at school) has taught me to account for this failure by concluding that I must have made a mistake. I must do the counting or the calculation, or both, again. This time I must try to count more carefully, or concentrate harder, or double-check my figures. Experience tells me that I will almost always be able to make the two numbers agree in the end (if not, I might explain the failure by concluding, for example, that I am too tired to work reliably). But when the numbers do eventually agree, as they generally will, what have I actually proved? Have I proved that beads (or other physical objects) are reliable things to count? Is counting, then, a practice that reflects the empirical properties of some physical objects (that they do not spontaneously vanish)? Have I proved that pencil and paper are reliable aids to calculation? Have I proved that my method of counting and/or my method of calculating is correct? Wittgenstein's conclusion is that I cannot decide which of these propositions I have "really" settled. All I can properly conclude is that the methods of counting and calculating that I am employing are good for my purposes.

We can now begin to see why Wittgenstein suggests that a calculation is like a measurement. The answer to a calculational problem is not given in advance. When I make a calculation I am doing an experiment to find out what answer I get. The product of 25 x 25, for example, is not already 625 before I make the calculation. It is true that, in this case, other people have reached this conclusion before me (and I might know that) but then I should say that I am doing an experiment to see if *I* can get that answer as well. If I routinely fail to agree with the majority of other calculators I should not conclude that I am in some profound sense wrong (although we find this explanation convenient) only that I cannot do what is generally accepted as "calculating" and that I would not make, for example, a good accountant. To return to the example above, if I routinely failed to get the answer 625 as the product of 25 x 25, I would simply conclude that calculating was not a useful practice for helping me work out how many beads I would need to put 25 in each of 25 boxes. It is because most people (though by no means all) can be taught a common practice of calculation that it is such a useful technique.[9]

We can very usefully push the analogy between calculation and measurement a little further. When a novel physical phenomenon is first generated and measured, it is often difficult for other experimenters to replicate the result. In order to overcome this problem, new standardized procedures and technologies are sometimes developed to make the new result easily reproducible.[10] The same is true for computational techniques. If it is important that a number of people agree on the outcome of a calculation, we can devise methods for making that agreement easier to reach. Suppose, for example, that we ask two people to calculate the product of 25 x 25 in their heads and to write down the answer. If they give different answers how do we know which, if either, is correct? We could adopt my commonsense approach above and set out 25 rows of 25 objects and count them (this assumes that we agree on how to count). However, this procedure is generally going to be impractical. More practically, we could develop a written method of calculating using pencil and paper. This enables us to reach agreement more easily by making the steps of the calculation visible and accessible to scrutiny.[11] Ironically, this practice will slow down the most gifted mental calculator but, crucially, it helps the majority of people to achieve a common level of computational competence.

We can now ask our two calculators to do their calculations on paper. If they still disagree we can now go through the calculations step by step (without recourse to counting objects) and decide who has given the correct answer. By developing a disciplined routine for calculating using the material technology of pencil and paper we have improved the reliability of the practice. Notice that this does not resolve our fundamental difficulties concerning the foundations of mathematics, it merely enables more people to do more complicated calculations more reliably. We might say that pencil and paper have now become part of the experimental apparatus. This is, of course, the import of the amusing epigraph at the beginning of this section. A clerk who calculates on paper is more valuable partly because his work is likely to be more reliable, but also because he provides a written record of his work that can be checked if a disagreement should arise. This example also points to further possibilities opened up by the introduction of disciplined techniques of calculation. As I noted above, such techniques enable far greater numbers of people to participate in the organized practice of calculating. This increase in the potential size of the calculating community is very important because it will more than compensate for the lost skill of the outstanding mental calculator. Furthermore, if the speed of the standardized calculator can also be increased by further development of calculating practice—through the introduction of mathematical tables or machines for example—then calculating can become increasingly reliable, widespread, and fast.

This brief excursion into the philosophy and sociology of mathematics has shown us that calculating is not merely a matter of following a series of logically determined steps. It is, rather, a human activity that is useful because most people can learn both its practice and to agree routinely about the outcome of calculations. What the above discussion also shows us is that the reliability and utility of complicated calculations, just like the reliability and utility of precision measurements, can be improved and extended by developing new methods and technologies of computation.

From these observations we can draw two conclusions. First, increasing the scope and reliability of techniques of calculation requires labor. If we wish, for whatever reason, reliably to undertake a large number of complex calculations, we will need to develop an appropriate technique. We could, for example, train human calculators in the most reliable methods of calculation and organize them into an efficient team. Alternatively, we might try to invent new computational techniques, either mathematical or mechanical, that would increase the power and speed of the individual calculator. Either way, increased speed and accuracy will involve increased work. This leads us to the second conclusion. These kinds of projects do not emerge in an historical or cultural vacuum. When the methods and uses of calculation change, we should look for an historical explanation of why that change is taking place and what resources are enabling it to occur. In the remainder of this essay I shall develop these points by looking at some examples of changing calculational practice. We shall see that, during the Victorian period, a range of resources were employed to enable rapid and reliable calculation to take place on a previously unknown scale. Without these resources, developed primarily for bureaucratic and commercial purposes, the exact sciences could not have achieved the definitive exactitude they now enjoy.

3. Mathematical Tables and Calculation in Victorian Britain

> Here is a comparison of two different times. In the [seventeenth century], the pupil was directed to perform a common subtraction with a voice-accompaniment of this kind: "7 from 4 I cannot, but add 10, 7 from 14 remains 7, set down 7 and carry 1; 8 and 1 which I carry is 9, 9 from 2 I cannot, &c." We have before us the announcement of the following table, undated, as open to inspection at the Crystal Palace, Sydenham, in two diagrams of 7ft. 2in. by 6ft. 6in. "The figure 9 involved into the 912th power, and the antecedent powers or involutions, containing upwards of 73,000 figures. Also the proofs of the above, containing upwards of 146,000 figures."
> *(Augustus De Morgan, 1861)*[12]

In his article on "Table" for *The English Cyclopaedia* of 1861, the mathematician Augustus De Morgan drew attention to an important shift in the practice of calculation that was taking place in mid-Victorian Britain. Where previously numbers had generally been computed as and when required, and by the user himself, they were now being mass produced for mass consumption. Admittedly, one or two specialist groups, astronomers and navigators for example, had long made use of mathematical tables to lighten the work of heavy calculation but, during the mid-nineteenth century, a veritable culture of calculation had emerged. By the 1860s, the favorite device for lessening the work of the computer, the mathematical table, had become an object whose dizzying rows of printed figures would fascinate the Victorian public. These tables displayed the limitless fecundity of numbers, and transformed them into a commodity that would bring the power of calculation within the reach of the ordinary citizen.[13] The centrality of tables of numbers and calculation to mid-Victorian commercial life was famously portrayed by Charles Dickens through the grim advocate of utilitarianism and political economy, Thomas Gradgrind. Gradgrind, who always had "a rule and a pair of scales, and the multiplication table" in his pocket, was a man of "facts and calculations" who based his life upon the principle that "two and two are four."[14]

De Morgan embellished his account of the remarkable table exhibited at the Crystal Palace by drawing attention to some of the astonishing feats of calculation that had recently been undertaken by European computers. These, he argued, were but a manifestation of a much deeper and more widespread obsession with numbers. In 1853, for example, William Shanks had calculated the value of π to 607 places of decimals.[15] Shanks was well aware that this immense labor would do little to enhance his reputation as a mathematician, but was content to establish himself as a "computer" and to demonstrate that "even in calculation" British mathematicians would "not allow themselves to be outstripped by their continental neighbours."[16] De Morgan reckoned that the mammoth efforts of men like Shanks did more than merely affirm "the capacity of this or that computer for labour and accuracy." He believed it to be indicative of a much deeper upsurge of interest in numbers in Victorian society. The example of π, he noted, was merely an indication of "a general increase in the power to calculate, and in the courage to face the labour."[17] As Shanks's work shows, that courage was generated, at least in part, by the belief that numerical precision was a matter of national pride.

Victorian interest in methods of calculation was not merely a strange fascination with numbers or the rules of arithmetic, it derived directly from the needs of commerce, state, and industry. From train journeys to life insurance, from navigation to international banking, an "avalanche of printed numbers" came to mediate between theory and practice, between supplier and consumer.[18] As Hacking has noted, however, the cultural transformation wrought by the publication of vast quantities of tabulated numerical data during the mid-nineteenth century was so all-embracing that it has remained strangely invisible to historians.[19] In much the same way that practical techniques of calculation are reckoned of secondary importance to the idealized logic of arithmetic, so the figures in a railway timetable are reckoned secondary to more substantial aspects of steam travel. But as A. H. Clough wrote in 1850:

> . . . the modern Hotspur
> Shrills not his trumpet of 'To Horse, To Horse!'
> But consults columns in a railway guide;
> A demigod of figures; our Achilles
> Of computation.[20]

The timetable was as important to the successful operation of a Victorian railway as was the technology of the steam locomotive. It is extremely important, then, that the tables of numbers used in Victorian Britain are not seen as mere referents to more substantial developments beyond the printed page. They should, rather, be seen as a form of currency that was as constitutive of those developments as were iron and steam. In this section I shall focus on the emergence of a division of labor between the mathematical physicist and the calculator during the nineteenth century. As we shall see, it was this division of labor, facilitated first and foremost by the mathematical table, that helped retain the link between increasingly complicated mathematical functions and increasingly precise physical measurements.

Prior to the nineteenth century, there were few mathematical tables published specifically for use in mathematical physics.[21] A mathematician who sought to generate numbers from an algebraic expression might have used standard tables of, say, logarithms, or persuaded a student to help with the work, but the production of such numbers was primarily his own responsibility. It was only in the 1780s that mathematical tables were identified as objects of potential national importance. The first state-funded project for the large-scale production of mathematical tables was begun by the French in 1784.[22] Inspired by Carnot, Prieur, and Brunet, a large team of computers was gathered together in order to calculate a new fourteen-figure table of logarithms as well as tables of the logarithms of sines, tangents, and other trigonometrical ratios, in relation to the centesimal division of the quadrant.[23]

In order to undertake the most extensive series of calculations ever executed, the calculators were divided by the director, Prony, into three groups. The first group consisted of "five or six of the most eminent mathematicians in France," including Legendre, their job being to devise the best method of calculating the tables and to carry out the calculation of "fundamental numbers."[24] These mathematicians developed the method of finite differences as the most efficient system for the production of the tables. The second group consisted of "seven or eight persons of considerable acquaintance with mathematics." Their job was to reduce the algebraic formulae produced by the first group into numbers that could be passed to the third group. This last group was made up of between seventy and eighty "ordinary computers" who undertook the bulk of the calculation. The tables took two years to complete, each figure being calculated twice (by independent computers) so that systematic errors could easily be detected. The completed manuscript of the tables filled some seventeen large folio volumes. Charles Babbage later claimed that Prony had devised and operated the finest example to date of the division of mental labor, each group of mathematicians possessing only that quantity of skill and knowledge required for the work that they were to undertake.[25] The preparation of the Cadastre Tables (as they were known) is interesting because it provides an excellent example of the way in which a government-sponsored bureaucracy employed hierarchical organization to solve a major computational problem. It is important to note, however, that the Cadastre tables were not produced to fulfill the needs of mathematicians or mathematical physicists, but to facilitate the complete implementation of the decimal system in France.[26] Indeed, the fact that they were never actually published suggests that the users of such tables did not consider their production vital.[27]

By the early years of the nineteenth century, the British government was also convinced of the national importance of accurate mathematical tables, especially for the purposes of navigation. In 1820, for example, a "distinguished member of the [London] Board of Longitude" was instructed to offer the French government £5000 towards the cost of a joint publication of the Cadastre Tables.[28] The project came to nothing.[29] Immediately following the failure, however, Charles Babbage and John Herschel were invited by the Royal Astronomical Society to form a committee to oversee the preparation of "certain tables" for the improvement of the Nautical Almanac.[30] Babbage and Herschel employed two (human) computers to prepare independent copies of the required tables. According to Babbage, it was during a meeting (in 1821 or 1822) to compare the two tables—at which many "discordances" were found—that he first expressed the wish to be able to "calculate by steam."[31] By 1823 he had built a small experimental model of the Difference Engine and opened negotiations with the British government in an effort to obtain funding for the construction of the real thing.

From the point of view of the present discussion, the most interesting aspect of Babbage's project is what it reveals about the importance and manufacture of mathematical tables in Britain during the 1820s and 1830s. As we have just seen, the first Difference Engine was not conceived of as an all-purpose computer, but as a one-off dedicated machine for the calculation and printing of mathematical tables. It was the tables, rather than the machine, that were to be the widely distributed aid to calculation. As late as 1832 Babbage described the Difference Engine as a direct substitute for the third group of Prony's computers, whose work, he believed, "may almost be termed mechanical."[32] Indeed, the Difference Engine was so called because it would calculate and print tables of numbers according to the method of finite differences developed by Prony's elite mathematicians.[33] Lardner's account of Babbage's "Calculating Engine," prepared under the direction of Babbage himself, also portrayed the practical importance of the machine solely in terms of mathematical tables. Those who, while admitting the "great ingenuity of the contrivance," he wrote, considered it "more in the light of a philosophical curiosity, than as an instrument for purposes practically useful" were quite wrong because they ignored the "extensive utility of those numerical tables which it is the purpose of the engine in question to produce."[34]

Lardner also provides us with an extremely useful overview of the range and relative utility of the mathematical tables then in use. He first mentioned multiplication tables that were of very widespread use in all forms of calculating and in the preparation of more specifically dedicated tables. Then came tables of trigonometrical ratios together with their squares, square roots, and other functions. These tables, often computed in very specific form, were widely employed by an enormous range of craftsmen and engineers. Next came tables of logarithms and the logarithms of any other tabulated functions that frequently needed to be multiplied (trigonometrical functions, for example). These tables, like the multiplication tables mentioned above, were of such widespread utility and use that their applications were too numerous to list. Lardner turned next to the wide range of tables whose importance, while in "no way inferior," were of a "more special nature." These were tables of interest, discount, and exchange, and tables of annuities widely used in banking, insurance, and commerce generally. Finally, Lardner noted that the discipline to which accurate mathematical tables were most indispensable was astronomy. He spared no efforts in spelling out the intimate link between astronomy and navigation, reminding the reader that accurate tables for the purposes of navigation were a necessity for the world's leading sea power.[35]

As is well known, Babbage's Difference Engine was abandoned during the early 1830s, but the production and publication of mathematical tables accelerated during the first half of the nineteenth century. De Morgan's article of 1861 reveals that an important characteristic of this period was the move toward providing cheap and handy tables for scientists, engineers, and those working in commerce. In 1861, for example, De Morgan himself, who was much concerned with improving the teaching of basic arithmetic in schools, published a table of three-figure logarithms. These tables were presented on one side of a card 7.5″ by 6″ and were intended partly as a useful means of introducing students to the use of logarithms in easing the labor of calculation. It was through the widespread use of such tables in industry, commerce, and the burgeoning primary and secondary education systems, that the culture of calculation spread so rapidly in mid-Victorian Britain.

De Morgan also devoted a section of his *Cyclopaedia* article to the tables that were required in "the higher mathematics." Most of these tables derived from developments in analysis during the late eighteenth and early nineteenth centuries. De Morgan drew attention to the whereabouts of tables of elliptic integrals, gamma and error functions, as well as tabulated values for some integrals that frequently needed to be evaluated for the purposes of mathematical physics. These tables were drawn from a strange assortment of sources and had generally been calculated only in the narrow range that suited the immediate purposes of the original calculator. The following example will illustrate the point. The most readily accessible tables of the values for the integral of $\exp(-x^2)$ were published in two forms drawn from two different sources. They had first been published (together with their logarithms) in Strassburg in 1799 by Kramp in a book on astronomical refraction, from which they had been reprinted in the *Encyclopaedia Metropolitana* in an article on "The Theory of Probability." This article also included a table of the same function in the form more commonly used in the theory of probabilities (that is, multiplied by $2/\sqrt{\pi}$) reprinted from the *Berlin Astronomisches Jahrbuch* for 1834. The latter table was also reprinted, with extensions, in the "Essay on Probabilities and Life Contingencies" in the *Cabinet Cyclopaedia* and in the article on "Probability" in the *Encyclopaedia Britannica.*

This example illustrates the several difficulties that would have confronted any mathematical physicist of this period who wished to compute a series of numerical values from an analytical solution to a problem in mathematical physics. First, it was often difficult to ascertain whether or not the required tables had already been computed by another mathematician. Furthermore, the fact that some functions were tabulated in connection with very specific applications made it easy to overlook a table even when it had been published. We have just seen, for example, that tables useful in the calculation of astronomical refraction might appear in an article on probabilities and life contingencies, and, of course, vice versa. Second, if the required tables had been calculated and could be located, it was quite possible that they would not cover the range of values required, would not be calculated to a sufficient number of decimal places, and would not be calculated for a small enough interval.[36] In each of these cases the user would have to interpolate or extrapolate from the values given in order to fulfill his own needs. Finally, if the tables had been copied and reprinted several times (as in the example above) it was highly probable that, in addition to the errors made in the preparation of the table, there would be numerous additional errors due to the typesetter. De Morgan remarked that in the production of mathematical tables, the printer's role was every bit as important as that of the most able computer. In a normal mathematical publication the logical unfolding of the argument would enable the attentive reader to spot and check a simple typographical error.[37] Once a mathematical table had been published, on the other hand, an error due to the misreading of a figure by the typesetter was indistinguishable from one due to the incompetence of the calculator—both errors being equally difficult to locate.

The problem of the unreliability of published tables was taken up by J.W.L. Glaisher during the early 1870s. He argued that it would be practically impossible to eliminate errors from mathematical tables while there existed no government body prepared to take responsibility for their publication and systematic correction.[38] Lists of errors found in the most accurate and widely used tables were published from time to time, but in a quite uncoordinated fashion. It was generally unclear, therefore, which errors, if any, had been corrected when a publisher produced a new edition of mathematical tables. Glaisher maintained that it was, in any case, quite unsatisfactory to leave responsibility for the accuracy of mathematical tables in the hands of commercial publishers. Publishers certainly found it expedient to claim that new editions of tables were published "with many errors corrected," but they found it equally expedient not to reveal which errors had been corrected so as not to render the old edition as valuable as the new one. Some commercial publishers even found it convenient to leave a few known errors in the tables so that they could easily detect and prosecute would-be pirates.

It was apparently during the late 1860s that the provision of mathematical tables began to become a serious problem both to mathematics and mathematical physicists. By 1871, the British Association (hereafter BA) had become sufficiently concerned to form a committee to report on the problem.[39] The Committee consisted of five members: Arthur Cayley, George Stokes, William Thomson, H.J.S. Smith, and J.W.L. Glaisher. Glaisher was the secretary of the Committee and it was he who, in practice, did most of the work. Although Glaisher had only recently graduated second wrangler in the Cambridge Mathematical Tripos, he was in fact ideally suited for the job of secretary of the Committee. Even as a child, Glaisher had been made acutely aware of the importance of mathematical tables through his father's work in astronomy and meteorology.[40] He regularly undertook computational work for his father who was himself deeply concerned with the production and accuracy of the mathematical tables used in astronomy.[41] Glaisher was not a typical student of the Mathematical Tripos. He had developed—presumably through his father—a range of interests in higher mathematics before he went to Cambridge, and instead of submitting himself completely to the grind of preparing for the annual examinations he published a number of original researches while still an undergraduate. He was particularly interested in the properties of the higher transcendental functions—especially elliptic integrals—and in their application in mathematical physics. He quickly realized that the utility of these functions would be severely restricted if it remained impossible to evaluate such equations numerically. In a postscript on elliptic functions at the end of his first report for the B.A., Glaisher wrote: "Apart from their interest and utility in a mathematical point of view, one of the most valuable uses of numerical tables is that they connect mathematics and physics, and enable the extension of the former to bear fruit practically in aiding the advance of the latter."[42]

In fact, by the time he was appointed to the B.A. Committee, Glaisher had already begun preparation of tables of exponentials, hyperbolic sines and cosines, elliptic functions, and Bessel functions.

The stated purposes of the Committee were twofold: first, to form as complete a catalogue as possible of existing mathematical tables; and, second, to reprint or, if necessary, to calculate, "tables which were necessary for the progress of the mathematical sciences."[43] The first report of the Committee, published in 1873, consisted principally in a massive bibliography (running to almost 150 pages) of all known mathematical tables that might be of interest to the mathematician or physicist. The tables were organized according to a scheme of classification drawn up by Cayley (see table 1). Cayley's classification gives a useful overview of the tables that the Committee considered relevant. The tables that were in most immediate need of tabulation for the purposes of mathematical physics were those of Section G, the transcendental functions.[44]

The first computational work undertaken by the Committee was the preparation of tables of Legendre functions and elliptic functions. The former were of major importance in potential theory, the latter in dynamics.[45] The actual work of computing and checking the tables invariably took much longer than initially anticipated by Glaisher. In the case of the Legendre functions, for example, he hoped that they would be completed and published by 1874; in practice they did not appear until 1879.[46] The preparation of the tables of elliptic functions was a mammoth task. These were, according to Glaisher, the most widely used transcendental functions in analysis after the elementary circular (sin, cos, etc.) and logarithmic functions, and such intense effort on the preparation of new tables had not been expended since the original calculation of tables of logarithms in the early seventeenth century.[47] Eight computers worked under the direction of Glaisher and his father for a period of three years before the basic calculations were completed.[48] The tables were printed and circulated privately, though, like the Cadastre Tables before them, never formally published.

The work on Legendre functions and elliptic functions constituted the

A. Auxiliary for non-logarithmic computations.
   1. Multiplication.
   2. Quarter-squares.
   3. Squares, cubes, and higher powers, and reciprocals.

B. Logarithmic and circular.
   4. Logarithms (Briggian) and antilogarithms (do.); addition and subtraction logarithms, &c.
   5. Circular functions (sines, cosines, &c.), natural, and lengths of circular arcs.
   6. Circular functions (sines, cosines, &c.), logarithmic.

C. Exponential.
   7. Hyperbolic logarithms.
   8. Do. antilogarithms (e^x) and h. l tan $(45° + \frac{1}{2}\phi)$, and hyperbolic sines, cosines, &c., natural and logarithmic.

D. Algebraic constants.
   9. Accurate integer or fractional values. Bernoulli's Nos., $\Delta^n 0^m$, &c. Binomial coefficients.
   10. Decimal values auxiliary to the calculation of series.

E. 11. Transcendental constants, e, π, γ, &c., and their powers and functions.

F. Arithmological.
   12. Divisors and prime numbers. Prime roots. The Canon arithmeticus &c.
   13. The Pellian equation.
   14. Partitions.
   15. Quadratic forms a^2+b^2, &c., and partition of numbers into squares, cubes, and biquadrates.
   16. Binary, ternary, &c. quadratic and higher forms.
   17. Complex theories.

G. Transcendental functions.
   18. Elliptic.
   19. Gamma.
   20. Sine-integral, cosine-integral, and exponential-integral.
   21. Bessel's and allied functions.
   22. Planetary coefficients for given $\frac{a}{a'}$.
   23. Logarithmic transcendental.
   24. Miscellaneous.

B 2

TABLE 1. Arthur Cayley's classification of mathematical tables for the British Association Committee of 1871. The preparation of tables under Section G was considered the most urgent for the advance of mathematical physics. J.W.L. Glaisher, "Report on Mathematical Tables," *British Association Report* 43 (1873), 1–175, 3.

bulk of the calculation undertaken by the first B.A. Committee which ceased to meet after 1883.[49] In 1888/89, however, a new B.A. Committee was convened specifically for the purpose of calculating and publishing functions useful in mathematical physics. Glaisher played a much-reduced role in the new Committee which was chaired by Lord Rayleigh. The Committee now considered the calculation and publication of Bessel functions to be the most pressing need. From his work in acoustics, Rayleigh was well aware of the need for tables of Bessel and allied functions. In his monumental book *The Theory of Sound* of 1877, he had shown that Bessel functions frequently arose as solutions to the equations of motion of vibrating systems. Since no table of Bessel functions had been published in Britain, Rayleigh had to reproduce appropriate sections from a table due to Lommel (1868) which, in turn, was due to Hansen (1843).[50]

The systematic computation of Bessel functions was supervised by the new secretary of the Committee, A. Lodge, and carried out by two computers. The work continued until after the turn of the century. Two interesting new developments in the Committee's calculation of the Bessel functions were the hiring of a "professional calculator" to help with the work, and the employment of a calculating machine to speed up their work and to improve its reliability.[51] Initially the Committee merely borrowed an Edmondson's calculator (from one Professor Mcleod) but, in 1893, they bought their own Edmondson machine.[52] The importance of the role quickly assumed by the machine may be estimated from the fact that its subsequent breakdown seriously hindered the computational work. When the machine had repeatedly to be returned to the maker, the Committee once again borrowed McLeod's machine on which "the greater part of the work [was] done."[53]

The use of calculating machines in the preparation of mathematical tables appears to have become the norm during the 1890s. The B.A. established a second Committee in 1896 (on which Glaisher also served) to oversee the preparation of tables of frequency distribution curves (and their logarithms) for use by statisticians and biologists. The moving force behind this Committee appears to have been Karl Pearson, who drew up the published reports. As with Glaisher's work on elliptic functions, the actual computation turned out to be "far more laborious than was initially anticipated."[54] Indeed, Pearson noted that had the true amount of labor required to produce accurate tables been anticipated in advance, it "would probably have sufficed to discourage any attempt to carry out the work." The work in question was carried out by five women using a Brunsviga calculator (see sect. 4).[55]

In 1910, when Glaisher wrote an article on "Mathematical Tables" for the *Encyclopaedia Britannica*, he noted that numerical tables were not merely a device for saving the labor of computation. They had also rendered practicable a "vast number of calculations and researches" that would otherwise have remained "beyond human reach."[56] The mathematicians of the nineteenth century had shown that an enormous range of problems in mathematical physics could be solved analytically in terms of transcendental functions, but these analytical solutions were of little practical value unless the terms in the solutions could be evaluated numerically. The systematic location, collection, and calculation of tables of these functions was begun by Glaisher and the B.A. in the 1870s. The fruits of their labor encouraged physicists to tackle a range of new and practical problems. By 1914 the B.A.'s tables of logarithms of Bessel functions were proving so popular with physicists that the Committee had to apply for a special grant to keep up with the demand.[57] It was the mass circulation of such tables of transcendental functions that enabled the power of nineteenth-century analysis to be brought to bear on a wide range of practical problems in physics in the early twentieth century.[58]

4. Brains of Steel

> The [arithmometer] asks for peculiar methods, and such as are not easily to be described. Herein there is room for skill and intelligence, so that hand and head may work together with mutual advantage.
> *(Hannyngton, 1871)*[59]

As will have become apparent in the discussion above, the term "computer" in Victorian Britain referred to any person engaged in the arduous task of calculating. It could apply equally to mathematicians and to clerks. At the beginning of the nineteenth century, men capable of undertaking unusual feats of mental arithmetic were both a novelty and a useful commodity. Their rare and remarkable skill could make them fairground freaks, practical calculators, or objects of scientific enquiry.[60] Zerah Colburn, for example, the illiterate son of an impoverished Vermont farmer, revealed a remarkable capacity for mental calculation around the age of six. Having achieved fame in the U.S. and Europe as a calculating prodigy, Colburn made astronomical calculations for the British Board of Longitude and was "examined" by several English mathematicians. Colburn's apparent ability to engage long-standing problems in arithmetic (at the age of eight or nine) by the "mere operation of his mind" led some mathematicians to conjecture that he had discovered, or innately possessed, some new arithmetical technique (perhaps similar to logarithms) unknown to mathematicians.[61] It gradually emerged, however, that the abilities of calculating prodigies, extraordinary though they were, could be explained without appeal to an unknown calculus. Once it was also recognized that ability in mental calculation did not signify broader intellectual ability, even in mathematics, "calculating boys" seem to have become a mere curiosity.

Indeed, with the increasing use of tables, the practice of mental calculation became not only uninteresting but positively frowned upon. A lone "lightning calculator" was of little practical use unless his work could be shown to be reliable. As we saw in the epigraph to section 2, Victorian clerks were expected to show their work so that it could be checked, but calculation on paper negated the very talent of the truly outstanding mental calculator. As the *Spectator* reported in 1878: "Whether it is from the cheapness and abundance of ready-reckoners, or the spread of education and the increase of ability to make use of logarithm tables, or contempt for the faculty, the lad who can multiply in the twinkling of an eye six figures by six is rare, and is in little request."[62]

The very utility of calculation in Victorian commerce meant that the rare gift of the outstanding mental calculator was discarded in favor of techniques that could be made more reliable and universal. In order to establish a still more robust technology of calculation, the late-Victorians turned to precision engineering.

We saw in section 3 that the problem of generating and printing accurate mathematical tables led Babbage to propose the construction of the Difference Engine in the early 1820s. The idea of delegating the computation and printing of such tables to a machine had obvious appeal but was fraught with problems. First, the construction of the Difference Engine was technologically demanding, making the project expensive and open-ended.[63] Second, the effective dedication of the machine to a single task—which could, in practice, be done by more conventional methods—made it very vulnerable to the withdrawal of funding by its only major sponsor, the British Government. Finally, the attempt to build a machine that would replace both Prony's computers and the printer's compositor in one go turned out to be overly ambitious. Even when more modest versions of the Difference Engine were successfully completed, they were barely financially viable and ran into further practical problems. Georg and Edvard Scheutz, for example, completed two production Difference Engines. One was sold to Dudley Observatory in the U.S., the other to the General Register Office in London. In practice the machines proved troublesome to use and could not approach the quality of printing achieved by conventional table manufacturers.[64]

The actual route by which mechanical devices were enrolled in computational work on a large scale was rather different. The first commercially successful calculating device, the arithmometer, was produced by the French insurance entrepreneur, Thomas de Colmar, in the early 1820s.[65] The arithmometer was manually operated and capable of the four basic arithmetic operations (+, −, x, ÷). The result of the computation made on the machine was not automatically printed, but read off by the operator. Unlike the large machines built for the production of tables, the arithmometer was relatively cheap to manufacture, portable, and applicable in any situation in which a large number of repetitive calculations needed to be undertaken. Furthermore, the modest commercial success of the machine during the mid-nineteenth century, enabled its manufacturers to make it increasingly reliable and adapted to the needs of the computer.

Although de Colmar's arithmometer was exhibited and awarded a medal at the Great Exhibition in 1851, the large-scale mechanization of computation in commerce appears to have begun only in the late 1860s.[66] The first public debate in Britain over the utility of the arithmometer to the computer took place in the pages of the *Assurance Magazine* (subsequently the *Journal of the Institute of Actuaries*) during the early 1870s. In 1865 the magazine published an article on the adaptation of some of the formulae used in the insurance industry for calculation on the arithmometer. The author of this short article, Major Hannyngton of Dublin, admitted that the formulae he had devised were not "the fittest for working by hand" but concluded that he was "persuaded that the days of handwork in the actuary's craft [were] coming to an end.[67] Hannyngton's article gave no account of the machine itself and must, therefore, have been directed at the then relatively small number of actuaries who already employed mechanical aids to calculation.

Four years later, in 1869, a second article appeared which gave a succinct account of the structure and operation of the arithmometer.[68] The fact that the article had been reprinted in translation from a German journal suggests that, while there was growing interest in the machine, it was still not easy to find authors in Britain who could give an authoritative account of its operation. Two years later, in 1871, Hannyngton gave a major address to the Institute of Actuaries on the design, operation, and application of the arithmometer in the work of the actuary. The talk was subsequently published in the *Journal of the Institute of Actuaries*, complete with a full technical description and illustration of the arithmometer reprinted from the *Engineer* magazine.[69] Hannyngton's paper effectively supplemented the manufacturer's instructions in two ways. First, it gave an alternative—and in Hannyngton's opinion superior—account of the methods for carrying out the basic arithmetical operations using the machine. Second, it explained in detail how the machine could best be applied to the work of the actuary. Hannyngton explained, for example, how the machine could be used after the fashion of a difference engine in the preparation of handy tables. He suggested that two computers, each using an arithmometer, could work together. The first computer would compute the differences and pass them to the second computer who would calculate the successive terms in the table by addition.

At the end of the paper, Hannyngton warned the audience that although in his experience the arithmometer invariably speeded up the work of the actuary, it was generally the case that new formulae needed to be invented in order that the labor of calculation be best suited to the computational routines demanded by the machine (see the epigraph at the beginning of this section). He also emphasized that these new methods could not at that time be explained or stated in any simple algorithm, but depended upon the skill of the computer in finding the most efficient division of labor between "hand and head."[70] Hannyngton's remarks remind us that changing the technology of calculation was not simply a process of mechanizing mental work; it constituted, rather, an effective reinvention of the practice of calculation.[71] As with the use of pencil and paper, the use of a calculating machine reorganizes the work of human beings according to a new discipline. In this case the work became the setting and reading of dials and the mechanical turning of a handle. The advantages of the new discipline were: first, that the calculator could work for longer periods without tiring; second, that, with practice, calculations could be done more rapidly; third, that teaching people to calculate reliably became easier; and fourth, that enrolling the machine in calculation meant that the practice could be progressively improved by the advances in machine technology. It must also be noted, however, that this new technology of calculation found a mass market only because of the increasing demand for repetitive calculation on a massive scale.[72] Furthermore, once the human calculators were accustomed to working with the machine, they were rendered effectively helpless without it.

These shortcomings were also appreciated at the time. Hannyngton's paper in the *Journal of the Institute of Actuaries* was followed by one which was much less sanguine about mechanical aids to calculation. This paper was based upon a lecture delivered by Edward Sang to the Actuarial Society of Edinburgh. Sang was an Edinburgh lawyer who had devoted a major part of his life to the calculation of mathematical tables. He was the first computer independently to recompute seven-figure logarithms for the numbers 20,000–200,000, and a keen advocate of the centesimal division of the quadrant. Sang argued that, with the exception of one or two special applications, the calculating machine was of little practical help to the computer. His chief objections to the use of the arithmometer were threefold. First, he believed that in copying the result of a calculation from the machine to paper, the computer was more likely to make an error than if he had made the calculation in his head. Second, he claimed that all of the shortcuts in computation that constituted the skill of the professional computer—including the use of tables—would be lost if a machine dictated the order of the work. Finally, he reckoned that if computers became reliant upon machines, they would be unable to cope when confronted by a calculation that could not be undertaken by machine. In conclusion he observed that " . . . on the whole, arithmeticians have not much to expect from the aid of calculating machines. A few tables, otherwise very easily made, comprise the whole extent of our expected benefits; and we must fall back upon the wholesome truth that we cannot delegate our intellectual functions, and say to a machine, to a formula, to a rule, or to a dogma, I am too lazy to think, do please think for me."[73]

Sang's remarks received little support from his readership. Even the generally even-handed editor of the journal appended a note to the paper expressing his own objections to the claims made by the author.[74] The editor noted that the main advantage of the arithmometer was not that it enabled individual calculations to be executed more quickly—though many were prepared to argue that it did—but that it reduced the work to a mechanical form that could be undertaken for long periods without mental fatigue (and the concomitant loss of speed and accuracy). The editor also invited a more thorough response from W.A.J. Hancock, an actuary and Secretary of the Patriotic Assurance Company, who was very experienced in the use of the arithmometer.[75] Hancock claimed that Sang's views on the shortcomings of the arithmometer derived largely from the fact that he had little experience in using one. Hancock then proceeded to answer Sang's objections in turn.

He dismissed Sang's first objection by noting that the work involved in any extended calculation gave the computer far more opportunity for error than did the entering and reading of figures on the arithmometer. To the second objection he responded that although some of the skill of the computer was lost when he worked with the arithmometer, various new forms of skill—of the kind discussed by Hannyngton—were developed. With practice, he added, these new skills both speeded up and improved the accuracy of computation. In response to Sang's final objection, Hancock noted that once the computer had learned to work with the machine he could, in principle, undertake any calculation using the same principles. "We do not expect the calculating machine to think for us," he concluded, "but to save the brain in doing mechanical work." The arithmometer converted the mental work of calculation into the mechanical work of setting pointers and turning a handle. Subsequent articles in the journal on specific applications of the arithmometer suggest that the machine became commonplace in the work of the actuary during the late 1870s and 1880s.[76]

The increasingly widespread use of arithmometers in actuarial work during the last quarter of the nineteenth century created a market which prompted several engineer-entrepreneurs to design and patent improved calculating machines.[77] The first arithmometer of English manufacture was built under licence by C. & E. Layton (who subsequently acquired the patents) and exhibited in London in 1883 (see fig. 1).[78] By 1914, a wide range of reliable machines was in use for statistical, actuarial, mathematical, astronomical, and laboratory research work (see fig. 2).[79] As an example of the development in design and manufacture of machines during this period I shall discuss the Brunsviga calculator mentioned in section 3 (see fig. 3). The basic mechanism for the Brunsviga was built in 1878 by the Swedish engineer W. T. Odhner.[80] Odhner took out a German patent for his machine in 1891 which was bought by the Braun-

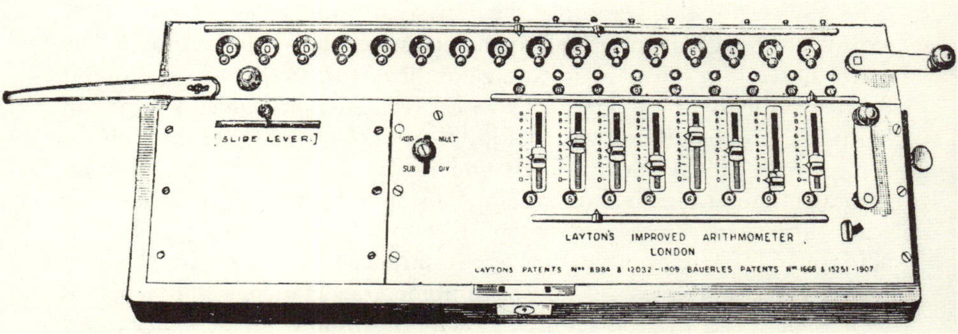

FIG. 1. Layton's Improved Arithmometer (ca. 1910). Based on the original de Colmar design, C. & E. Layton's arithmometer was the first of English manufacture. The design was continually updated from the early 1880s until well after World War I. E. M. Horsburgh, ed., *Modern Instruments and Methods of Calculation: A Handbook of the Napier Tercentenary Exhibition* (London, 1914), 103.

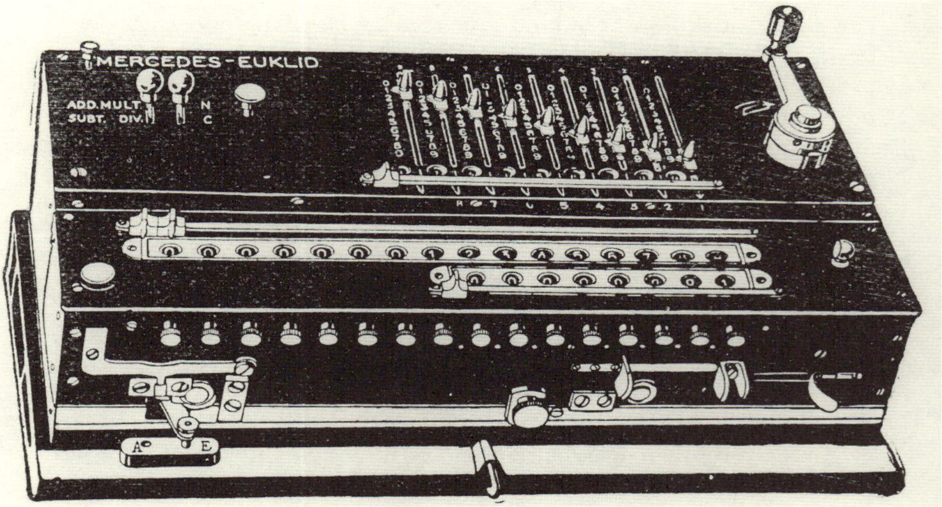

FIG. 2. The Mercedes-Euclid. Also based on the de Colmar design, the "electrically-driven" Mercedes-Euclid was the German equivalent of Layton's Improved Arithmometer. The machine was designed to minimize the number of operations made by the human operator (especially in division) so as to reduce fatigue and error. E. M. Horsburgh, ed., *Modern Instruments and Methods of Calculation: A Handbook of the Napier Tercentenary Exhibition* (London, 1914), ii.

Brunsviga Brains of Steel **Calculators**

The most Perfect, Serviceable, Reliable Calculator for every description of Scientific or Technical Calculations, including

Statistical, Actuarial, Mathematical, Astronomical, Laboratorial or Research Work.

The Smallest Machine with the Largest Capacity ever constructed; manufactured of the best Steel with mathematical accuracy.

Brunsviga Standard Calculators have more than 20 years' record for quality and for services rendered to the Scientific, Commercial, and Official World. Every Calculation registered, and an absolute check given to the Operator.

Trinks-Brunsviga Calculator

Modern Miniature Machines

Rapid Action, Silent Running, Weight, 9 lbs. only, Automatic Carriage.

Trinks-Triplex, Three Machines in One

with 20 figures tens carrying capacity.

$$\frac{336{\cdot}7275}{14{\cdot}25} \times 65{\cdot}37 = 1544{\cdot}6931$$

This Rule-of-Three Calculation worked out in a Single Operation: every Factor, Dividend, Divisor, Quotient, and total Result, being visible simultaneously.

Free Demonstrations given Anywhere upon Appointment.

GRIMME, NATALIS & CO., LTD.,

SOLE MANUFACTURERS,

LONDON, E.C., 46, Cannon St. Tel.: City 8120. **MANCHESTER, 38, Deansgate.** Tel.: City 6485.

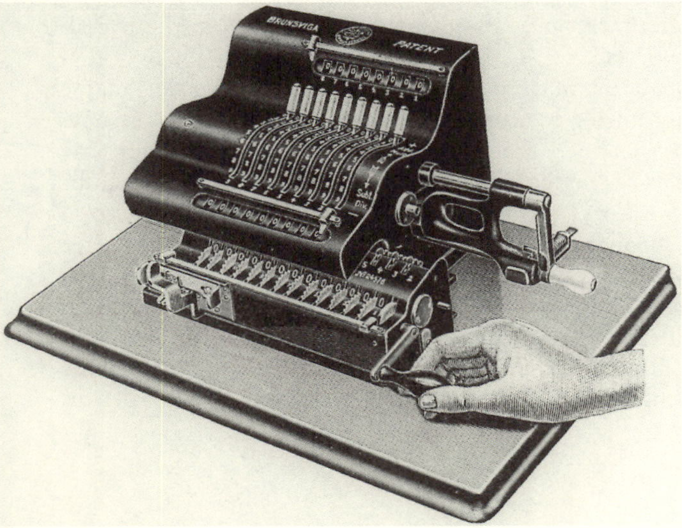

FIG. 3. The Brunsviga employed a different mechanism from the de Colmar–type machines and demanded a different working routine of the operator. The important features contested by the manufacturers were speed, accuracy, weight, and noise in operation. The disembodied hand suggests ease of operation and reveals the size of the machine. E. M. Horsburgh, ed., *Modern Instruments and Methods of Calculation: A Handbook of the Napier Tercentenary Exhibition* (London, 1914), v (advertisement), 87 (illustration).

schweig-based company, Grimme, Natalis & Co., in March 1892. During the following twenty years the technical and managing director of the company, F. Trinks, took out 130 German patents, 300 patents in other countries, and 220 German registered designs, in improvements to perfect the operation of the calculator.[81] The Brunsviga calculator was first introduced into Britain at the Oxford meeting of the B.A. in 1894. This machine was the 123rd manufactured and continued in reliable daily use until at least 1914. By 1912, Brunsviga had manufactured and sold twenty thousand machines.

Bearing in mind that Brunsviga was only one of many companies successfully selling calculating machines, it is clear that the mechanization of computation was well under way in the commercial world by the early years of the twentieth century. The extent to which mechanized computational procedures were adopted in the sciences is (at present) more difficult to estimate. As we have just seen, the arithmometer began to be used in the preparation of tables for actuarial work during the late 1870s and early 1880s and, as we saw in section 3, calculating machines were commonly employed in the preparation of tables for mathematical physics from the early 1890s. Arithmometers were also used in some astronomical observatories in the 1870s, although we currently have little sense of how widespread this practice was.[82] Two English calculating machines exhibited at the International Inventions Exhibition in South Kensington (London) in 1885 attracted so much attention from visitors that C. V. Boys was invited by the Society of Arts to write a major article on mechanical computation.[83] According to Boys, the arithmometer was already (1886) so well known and widely used in Britain that more than a brief description of the machine would have been superfluous.[84]

By the turn of the century, calculating machines were being used routinely in laboratories, and being advertised directly to scientists in popular scientific journals. The American "Comptometer," for example, was launched in Britain through an advertising campaign in the British scientific press in 1901 (see fig. 4).[85] The Comptometer was the first keyboard-operated calculating machine, and would quickly become a major competitor of arithmometer-type machines. Boys, who reviewed the new machine for *Nature*, claimed that, on balance, the Comptometer was "not as convenient as the arithmometer for reducing and computing observations in the laboratory" partly because it made an aggravating noise "like a typewriter through a megaphone."[86] In fact, the Comptometer was a much less sophisticated device than the arithmometer and the speed of its operation derived almost completely from the speed with which an experienced computer could operate the keyboard. By the outbreak of World War I, astronomers had also become the target of advertisements for calculating machines. The Royal Astronomical Society recorded in

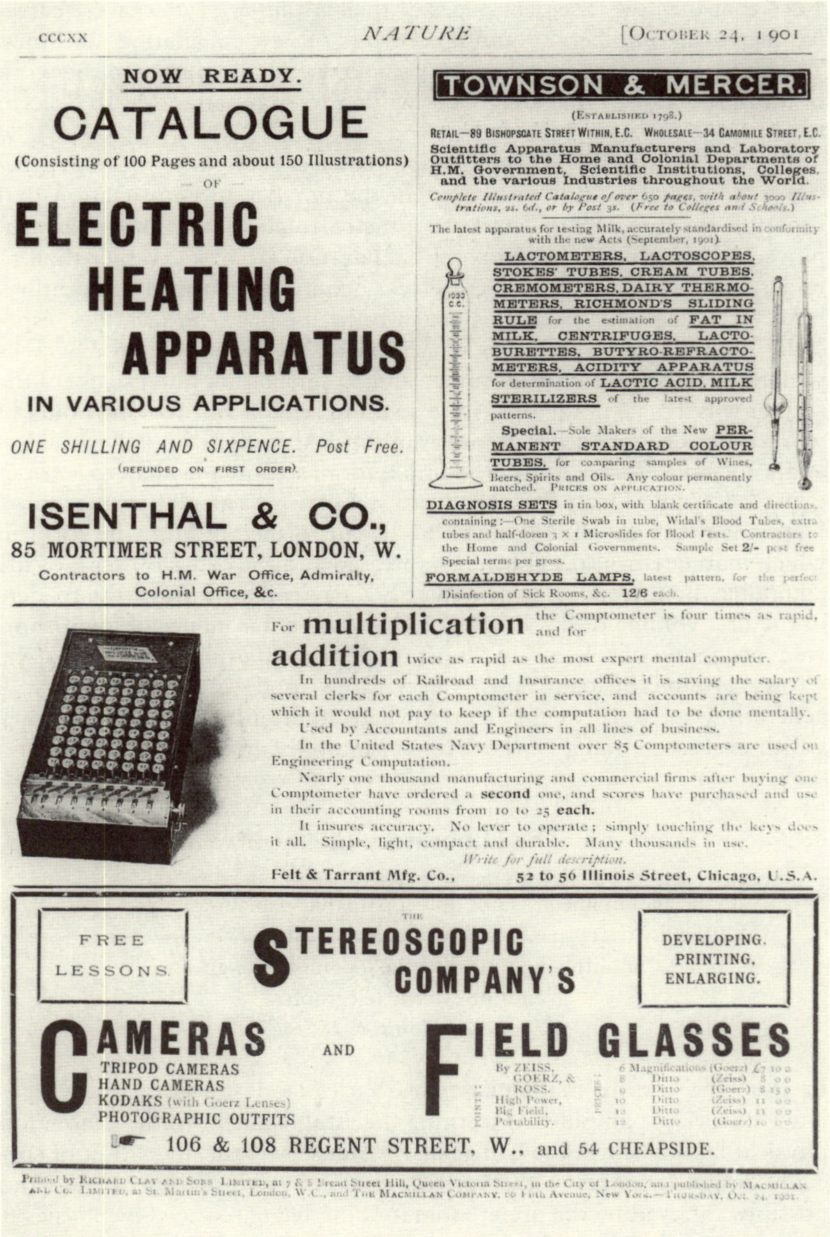

NOW READY.

CATALOGUE

(Consisting of 100 Pages and about 150 Illustrations)

— OF —

ELECTRIC HEATING APPARATUS

IN VARIOUS APPLICATIONS.

ONE SHILLING AND SIXPENCE. Post Free.

(REFUNDED ON FIRST ORDER).

ISENTHAL & CO.,

85 MORTIMER STREET, LONDON, W.

Contractors to H.M. War Office, Admiralty, Colonial Office, &c.

TOWNSON & MERCER.

(ESTABLISHED 1798.)

RETAIL—89 BISHOPSGATE STREET WITHIN, E.C. WHOLESALE—34 CAMOMILE STREET, E.C.

Scientific Apparatus Manufacturers and Laboratory Outfitters to the Home and Colonial Departments of H.M. Government, Scientific Institutions, Colleges, and the various Industries throughout the World.

Complete Illustrated Catalogue of over 650 *pages, with about* 3000 *Illustrations,* 2s. 6d., *or by Post* 3s. *(Free to Colleges and Schools.)*

The latest apparatus for testing Milk, accurately standardised in conformity with the new Acts (September, 1901).

LACTOMETERS, LACTOSCOPES, STOKES' TUBES, CREAM TUBES, CREMOMETERS, DAIRY THERMOMETERS, RICHMOND'S SLIDING RULE for the estimation of **FAT IN MILK, CENTRIFUGES, LACTO-BURETTES, BUTYRO-REFRACTOMETERS, ACIDITY APPARATUS** for determination of **LACTIC ACID, MILK STERILIZERS** of the latest approved patterns.

Special.—Sole Makers of the New **PERMANENT STANDARD COLOUR TUBES,** for comparing samples of Wines, Beers, Spirits and Oils. Any colour permanently matched. PRICES ON APPLICATION.

DIAGNOSIS SETS in tin box, with blank certificate and directions, containing:—One Sterile Swab in tube, Widal's Blood Tubes, extra tubes and half-dozen 3 × 1 Microslides for Blood Tests. Contractors to the Home and Colonial Governments. Sample Set **2/-** post free Special terms per gross.

FORMALDEHYDE LAMPS, latest pattern, for the perfect Disinfection of Sick Rooms, &c. **12/6** each.

For **multiplication** the Comptometer is four times as rapid, and for **addition** twice as rapid as the most expert mental computer.

In hundreds of Railroad and Insurance offices it is saving the salary of several clerks for each Comptometer in service, and accounts are being kept which it would not pay to keep if the computation had to be done mentally.

Used by Accountants and Engineers in all lines of business.

In the United States Navy Department over 85 Comptometers are used on Engineering Computation.

Nearly one thousand manufacturing and commercial firms after buying one Comptometer have ordered a **second** one, and scores have purchased and use in their accounting rooms from 10 to 25 **each.**

It insures accuracy. No lever to operate; simply touching the keys does it all. Simple, light, compact and durable. Many thousands in use.

Write for full description.

Felt & Tarrant Mfg. Co., **52 to 56 Illinois Street, Chicago, U.S.A.**

FREE LESSONS.

THE STEREOSCOPIC COMPANY'S

DEVELOPING, PRINTING, ENLARGING.

CAMERAS

TRIPOD CAMERAS
HAND CAMERAS
KODAKS (with Goerz Lenses)
PHOTOGRAPHIC OUTFITS

AND

FIELD GLASSES

By ZEISS, GOERZ, & ROSS.

POINTS: High Power, Big Field, Portability.

PRICES:			
6 Magnifications		(Goerz)	£7 10 0
8	Ditto	(Zeiss)	8 0 0
9	Ditto	(Goerz)	8 15 0
10	Ditto	(Zeiss)	11 0 0
12	Ditto	(Zeiss)	11 0 0
12	Ditto	(Goerz)	10 0 0

106 & 108 REGENT STREET, W., and 54 CHEAPSIDE.

Printed by RICHARD CLAY AND SONS LIMITED, at 7 & 8 Bread Street Hill, Queen Victoria Street, in the City of London, and published by MACMILLAN AND CO. LIMITED, at St. Martin's Street, London, W.C., and THE MACMILLAN COMPANY, 66 Fifth Avenue, New York.—THURSDAY, Oct. 24, 1901.

FIG. 4. The Comptometer. The advertisement shows the calculating machine beginning to take its place alongside other laboratory apparatus as a standard scientific instrument. *Nature*, 24 Oct. 1901, cccxx.

1914 that the advertisement pages of their *Monthly Notices* were carrying increasing numbers of announcements of new calculating machines. They added that these machines "form so useful an adjunct in the work of an observatory" that they were willing to draw attention to "recent modifications."[87]

Boys's 1901 review of the Comptometer is especially interesting because it highlights another change in the practice of computation that began in the late 1890s. Until this time calculating machines were employed by scientists largely for the preparation of tables, rather than to undertake the calculations themselves. Around the turn of the century, however, calculating machines began to be used *instead* of tables, especially for the basic operations of arithmetic. Boys considered machines preferable to tables of logarithms as aids to calculation because use of the former was less mentally tiring and so more reliable during long hours of work.[88] He also stressed that just as in the past computers had learned to adapt formulae for logarithmic computation, so now they had to learn to express formulae in forms that were suitable for calculating machines. Here again the calculating machine began to dictate both the form in which mathematical results were expressed and the order of work of the computer. By the end of World War I, Karl Pearson warned the readers of the first volume of his *Tracts for Computers* that tables of logarithms were becoming obsolete.[89] In "most modern computing laboratories" he wrote, "a table of logarithms is very rarely used—and when used it is generally one to 10 or 14 figures where multiplications are necessary which exceed the range of the ordinary multiplying machine."[90]

5. The Laboratory of Theory

> The arithmometer is not only a palpable evidence of a great difficulty overcome, it is an element of wealth, a new means of multiplying time, like the locomotive engine and the electric telegraph.
>
> *(Insurance Monitor, 1856)*[91]

In 1868 the Oxford mathematician Charles Dodgson wrote and circulated a satirical pamphlet concerning the establishment of the Clarendon physical laboratory in the University.[92] Amused by the allocation of specially designed spaces to the various needs of exact measurement—and possibly annoyed by the preferential funding being received by physics—he suggested, half mockingly, that the Trustees might also consider the establishment of a mathematical laboratory. This fanciful laboratory was to consist of a series of dedicated spaces for the practice of mathematics, including a "very large room for calculating Greatest Common Measure . . . a piece of open ground for keeping roots and practising their extraction . . . [and] a narrow strip of ground, railed off and carefully levelled, for investigating the properties of Asymptotes, and testing practically whether Parallel Lines meet or not." The naive reader was left in no doubt that the author's remarks were intended satirically by the observations that it was advisable to "keep square roots by themselves, as their corners were apt to damage others" and that the ground set aside to investigate the properties of parallel lines "should reach, to use the expressive language of Euclid, 'ever so far.' "[93]

Dodgeson's humorous parody of developments taking place in laboratory science during the late 1860s, underlines an important difference between the physics and mathematics of the period. Physics was an emergent discipline, which brought together analytical mechanics, experimental natural philosophy, and precision measurement. The discipline placed observation and exact measurement at its foundation and, thanks to its perceived virtues of technical exactitude and practical utility, came to occupy an important role in secondary and tertiary education in Victorian Britain.[94] Advanced mathematics, by contrast, remained an esoteric discipline. It too played an important role in the educational system of nineteenth-century Britain, but as part of the liberal education received by the professional Victorian gentleman rather than as part of the vocational training of the future scientist or industrialist. In Cambridge, for example, where the elite Mathematical Tripos stood at the very heart of the liberal education, students were examined and ranked strictly according to their ability rapidly to reproduce elegant solutions to difficult problems in higher analysis. On no occasion, however, were they expected to compute (or even to consider) the numerical value that might be derived from the tidy algebraic expression that marked the endpoint of an examiner's question. At Cambridge, the timeless truth and relentless logic of dynamics and analysis were not to be sullied by the mechanical toil and necessary approximation of numerical calculation.

Dodgson's fantastic mathematical laboratory expressed the flight of fancy required even to imagine such a place in an elite academic institution in the 1860s. The Cambridge-trained mathematician did not generally concern himself with what might be termed the experimental aspects of his discipline. If a differential equation or integral could not be evaluated analytically, then it remained a research problem for future mathematicians. There seemed little purpose in devising numerical methods and mechanical instruments to extract, by force of uninspired labor, a solution that ought properly to be obtained by analytic elegance and intellectual ingenuity. But for those concerned primarily with the application of mathematics to practical problems in accountancy, statistics, physics, and astronomy, methods of approximation were always required if equations were to yield useful results. Indeed, had Dodgson studied the plans for the proposed Clarendon laboratory more carefully, he might have noticed the "calculation room," where a range of devices was on hand for solving recalcitrant equations and generating numerical data. The Cavendish Laboratory in Cambridge also contained a calculating room equipped with a range of drawing instruments and planimeters.[95] In section 3 we saw how the application of Bessel functions, Legendre functions, and elliptic integrals to problems in physics generated a need for the provision of mathematical tables. Without reliable and accessible tables of these functions, physicists would have been severely hampered in their attempts to utilize higher analysis in the service of physics. Likewise it was the practical mathematical problems facing actuaries and statisticians in the latter nineteenth century that led them to take the lead in developing numerical methods.[96]

By 1913 the techniques used to apply mathematics to a range of practical problems had become so numerous and commonplace that E. T. Whittaker was able to found a mathematical laboratory in Edinburgh University. Whittaker had been second wrangler in the Cambridge Mathematical Tripos of 1895 but, like Glaisher, was by no means typical of its graduates. From the beginning of his undergraduate career in Cambridge, Whittaker was interested in the theory of functions, rather than their use in the solution of special problems in analytical dynamics. In his third year he abandoned his mathematics coach (R. R. Webb) and the cram for the Tripos examination in order to work with Cambridge's leading analyst A. R. Forsyth.[97] Whittaker subsequently became a mathematics lecturer at Trinity College. The courses that he gave to Trinity students later formed the substance of his classic textbook *A Course in Modern Analysis*, the first book in Britain to present the functions of a complex variable at undergraduate level.[98]

In 1906 Whittaker was appointed professor of mathematics at Trinity College Dublin, with the title of Astronomer Royal of Ireland. It was in connection with his work in astronomy that he first became interested in numerical mathematics. When he moved to Edinburgh in 1912 as professor of mathematics he immediately made plans to open a special laboratory devoted to the investigation of mathematical problems that could not be solved analytically. The announcement of the opening of the laboratory generated enormous interest in the British mathematics community, and an initial colloquium organized by Whittaker in August 1913 was attended by more than eighty mathematicians. The topics to be studied in the laboratory—both at student and research level—included interpolation, the method of least squares, practical Fourier analysis, the numerical evaluation of definite integrals, the numerical solution of differential equations and the construction of tables of functions not previously tabulated. This was the first time that these topics had been systematically investigated and taught at any university in the world.

Whittaker's mathematical laboratory had much more in common with the physical laboratory than Dodgson's spoof might have led one to expect. The rooms contained a wide variety of instruments for calculation. Various forms of mechanical integrator were available for solving integrals for which no analytical solution was available. There was also a wide range of differentiating machines and harmonic analyzers for producing analogue solutions to differential equations and resolving periodic functions into harmonic components, respectively. Most central to the work of the mathematical laboratory, however, was the development of numerical methods. Experience showed that problems that were accessible to solution by both analogue and numerical methods were generally best tackled by the latter. Whittaker developed a collection of new techniques for solving problems numerically and worked out the most efficient routines for undertaking the numerical work. Each computer worked at a specially designed desk which was described in detail in one of the early publications of the laboratory.

> Those [computer's desks] used in the mathematical laboratory of the University of Edinburgh are 3′ 0″ wide, 1′ 9″ from front to back, and 2′ 6″ high. They contain a locker, in which computing paper can be kept without being folded, and a cupboard for books, and are fitted with a strong adjustable book-rest. Thus the computer can command a large space and utilize it for books, papers, drawing-board, arithmometer, or instruments. Each desk is supplied with a copy of Barlow's tables . . . a copy of Crelle's multiplication table . . . and with tables giving the values of the trigonometric functions and logarithms. These may, of course, be supplemented by a slide rule, or any of the various calculating machines now in use, and such books of tables as bear particularly on the subject in question.[99]

In addition to the special desks, mathematical tables, calculating machines, and instruments, the procedure to be followed during each calculation was governed by the use of a special printed form. If, for example, a collection of data points that exhibited periodicity was to be represented by an analytical function consisting of sinusoidal components (i.e., Fourier analyzed), the period was divided into twenty-four standard intervals and then analyzed using a standard form containing boxes for each stage of the calculation. The form ensured that the computer recorded each of the steps of his calculation and undertook a self-check (also recorded) to ensure that he had not made any major computational

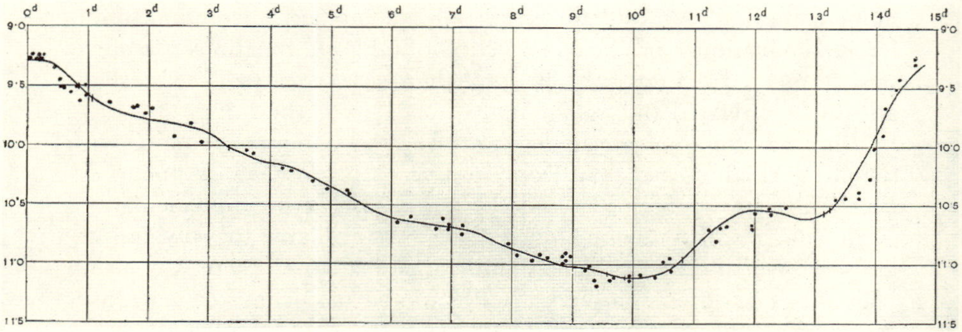

FIG 5. Manufacturing an empirical law. The points displayed on the chart show the measured variations in magnitude of the star RW Cassiopeiae during a period of two weeks. The standard form in Table 2 shows the regimen of calculation used to generate an analytical function representing the variation in magnitude. This "empirical law" is then illustrated as the continuous line on the chart. E. T. Whittaker and G. Robinson, *The Calculus of Observations: A Treatise on Numerical Mathematics* (London, 1924).

errors. Furthermore, any extra work undertaken by the computer had to be done according to strict rules. Special computing paper, 26″ x 16″ divided by faint ruling into quarter-inch squares, was deemed "essential," the computer being taught to copy down numbers two digits at a time in ink (never pencil). This procedure was designed to combine accuracy and speed with the minimum of eyestrain and mental fatigue. The sample form reproduced in figure 5 shows some observations of the variable star RW Cassiopeiae made by Whittaker himself while working in Dublin as Astronomer Royal. Having expressed the periodic changes in magnitude of the star as due to a series of underlying sinusoidal effects—that is, having produced an empirical law—Whittaker attempted to explain the origin of the effect in terms of stellar physics (table 2).[100]

The important point to notice is that the production of the empirical law from the raw data (which could then be presented to the astrophysicist as a "natural" phenomenon awaiting solution) was as much mediated by machines, tables (both wooden and paper), skilled workers, and standardized working practices, as was the collection of the data in the observatory in the first place. Mathematical physicists were likewise reliant upon the same technology of calculation to generate numbers from elliptic integrals, Bessel functions, or Legendre polynomials, and to provide mechanical or numerical solutions to integrals and differential equations that could not be solved analytically. The opening of Whittaker's mathematical laboratory in 1913 marked the formal recognition of the fundamental importance of the computation industry in late Victorian and Edwardian physical science. The students and textbooks that Whit-

UNIVERSITY OF EDINBURGH—MATHEMATICAL LABORATORY

Harmonic Analysis of the Light-curve of the Variable Star RW Cassiopeiae. Period 14·81 Days.

θ.		Days.	m.	$100(m-9)$.
0°	u_0	0·00	9·20	20
15°	u_1	0·62	9·52	52
30°	u_2	1·23	9·64	64
45°	u_3	1·85	9·72	72
60°	u_4	2·47	9·89	89
75°	u_5	3·08	9·98	98
90°	u_6	3·70	10·05	105
105°	u_7	4·32	10·21	121
120°	u_8	4·94	10·36	136
135°	u_9	5·55	10·48	148
150°	u_{10}	6·17	10·62	162
165°	u_{11}	6·79	10·67	167
180°	u_{12}	7·41	10·75	175
195°	u_{13}	8·02	10·88	188
210°	u_{14}	8·64	10·95	195
225°	u_{15}	9·26	11·10	210
240°	u_{16}	9·88	11·12	212
255°	u_{17}	10·49	11·03	203
270°	u_{18}	11·11	10·80	180
285°	u_{19}	11·73	10·64	164
300°	u_{20}	12·35	10·53	153
315°	u_{21}	12·96	10·53	153
330°	u_{22}	13·58	10·42	142
345°	u_{23}	14·19	9·66	66

(u_0 to u_{12})	20	52	64	72	89	98	105	121	136	148	162	167	175
(u_{23} to u_{13})		66	142	153	153	164	180	203	212	210	195	188	
Sums (v_0 to v_{12})	20	118	206	225	242	262	285	324	348	358	357	355	175
Diffs. (w_1 to w_{11})		−14	−78	−81	−64	−66	−75	−82	−76	−62	−33	−21	

(v_0 to v_6)	20	118	206	225	242	262	285
(v_{12} to v_7)	175	355	357	358	348	324	
Sums (p_0 to p_6)	195	473	563	583	590	586	285
Diffs. (q_0 to q_5)	−155	−237	−151	−133	−106	−62	

(w_1 to w_6)	−14	−78	−81	−64	−66	−75
(w_{11} to w_7)	−21	−33	−62	−76	−82	
Sums (r_1 to r_6)	−35	−111	−143	−140	−148	−75
Diffs. (s_1 to s_5)	7	−45	−19	12	16	

(s_1 to s_3)	7	−45	−19
(s_5, s_4)	16	12	
Sums (k_1 to k_3)	23	−33	−19
Diffs. (n_1, n_2)	−9	−57	

(p_0 to p_3)	195	473	563	583
(p_6 to p_4)	285	586	590	
Sums (l_0 to l_3)	480	1059	1153	583
Diffs. (m_0 to m_2)	−90	−113	−27	

	$l_1=1059$	$l_2=1153$	$m_2=-27$	$k_1=23$
(×½)	$l_1'=529·5$	$l_2'=576·5$	$m_2'=-13·5$	$k_1'=11·5$

	$m_1=-113$	$k_2=-33$	$n_1=-9$	$n_2=-57$
(×0·866)	$m_1'=-97·9$	$k_2'=-28·6$	$n_1'=-7·8$	$n_2'=-49·4$

	$q_1=-237$	$q_3=-133$	$q_5=-62$	$r_1=-35$	$r_3=-143$	$r_5=-148$
(×0·707)	$q_1'=-167·6$	$q_3'=-94$	$q_5'=-43·8$	$r_1'=-24·7$	$r_3'=-101$	$r_5'=-104·6$

	$q_2=-151$	$t_1=-211·4$	$h_1=-129·3$	$r_4=-140$
(×0·866)	$q_2'=-130·8$	$t_1'=-183·1$	$h_1'=-112·0$	$r_4'=-121·2$

	$l_0=480$ $l_1=1059$ $l_2=1153$ $l_3=583$	$l_0=480$ $l_2'=576·5$ $l_1'=529·5$ $l_3=583$	$l_0=480$ $l_1'=529·5$ $l_3=583$ $l_2'=576·5$	$m_0=-90$ $m_1'=-97·9$ $m_2'=-13·5$	$k_1'=11·5$ $k_2'=-28·6$ $k_3=-19$
Sum of 1st column	1633	1009·5	1063	−103·5	−7·5
Sum of 2nd column	1642	1159·5	1106	−97·9	−28·6
Sum	$3275=24a_0$			$-201·4=12a_2$	$-36·1=12b_2$
Diff.	$-9=24a_{12}$	$-150=12a_4$	$-43=12a_8$	$-5·6=12a_{10}$	$21·1=12b_{10}$

	$m_0=-90$ $m_2=-27$	$k_1=23$ $k_3=-19$	$n_1'=-7·8$ $n_2'=-49·4$
Sum			$-57·2=12b_4$
Diff.	$-63=12a_6$	$42=12b_6$	$41·6=12b_8$

	$q_1'=-167·6$ $q_5'=-43·8$	$r_1'=-24·7$ $r_5'=-104·6$
Sum	$-211·4=t_1$	$-129·3=h_1$
Diff.	$-123·8=t_2$	$79·9=h_2$

	$q_4=-106$	$t_2=-123·8$	$h_2=79·9$	$r_2=-111$
(×½)	$q_4'=-53$	$t_2'=-61·9$	$h_2'=40·0$	$r_2'=-55·5$

	$t_1'=-183·1$ $t_2'=-61·9$	$q_4'=-53$ $q_2'=-130·8$	$r_3'=-101$ $h_2'=40$	$r_2'=-55·5$ $r_6=-75$
Sum	$-245=f_1$	$-183·8=e_1$		$-130·5=j_2$
Diff.	$-121·2=f_2$	$77·8=e_2$	$-141=j_1$	

	$q_0=-155$ $q_3'=-94$ $e_1=-183·8$ $f_1=-245$	$q_0=-155$ $t_2=-123·8$ $-q_4=106$ $-q_3'=94$	$q_0=-155$ $f_2=-121·2$ $e_2=77·8$ $-q_3'=94$	$j_1=-141$ $j_2=-130·5$ $h_1'=-112$ $r_4'=-121·2$	$h_2=79·9$ $r_2=-111$ $r_3'=-101$ $-r_6=75$	$h_1'=-112$ $j_2=-130·5$ $-j_1=141$ $-r_4'=121·2$
Sum of 1st col.	−338·8	−49·0	−77·2	−253	−21·1	29
Sum of 2nd col.	−339·0	−29·8	−27·2	−251·7	−36	−9·3
Sum	$-677·8=12a_1$	$-78·8=12a_3$	$-101·4=12a_5$	$-504·7=12b_1$	$-57·1=12b_3$	$19·7=12b_5$
Diff.	$0·2=12a_{11}$	$-19·2=12a_9$	$-50·0=12a_7$	$-1·3=12b_{11}$	$14·9=12b_9$	$38·3=12b_7$

CHECKS: $u_0=a_0+a_1+a_2+a_3+a_4+a_5+a_6+a_7+a_8+a_9+a_{10}+a_{11}+a_{12}$.

$\frac{1}{2}(u_1-u_{23})=·259\,(b_1+b_{11})+\frac{1}{2}(b_2+b_{10})+·707\,(b_3+b_9)+·866\,(b_4+b_8)+·966\,(b_5+b_7)+b_6$.

RESULT: $u=a_0+a_1\cos\theta+a_2\cos 2\theta+a_3\cos 3\theta+a_4\cos 4\theta+a_5\cos 5\theta+a_6\cos 6\theta+a_7\cos 7\theta+a_8\cos 8\theta+a_9\cos 9\theta+a_{10}\cos 10\theta+a_{11}\cos 11\theta+a_{12}\cos 12\theta$
$+b_1\sin\theta+b_2\sin 2\theta+b_3\sin 3\theta+b_4\sin 4\theta+b_5\sin 5\theta+b_6\sin 6\theta+b_7\sin 7\theta+b_8\sin 8\theta+b_9\sin 9\theta+b_{10}\sin 10\theta+b_{11}\sin 11\theta$.

$m=9+0·01u=10·36-0·565\cos\theta-0·168\cos 2\theta-0·066\cos 3\theta-0·125\cos 4\theta-0·087\cos 5\theta-0·053\cos 6\theta-0·042\cos 7\theta-0·036\cos 8\theta-0·016\cos 9\theta-0·005\cos 10\theta-0·004\cos 12\theta$
$-0·420\sin\theta-0·030\sin 2\theta-0·048\sin 3\theta-0·048\sin 4\theta+0·016\sin 5\theta+0·035\sin 6\theta+0·032\sin 7\theta+0·035\sin 8\theta+0·012\sin 9\theta+0·018\sin 10\theta-0·001\sin 11\theta$
$=10^{m}·36+0·70\sin(\theta+233°)+0·17\sin(2\theta+260°)+0·08\sin(3\theta+234°)+0·13\sin(4\theta+249°)+0·09\sin(5\theta+250°)+0·06\sin(6\theta+304°)$.

To face page 280.

TABLE 2.

taker's school produced, subsequently took his methods into universities and technical colleges throughout Britain.

I have concluded my historical survey in Whittakers's laboratory because it represents an important change in the role of advanced mathematics in British society. Fifty years before its foundation, advanced mathematics was widely seen in Britain as a static and esoteric discipline, whose principal use was to sharpen the mental faculties of the intellectual elite. Fifty years after its foundation, numerical methods and calculation using digital computers were at the forefront of mathematical research. Whittaker's mathematical laboratory also makes explicit the link between calculating and precision measurement. Researchers in the laboratory brought a range of calculating techniques to bear on important problems in an effort to establish which techniques were the fastest and the most reliable. They experimented with numerical methods, analogue and digital calculating devices, mathematical tables, and a range of disciplined practices to enable the human calculator to work with maximum accuracy and minimum fatigue. The problem of generating numbers from analytical expressions, and vice versa, had now become a specialized research activity, with its own methods and technology.

Whittaker's mathematical laboratory also makes the relationship between "precision calculation" and "precision measurement" explicit in two further ways. First, it is surely not coincidental that the problem of computation identified by the B.A. Committee in 1871 arose just as the physical laboratory was becoming established as the key site for precision measurement in physics. As the existence of calculating rooms in the Cavendish and the Clarendon suggests, the problem of forging agreement between precision measurement and theoretical prediction became increasingly acute during the 1860s and 1870s. We can at least conjecture, therefore, that the need for new mathematical tables and other aids to calculation arose in the 1870s and 1880s as new and highly mathematical theories of electromagnetism and heat were accommodated within these new laboratories. Viewed in this light, Whittaker's work in Edinburgh appears as a transformation of the "calculation room" into a separate laboratory in its own right. This brings me to the second point. The technicians, workshops, machines, instruments, and disciplined working practices that sustained the daily life of physical laboratories, remind us that such institutions were the products of a wider industrial and manufacturing culture.[101] The mathematical laboratory must be similarly understood. It too was made up of skilled workers, machines, and other devices, all carefully coordinated to facilitate new kinds of calculation. But, as the epigraph at the beginning of this section makes clear, the devices inside the laboratory were as much a part of the wider Victorian economy as were steam engines and the electric telegraph. Without the printing and precision engineering industries, the bureaucratic and commercial demand for reliable calculating practices, and the willingness of a major university to support the setting up of a "laboratory" in this novel area of research, there would have been no laboratory for theory.

6. Conclusion

I began this essay with the suggestion that calculation and precision measurement are not wholly dissimilar activities. I argued that the practice of calculating is not dictated by the inherent properties of either numbers or physical objects. Calculating is a human activity in which, for example, the behavior of physical objects can be successfully predicted through symbolic representation and experimentation. How many beads do I need in order to place 25 in each of 25 boxes? The answer is 625; I do not need to count them. But, as with other forms of experimental activity, the speed, reliability, and scale of the calculating enterprise will depend upon the resources brought to the task. Mental calculation, for example, is a fragile skill that is distributed very unevenly among human beings; it could not act as a substitute for machine-based calculation in a highly numerate society like our own. We can improve people's ability for mental calculation up to a point—by making children learn their tables for example—but it is more effective to invent new ways of calculating that require only the skills that most people can easily master. In late nineteenth-century Britain, the practice of calculating was reorganized, like so many aspects of Victorian culture, according to the rhythm of machines. It was, moreover, the mass production and distribution of such machines that eventually generated an international community with a common ability to calculate on a new scale.

One of the remarkable characteristics of the modern western world is the extraordinary range of devices it employs to facilitate numeracy. As Crump has shown, most human cultures, both past and present, make use of what we would recognize as mathematics, yet no other society has developed anything like our facility for calculation.[102] That facility is, moreover, vital to the political, economic, and military power wielded by the modern industrial nations. The teaching and examining of numeracy is required by law in these nations and a significant part of their industrial effort is devoted to the technology of computation. It is, however, the very taken-for-granted nature of numeracy in the modern world that obscures the manufactured nature of the competency. When a science student punches the buttons on an electronic calculator to extract a numerical answer from an algebraic expression, the answer appears instantly to ten or more significant figures.[103] Science teachers frequently have to remind students that most of these figures are practically meaningless because they greatly transcend the empirical accuracy of the physical data to hand. The redundancy of precision built into commercial calculators thus reinforces the notion that numerical precision springs effortlessly from a theory, only to be reined back by the limitations of exact measurement. Computation often appears so cheap at the point of consumption that its seemingly miraculous precision can be freely attributed to the theory from which it was generated. In reality, of course, the power of the electronic calculator originates in the disciplined skill of the user and a technology developed to handle much larger problems in data management.[104] The massive information technology industry was developed to serve the needs not of the pure scientist or science student, but of government bureaucracy, business, and the military; and these latter consumers would not look upon calculation as a cheap commodity.

Whether we are trying to produce algebraic expressions from empirical data (phenomenological laws) or use them to generate numbers (predictions), a technology of calculation will always be implicated. And this brings me back to my remarks at the beginning of this essay. The sanctity with which we treat physical laws (expressed in mathematical form) depends upon which parts of scientific practice we choose to privilege. Traditionally, as Cartwright and Hacking have argued, physicists privilege what they call "laws of nature," and play down the role of other assumptions required to produce testable predictions. This is the theoretical hierarchy by which physicists conceptualize their world. But from a philosopher's perspective, the additional assumptions and approximations necessary to make a theory work can also be granted the same status as the laws themselves. The crux of my argument can be similarly stated. As *historians* of science we are not bound to acknowledge the same hierarchy in scientific practice as do scientists themselves. Indeed, if we want to give an *historical* explanation of the exactitude of the exact sciences we cannot limit our analysis to assessments of the partial truth or falsity of laws of nature, even when supplemented by the auxiliary assumptions and approximations required to make testable predictions. We must attribute eqivalent status to all aspects of scientific practice, including such seemingly mundane techniques as computation. As I have tried to show in this essay, once we see the practice of calculating as constitutive of the robust link between mathematical theory and experimental data, the social-historical origins of exactitude in the exact sciences become researchable.

I would like to thank David Edgerton, Stephen Johnston, Simon Schaffer, and Alan Yoshioka for reading an earlier draft of this essay and making a number of very helpful comments and suggestions. I would also like to thank Stephen Johnston for his generosity in giving me access to his unpublished research on the history of nineteenth-century calculating machines.

Notes to Chapter Twelve

1. J.W.L. Glaisher, "Mathematical Tables," *Encyclopaedia Britannica* (1910/11), 326.

2. N. Cartwright, *How the Laws of Physics Lie* (Oxford, 1983), 100–127.

3. A phenomenological law is one derived directly from empirical data. See Cartwright *How the Laws of Physics Lie* (n. 2), 1–4.

4. I. Hacking, *Representing and Intervening* (Cambridge, 1983), 215–19. Hacking argues that the canonical demonstrations given in physics textbooks provide a "semantic bridge" that lends plausibility to the steps in the mathematical argument (215). Cartwright and Hacking's work presents a major challenge to falsificationists, such as Popper, who claim that laws of nature can be straightforwardly tested by experiment.

5. On the importance of knowing how to apply mathematical laws and principles, see A. Warwick, "Cambridge Mathematics and Cavendish Physics: Cunningham, Campbell and Einstein's Relativity. Part I: The uses of theory," *Studies in the History and Philosophy of Science* 23 (1992), 625–56.

6. A. Cunningham-Robertson, "The late Mr. G. P.Bidder," *Spectator* 51 (1878), 1634–35, 1634.

7. Wittgenstein discusses the relationship between experimenting and calculating most directly in L. Wittgenstein, *Remarks on the Foundations of Mathematics*, ed. G. E. Wright, R. Rhees, and G.E.M. Anscombe. Trans. G.E.M. Anscombe (Oxford, rev. ed. 1978), 94–111.

8. For an excellent discussion of the strategies used by philosophers to try to secure the epistemological foundations of mathematics, and their shortcomings, see P. Kitcher, *The Nature of Mathematical Knowledge* (Oxford, 1984).

9. For discussions of how mathematical practice is stabilized in social communities, see D. Bloor, *Wittgenstein: A Social Theory of Knowledge* (London, 1983), and H. M. Collins, *Changing Order: Replication and Induction in Scientific Practice* (Sage, 1985).

10. On replication and its problems in physical science, see H. M. Collins, *Changing Order* (n. 9).

11. I am not claiming that the pencil-and-paper calculation makes the mental process of calculation visible. Calculating using pencil and paper is rather a new kind of calculating that takes place in recorded steps.

12. A. De Morgan, "Table," *The English Cyclopaedia: Conducted by Charles Knight: Arts and Sciences*, vol. 7 (London, 1861), 1006. De Morgan actually refers to the "age of Cocker." *Cockers Arithmetick* (John Hawkins, 1678) was a standard text on arithmetic during the late seventeenth and early eighteenth centuries.

13. On mid-Victorian exhibitions as public displays of consumer culture, see

T. Richards, *The Commodity Culture of Victorian England: Advertising and Spectacle, 1851–1914* (Verso, 1990), chap. 1.

14. C. Dickens, *Hard Times* (Oxford, 1989), 3. The book was written in 1854.

15. De Morgan claimed that Shanks's calculation constituted a table in its own right. Shanks had, for example, been required in the course of his calculation to compute the value of the reciprocal of 601.5 raised to the power 601. The description of this calculation alone ran to 87 pages. By 1873, Shanks had computed π to more than 700 places of decimals, a feat that was not bettered until after 1945.

16. W. Shanks, *Contributions to Mathematics: Comprising Chiefly the Rectification of the Circle to 607 Places of Decimals* (London, 1853), iii. During the 1840s the German computers Dase and Clausen had calculated the value of π to 205 and 250 places of decimals, respectively. See J.W.L. Glaisher, "Report on Mathematical Tables," *British Association Report* (hereafter *B.A. Report*), 43 (1873), 1–175, 122. Shanks's work was reckoned by some in the 1880s to stand as the most astounding monument to calculation, and to refute the notion that the Germans were more able than the British at methodical calculation. See, for example, S. Lupton, "The "Art of Computation for the Purposes of Science," *Nature* 37 (1888), 237–39, 262–63, 263.

17. A. De Morgan, "Table" (n. 12), 1006.

18. I. Hacking, *The Taming of Chance* (Cambridge, 1990), 2.

19. Ibid., 3. I have been unable to locate any study which addresses the rise of tabulated data directly. Glaisher's catalogue (see n. 21) of all known printed mathematical tables to 1871 shows that the production of such tables took place at a fairly constant rate from the early seventeenth to the late eighteenth centuries, but accelerated dramatically after 1815. This catalogue also shows that almost half of the total number of tables recorded were made in Britain and Ireland.

20. A. Briggs, "The Imaginary Response of the Victorians to New Technology: The Case of the Railways," in C. Wrigley and J. Shepherd, eds., *On the Move* (Hambledon, 1991), 58–75.

21. See Glaisher, "Report on Mathematical Tables" (n. 16).

22. Ibid.; and A. Hyman, *Charles Babbage: Pioneer of the Computer* (Oxford, 1982), 61–62.

23. The centesimal division (that is, the division into 100 rather than 90 degrees) of the quadrant was intended to facilitate the full introduction of the decimal system then recently adopted in France.

24. C. Babbage, *On the Economy of Machinery and Manufactures* (London, 1832), 156; and Glaisher, "Report on Mathematical Tables" (n. 16), 56–57.

25. For Babbage's account of Prony as a factory manager, see *On the Economy of Machinery and Manufactures* (n. 24), 157.

26. The production of an accurate and standardized set of tables would, of course, have been welcomed in many spheres, but mathematical physicists do not appear to have seen this as a pressing need during this period.

27. It is unclear exactly why the publication was halted. Babbage comments merely that "the sudden fall of the assignats rendered it impossible for Didot [the

printer] to fulfil his contract with the government." See A. Hyman, *Charles Babbage: Pioneer of the Computer* (n. 22), 62.

28. Glaisher, "Report on Mathematical Tables" (n. 16), 57. E. M. Horsburgh, ed., *Modern Instruments and Methods of Calculation: A Handbook of the Napier Tercentenary Exhibition* (London, 1914), 40.

29. There is considerable disagreement over why the Anglo-French project to publish the Cadastre Tables came to nothing. DeMorgan claims that the proposal was declined by the French. Sang suggests that the English Commissioners were "dissatisfied with the soundness of the calculations." Glaisher points out that the centesimal division of the quadrant employed in the tables rendered all but the logarithms practically useless in Britain. See Horsburgh, *Modern Instruments and Methods of Calculation* (n. 28), 40; and Glaisher, "Report on Mathematical Tables" (n. 16), 57, 64.

30. A. Hyman, *Charles Babbage: Pioneer of the Computer* (n. 22), 49.

31. Ibid., 49.

32. C. Babbage, *On the Economy of Machinery and Manufactures*, (n. 24), Art. 241 *et seq*.

33. See A. Hyman, *Charles Babbage: Pioneer of the Computer* (n. 22), 48. Babbage described to Davy in a letter how the Analytical Engine might be deployed if such a mammoth calculation were repeated. See A. Hyman, *Science and Reform: Selected works of Charles Babbage* (Cambridge, 1989), 48.

34. See A. Hyman, *Science and Reform: Selected works of Charles Babbage*, (n. 33), 53.

35. Ibid., 66.

36. The "interval" is the incremental change between the values of the independent variable for which the dependent variable is tabulated. If, for example, sin θ was tabulated for $\theta = 10$, $\theta = 20$, etc., the interval would be 10 degrees.

37. See De Morgan, "Table" (n. 12), 978; and Glaisher, "Report on Mathematical Tables" (n. 16), 13.

38. J.W.L. Glaisher, "On the Progress to Accuracy of Logarithmic Tables," *Monthly Notices of the Royal Astronomical Society* 33 (1872/3), 339–40. Glaisher expressed some disappointment that neither Greenwich Observatory nor the late Board of Longitude had seen fit to take on the responsibility.

39. Glaisher did not believe that the B.A. Committee would be able to coordinate table production because its life would be relatively short. In practice, however, it continued to meet and to oversee the publication of tables until 1948 when it handed the job over to the Royal Society.

40. J.W.L. Glaisher's father, James Glaisher, was George Airy's assistant at the Cambridge Observatory and subsequently followed Airy to Greenwich Observatory. At Greenwich, Glaisher was placed in charge of observations in meteorology and magnetism.

41. A. R. Forsyth, "James Whitbread Lee Glaisher, 1848–1928," *Proceedings of the Royal Society* 126 (1929/30), i–ix, ix.

42. Glaisher, "Report on Mathematical Tables" (n. 16), 172.

43. Ibid., 3.

44. The transcendental functions in this case are those containing trigonometric, logarithmic, and exponential functions.

45. See I. Todhunter, *An Elementary Treatise on Laplace's Functions, Lame's Functions, and Bessel's Functions* (London, 1875); and N. M. Ferrers, *An elementary treatise on spherical harmonics and subjects connected with them* (London, 1877).

46. *British Association Report* 48 (1879), 46–57.

47. Glaisher, "Report on Mathematical Tables" (n. 16), 172. Glaisher excluded the preparation of the Cadastre Tables as they had recalculated extant tables.

48. See J.W.L. Glaisher, *Messenger of Mathematics* 6 (1876), 111–12.

49. The rest, on factors and prime numbers, was due to James Glaisher.

50. J.W.S. Rayleigh, *The Theory of Sound* (Macmillan, 1877), vol. 1, 321. E. Lommel, *Studien ueber Bessel'schen Functionen* (Leipzig, 1868), 127; P. A. Hansen, *Ermittelung der Absoluten Stoerungen in Ellipsen von beliebiger Excentricitaet und Neigung, Teil I* (Gotha, 1843).

51. The committee applied for a grant of £15 to enable them to employ the professional calculator; see *B.A. Report* 61 (1891), 129.

52. See *B.A. Report* 63 (1893), 227–79, 228. The "Edmondson calculator" was designed and patented by Joseph Edmondson in 1883. The machine was a modified form of the Thomas arithmometer; see C. V. Boys, "Calculating Machines," *Journal of the Society of Arts* 34 (1886), 376–89, 378–80.

53. See *B.A. Report* 63 (n. 52), 228.

54. See *B.A. Report* 69 (1899), 65–120, 68.

55. Ibid., 68. The women were working at University College (London) with Pearson.

56. Glaisher, "Mathematical Tables," *Encyclopaedia Britannica* (1910/11), 326.

57. See *B.A. Report* 84 (1914), 75–102, 75. The committee was now receiving an annual grant of £30 to help pay for the calculation and proposed to publish the tables separately from the annual *B.A. Reports*. See also *B.A. Report* 83 (1913), 87–170, 87.

58. In 1914 the B.A. Committee was officially named "The Committee for the Calculation of Mathematical Tables"; see *B.A. Report* 83 (n. 57), 87. On the subsequent work of the Committee see *Mathematical Tables*, Prepared by the Committee for the Calculation of Mathematical Tables, vol. 1 (London, 1931), iii–iv. For an overview of scientific computing in Britain after World War II, including the work of the B.A. Committee, see M. Croarken, *Early Scientific Computing in Britain* (Oxford, 1990).

59. M. Hannyngton, "On the Use of M. Thomas de Colmar's Arithmometer," *Journal of the Institute of Actuaries* 16 (1871), 244–53, 253.

60. For an interesting discussion of remarkable mental calculators, see S. B. Smith, *The Great Mental Calculators* (Columbia University Press, 1983).

61. "On Some Strange Mental Feats," *Cornhill Magazine* 32 (1875), 157–75, 161.

62. "Calculating Boys," *Spectator* 51 (1878), 1208–9, 1208.

63. When the British Government withdrew from the project in 1832 they had invested £17,470, a large sum by the standards of the day. See D. Swade, *Charles Babbage and his Calculating Engines* (Science Museum: London, 1991), 18.

64. Ibid., 18–20. The layout and quality of print of tables that were to be used for long hours was of crucial importance. See A. De Morgan, "Table" (n. 12), 977–78.

65. On the early history of the arithmometer, see Palfreman and Swade (1991), 22; C. G. Chase, "History of Mechanical Computing Machinery," *Annals of the History of Computing* 2 (1980), 198–226, 204.

66. Boys, "Calculating Machines" (n. 52), 377.

67. M. Hannyngton, "On the Adaptation of Assurance Formulae to the Arithmometer of M. Thomas," *The Assurance Magazine* 12 (1866), 184.

68. "Dr. Zillmer on the Arithmometer," *Journal of the Institute of Actuaries* 15 (1869), 25, 26–27. Zillmer was president of the German Life Insurance Institute, and used the arithmometer extensively in his own work. See *Journal of the Institute of Actuaries* 16 (1871), 265.

69. Hannygton, "On the use of M. Thomas de Colmar's Arithmometer" (n. 59). The full technical specification of the arithmometer given in the *Engineer* 29 (1870), 319, was probably the first full account of the machine published in Britain (apart from the 1851 English Patent, 1851/13,504).

70. Hannygton, "On the use of M. Thomas de Colmar's Arithmometer" (n. 59), 253.

71. For an account of why we should not think of a calculating machine as a substitute for the human mind, see H. M. Collins, *Artificial Experts: Social Knowledge and Intelligent Machines* (MIT Press, 1990), chap. 4. A calculating machine does not "do arithmetic" any more than does the paper and pencil I use to work out a multiplication problem.

72. On the rapid increase in demand for commercial calculation in the insurance industry after 1850, see M. Campbell-Kelly, "Large-scale data processing in the Prudential, 1850–1930," *Accounting Business and Financial History* 2 (1992), 117–39.

73. E. Sang, "On Mechanical Aids to Calculation," *Journal of the Institute of Actuaries* (1871), 253–65, 265.

74. Ibid., 265.

75. Ibid., 265–69.

76. See, for example, P. Gray, "On the Arithmometer of M. Thomas (de Colmar), and its applications to the Construction of Life Contingency Tables," *Journal of the Institute of Actuaries* 17 (1873), 249–66; and D. Carment, "On the application of the Arithmometer to the Calculation of Tables of the Values of Endowment Assurance Policies," *Journal of the Institute of Actuaries* 22 (1881), 368–80. For the case of the Prudential, see Campbell-Kelly, "Large-scale data processing in the Prudential, 1850–1930" (n. 72).

77. See Chase, "History of Mechanical Computing Machinery" (n. 65).

78. See Horsburgh, *Modern Instruments and Methods of Calculation* (n. 28), 102.

79. For a technical description of the leading machines on sale in 1914, see Horsburgh, *Modern Instruments and Methods of Calculation* (n. 28), 69–135. This overview of calculating machines was originally prepared by F.J.W. Whipple to accompany an exhibition of machines at the Fifth International Congress of Mathematicians held in Cambridge in 1912.

80. See Chase, "History of Mechanical Computing Machinery" (n. 65), 205–7; and Horsburgh, *Modern Instruments and Methods of Calculation* (n. 28), 84. The St. Louis–based engineer F. S. Baldwin had designed and built a similar calculating machine in the early 1870s, but it was unclear whether this was the source of Odhner's design.

81. So fundamental were many of Trinks's improvements that, by 1914, the machine was known as the "Trinks-Brunswiga" calculator. See Horsburgh, *Modern Instruments and Methods of Calculation* (n. 28), 85.

82. Couch Adams, for example, was using an arithmometer in the Cambridge Observatory by the mid-1870s. See "The Late Mr. Thomas Fowler," *Transactions of the Devonshire Association for the Advancement of Science, Literature and the Arts* 7 (1875), 171–78, 174.

83. The machines in question were "Layton's Arithmometer" and "Edmondson's Calculator." See Boys, "Calculating Machines" (n. 52).

84. Ibid. We do not yet know exactly when and to what extent calculating machines were used for scientific work. The account given in the *Engineer* (n. 69) claims that the machines were used in government offices, observatories, insurance offices, and by civil engineers. My limited research leads me to believe that they were not widely used, even in astronomical work, until the 1890s.

85. The Comptometer was first manufactured in 1887 and was used mainly in the commercial world. The manufacturers may well have seen laboratories as a potentially new market around the turn of the century. See Horsburgh, *Modern Instruments and Methods of Calculation* (n. 28), 98. It was also in 1901 that Brunsviga began to advertise directly to a scientific market. See Croarken, *Early Scientific Computing in Britain* (n. 58), 15.

86. C. V. Boys, "The Comptometer," *Nature* 64 (1901), 265–68, 268.

87. Anon, *Monthly Notices of the Royal Astronomical Society* 37 (1914), 226.

88. For an example of the different forms of expression suited to logarithmic and machine calculation, see Boys, "The Comptometer" (n. 86), 268.

89. K. Pearson, ed., *Tracts for Computers* (Cambridge, 1919), 1.

90. Ibid.

91. Anon, *Post Magazine and Insurance Monitor*, 26 November 1856, 371.

92. Dodgson is better known as Lewis Carrol, the name under which he published the "Alice" books.

93. C. Dodgson, "The Offer of the Clarendon Trustees," in *The Complete Works of Lewis Caroll with an Introduction by Alexander Woollcott* (London, 1939), 1010.

94. See G. Gooday, "Precision measurement and the genesis of physics teaching laboratories in Victorian Britain," *British Journal for the History of Science* 23 (1990), 25–51.

95. See Maxwell's accounts for the Laboratory in the University Library Archive MSS Add 7655/vj/3 and 7655/vj/4.

96. The sources cited in E. T. Whittaker and G. Robinson, *The Calculus of Observations: A Treatise on Numerical Mathematics* (London, 1924), afford some insight to the practical origins of numerical methods in the nineteenth cen-

tury. Whittaker and Robinson also offer alternative formulae for those with access to calculating machines; see, for example, 14.

97. Forsyth had just completed the first advanced teaching text in Britain on functional analysis. See A. R. Forsyth, *Theory of Functions of a Complex Variable* (Cambridge, 1893).

98. E. T. Whittaker, *A Course in Modern Analysis* (Cambridge, 1902).

99. D. Gibb, *A Course in Interpolation and Numerical Integration* (London, 1915), 1.

100. See Whittaker and Martin, *Monthly Notices of the Royal Astronomical Society* 71 (1911), 511–16.

101. For excellent discussions of the Cavendish Laboratory as a product of late nineteenth-century British culture, see J. Bennett et al., eds., *Empires of Physics* (Whipple Museum Cambridge, 1993), 3–38; and S. Schaffer, "Late Victorian Metrology and its Instrumentation: a Manufactory for Ohms," in S. Cozzens and R. Bud, eds., *Invisible Connections* (SPIE Press, 1992).

102. T. Crump, *The Anthropology of Numbers* (Cambridge, 1990).

103. I am not suggesting that using a pocket calculator is an unskilled activity. What I am suggesting is that the work of calculation has to be reorganized to bring more rapid and complex calculation within the ability of most people. For an excellent discussion of the skill involved in using an electronic calculator, see H. M. Collins, *Artificial Experts: Social Knowledge and Intelligent Machines* (n. 71), chap. 5.

104. Norberg notes that histories of modern computing tend to focus on theoretical innovations in scientific computing, especially on the development of digital computers during and after World War II. The technologically more important development of data management in business and government is largely ignored. See A. L. Norberg, "High-Technology Calculation in the Early 20th Century: Punched Card Machinery in Business and Government," *Technology and Culture* 31 (1990), 753–79.

THIRTEEN

PRECISION: AGENT OF UNITY AND PRODUCT OF AGREEMENT

PART III—"TODAY PRECISION MUST BE COMMONPLACE"

M. Norton Wise

A Qualitative Change in the Numbers

THE PAPERS in this section relate the arrival, by the end of the nineteenth century, of high-quality voltmeters, diffraction gratings, and mechanical calculators, and at a relatively low cost. They transformed laboratory life. Ordinary college students in science and engineering (whose numbers were exploding) could now carry out experiments and calculations with an exactitude that the most talented natural philosophers and mathematicians of 1850 could not envisage. By that date, James Joule, with an exorbitant expenditure of time and skill, had measured the mechanical equivalent of heat by electrical heating of water as about 772 ft-lb/°F per pound of water.[1] By the end of the century, students using commercially "black-boxed" ammeters and voltmeters regularly obtained the more accurate 779 ft-lb in ten minutes. For twenty dollars, anyone interested could obtain a low-end Rowland grating with which to locate spectral lines undetectable twenty years earlier. And mathematical physicists, as well as employees of insurance companies, made their calculations using tables accurate to eight or more figures produced by mechanical calculators.

These are but three examples of an explosion in everyday precision. They represent the new resource base of skills, materials, and common knowledge available to dedicated artists with more time and money. With five years of work and the support of the British Association, Griffiths had refined the heat equivalent by two orders of magnitude to 779.77; Rowland's best gratings specified spectral lines reliably to 10^{-11} cm, and multiplied twentyfold the number of lines Ångström had published for the solar spectrum in 1869; mathematical physicists translated complex equations into precise numbers using machine-calculated tables of special mathematical functions (unavailable fifty years earlier) given to ten or more places. These and many similar refinements, typically of two orders of magnitude, essentially set the agenda of precision for the next half-century, which saw the spread of such standards to every area of science and technology. And that agenda seems to characterize the values of "modernism" as well as any other single feature.

When in 1944, Bausch and Lomb advertised their wares on the cover of Science magazine under the slogan, "Today Precision Must be Commonplace," they were referring not only to the hardware of war that they actually depicted (fig. 1).[2] Rather, they were associating the urgency and prestige of military power with their own place in modern scientific culture, with "production line accuracy" of a ten-thousandth of an inch. Wrapping science, industry, and the state around their Contour Measuring Projector, they identified its capacity with cultural dominance in education, research, industry, and war, and simultaneously with the virtues of the eyeglasses they could provide to every far-sighted citizen who wanted to participate in the coming victory. Precision had indeed become a necessity of modern life. Bausch and Lomb would do for the peacetime laboratory what they were doing for American fighting men against their wartime enemies.

One of the most interesting (if also in retrospect the most obvious) features of the papers here and in the rest of the volume is their attention to the fact that attaining precision in the sciences has always depended on the existing capacities and motivations of machinists, engineers, and other technicians in the commercial world. The signal of this dependence for the late nineteenth century was surely the underlying transformation in the precision of machining and manufacturing. In 1850 machinists worked to a tolerance of about one-hundredth of an inch. By 1900, when every machinist used micrometer calipers developed in the 1880s, they regularly maintained 0.001 in. and could manage 0.0001 when necessary. At that time Pratt and Whitney sold throughout the world measuring devices based on the Rogers-Bond comparator that were accurate to a hundred-thousandth of an inch, read by microscope. These commercial devices were used to produce and inspect the gauges employed in precision manufacturing.[3]

The fifty-year difference in machining precision marks the arrival of interchangeable parts manufacturing and the beginnings of a massive change in the technologies of everyday life: first rifles, then sewing machines, typewriters, automobiles, machine guns, and airplanes.[4] But wide distribution of laboratory voltmeters, diffraction gratings, and calculating machines depended on the same underlying transformation in tolerances, as well as refinement in the quality control of materials like copper, steel, and magnetic alloys. As Sweetnam observes, the collabora-

SCIENCE

New Series Vol. 99, No. 2564 | Friday, February 18, 1944 | Subscription, $6.00 Single Copies, .15

Bausch & Lomb Contour Measuring Projector

Today Precision *Must* Be Commonplace

American fighting men on our fighting fronts depend upon production line accuracy . . . for ten-thousandths of an inch variation on the production line can mean the difference between a hit or a miss on the battleline.

The Bausch & Lomb Contour Measuring Projector makes such accuracy possible on the fastest moving production lines, because it takes many vital inspection jobs "off the surface plate" and eliminates the tedious, time-consuming computations of the "sine bar." Inspections for accuracy become routine jobs.

Throwing an accurate, sharply defined shadow image of the object under examination on a translucent screen, the B&L Contour Projector permits exact measurements or comparison with an enlarged template drawing at magnifications great enough for easy and accurate dimensioning.

Here again is a Bausch & Lomb peacetime development that serves America at War. The B&L Contour Measuring Projector is helping speed production of fighting tools for our fighting men.

For Bausch & Lomb Instruments essential to Victory—priorities govern delivery schedules.

BAUSCH & LOMB
OPTICAL CO. • ROCHESTER, NEW YORK
ESTABLISHED 1853

AN AMERICAN SCIENTIFIC INSTITUTION PRODUCING OPTICAL GLASS AND INSTRUMENTS FOR MILITARY USE, EDUCATION, RESEARCH, INDUSTRY AND EYESIGHT CORRECTION

Science: published weekly by The Science Press, Lancaster, Pa.

FIG. 1. Wartime advertisement for the Bausch & Lomb Contour Measuring Projector. (Ref. 2)

tion between the Harvard astronomer William A. Rogers and George M. Bond of Pratt and Whitney, which resulted in the Rogers-Bond comparator, had precursors in Rogers's collaborations with the instrument makers Buff and Berger and with the Waltham Instrument Company in producing ruling engines for plane diffraction gratings, which Rogers was producing by 1880. Results obtained with gratings like these, compared with results from Ångström's grating and from Kirchhoff's set of four prisms, epitomize the qualitative change which had already occurred in the entire context for precision based on machining (fig. 2).[5] We need to know a great deal more about such relationships between commercial and scientific instruments. The present papers, nevertheless, warrant the general view that the larger culture of precision was as much the essential resource base for the best numbers scientists could produce as it was a product of scientific achievement.

Natural philosophers had a pressing need for tables of special functions, but they notably failed to produce them by their own efforts, Warwick shows, until calculating machines became routinely available in insurance offices, which was a matter of commercial demand and precision manufacturing. Gooday observes similar factors driving perfection of the now standard ammeters and voltmeters of the laboratory. Every electrical engineer required accurate and portable meters by the 1880s, just as every machinist required micrometer calipers by that date. Development and distribution of both sorts of instruments depended in the same way on this demand and on the new manufacturing techniques that could satisfy it. Finally, if commercial demand in the usual sense did not directly motivate Rowland's grating operation, his ruling engine still depended critically on the drive screw that he was able to produce by obsessive attention to the procedures used to make the drive screws for graduating machine-shop scales and for gauging gauges.

In fact, Deborah Warner has shown that the entire history of ruling engines for gratings is a story of the drive screw and of successive collaborations between grating makers and machinists, starting with the first gratings made by John Barton, Controller of the Mint in London, who enjoyed a competition to make the perfect drive screw with his close friend Henry Maudslay, who is as famed for his interchangeable machine screws and nuts as he is for his comparator, the "Companion of the Bench," or "Lord Chancellor." Barton began with a dividing engine inherited through his wife from none other than John "Longitude" Harrison, of eighteenth-century chronometer fame, to which he presumably affixed a new drive screw. Ångström, Warner says, owed his gratings to the dividing engine of the Pomeranian instrument maker F. A. Nobert. Similarly, Otto Sibum shows that Joule owed the all-important division of his thermometer scales at mid-century to the Manchester instrument

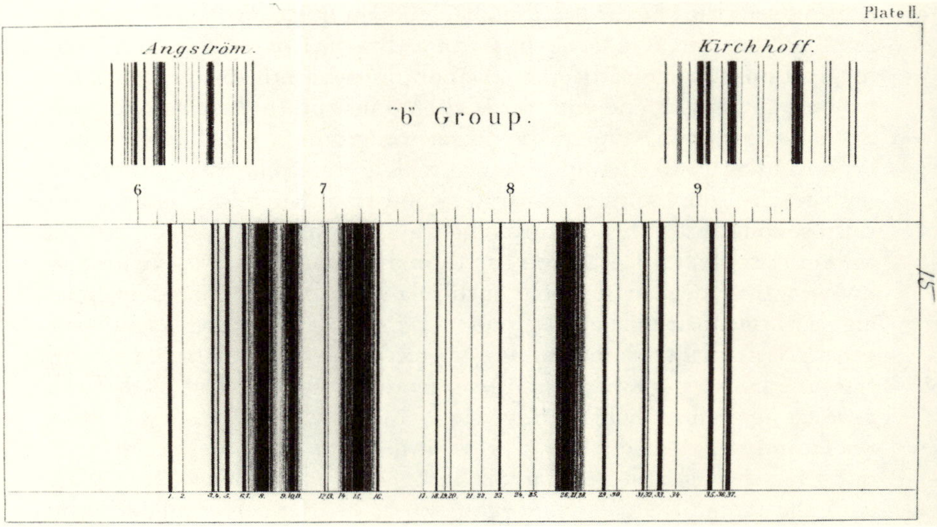

FIG. 2. A portion of the solar spectrum taken with a Rogers plane grating as compared with spectra by Ångström and Kirchhoff. (Ref. 5)

maker J. B. Dancer, who employed a "traveling microscope" controlled by a drive screw and graduated disc that in principle allowed readings of 1/4000 inch.[6] This ubiquitous dependence of scientific precision on the material interests and technical capacities of nonscientists, has a strange corollary: ambivalence toward, if not outright distrust of, technicians.

Trust and Distrust

When René Descartes in the late 1620s wanted to make hyperbolic lenses he had to find an instrument maker who could construct the lens-grinding machine that he envisaged. Thus began a tortured interaction with Jean Ferrier, a mechanic of considerable talent who found Descarte's ideas for such a machine less brilliant than did the great philosopher himself, who seemed to Ferrier to have little appreciation of the properties of materials or of working machines. No such lens-grinding engine was ever constructed in the seventeenth century. Descartes and other natural philosophers put it down to the failings of mechanics, whom they continually hoped to replace with machines.[7] From then until now, natural philosophers have alternately praised and blamed the skills of craftsmen and engineers, mythologizing their artistry and damning their incompetence.

The problems have an ironic class basis. From the perspective of higher-status natural philosophers, technicians do what anyone could in principle do. And yet what anyone could supposedly do turns out to be what only a few highly skilled individuals, possessing detailed knowledge of materials and techniques, can actually accomplish. As Mario Biagioli has recently shown, when mathematicians were still technicians Galileo moved outside the universities to the Medici court to acquire the higher standing of "philosopher" and to begin to make mathematical argument relevant to natural philosophy, while effacing his own monetary interests and the utilitarian aspects of his work. Still in the eighteenth century, mathematicians labored at the Royal Society under the image of "tool users" rather than thinkers.[8] Not until the end of the nineteenth century did mathematical expression by itself attain high status among natural philosophers, ultimately as the very foundation of "modern" physics. (Its formerly suspect boundary position has now been taken over by computer science, halfway between proper science and practical engineering, which in turn is rapidly becoming the foundation of "postmodern" science.)

This long-standing bias against the practical has always been in part a matter of morals, of the high and the low, of ideality versus materiality, and purity versus corruption. So scientists, while constantly dependent on mechanical and material resources to accomplish their ends, have constantly tried to distance themselves from the merely practical in order to maintain their high moral ground. In Britain, as several previous papers have stressed, this ground was typically that of the (originally land-based) gentleman. Natural philosophers at the end of the nineteenth century, Gooday argues, guarded that ground as carefully as had Joseph Banks in the eighteenth. The responsible, independent individual relied on his personal judgment and control of circumstances both to attain truth and to act in new situations. He placed his trust in neither mechanisms nor mechanics whose mysterious actions he could not directly oversee.

Several natural philosophers in Gooday's account went so far as to argue that attaining a poor result through self reliance was more valuable in acquiring proper character than attaining a precise one through direct-reading instruments, like a mechanic tending a machine. Warwick puts an interesting twist on this rhetoric of the gentleman when he shows that it could cut two ways. While one noted actuary and calculator denounced the arithmometer for destroying the intellectual capacity of arithmeticians—arguing in essence that there could be no "manufactory" of arithmetical thinking, just as their could be no manufactory of ohms at Cambridge (Schaffer), nor of actuarial tables for the Society of Actuaries (Porter), nor of mechanical equivalents for natural philosophers (Gooday)—another turned the argument back on him by asserting that arithmetic was merely mechanical work anyway and that the arithmometer saved high-class brains for high-class work, namely, thinking and directing rather than performing mechanical work. On this basis too, it seems, the commercial ammeters and voltmeters earlier decried by natural philosophers entered their laboratories in the first decade of the twentieth century, as did calculating machines.

Class identity, however, is not the only basis for the conflict over practical values. The issues may perhaps be put more generally in a positive form. Precision instruments are packaged trust. And the trust they carry is a social accomplishment. We never calibrate our own instruments against the external world. Even if we check to see that a Fahrenheit thermometer reads 212° when placed in boiling water at a standard pressure, we must check the pressure using a barometer that in turn requires calibration. But checking a single point barely counts in any case. Obtaining precision in a thermometer requires rigorous control of the uniformity and stability of the materials, of the manufacturing process, and of all aspects of handling, as well as calibration of the entire scale. No one is able to do all this on their own. To get started we normally find another expert to do the calibration, and we trust their work. But even that expert can only check results against other instruments, standards, and theoretical calculations whose reliability is guaranteed by other people, and so on. It is this potentially infinite chain of trusted calibrations, both theoretical and experimental, which objectifies the external world, establishing the legitimacy of amps, ohms, logarithms, and life tables. Where trust is weak, the chain is weak, and measurements and calculations are compromised. Famously, an early UNIVAC computer set up in 1952 for Walter Cronkite's election coverage on national television predicted an Eisenhower landslide with only 7 percent of the vote tallied; but so few people—including its programmers—trusted the computer, whose result contradicted all polls, that its original prediction was suppressed.[9] Which is not to say that election forecasters should have trusted the computer. They should not have because the network of trust required to support it had not been established. Much the same could be said of early calculating machines and of Ayrton and Perry's electrical meters. They were not trustworthy because, put narrowly, not enough of the "right class" of people were prepared to stand behind them, or put more generally, not enough trustworthy people, instruments, and standards had established their authority. We cannot say simply that the instruments were, in themselves, unstable and inconsistent, because their stability and consistency could only be established within a wider system of agreement.

For Rowland in the United States, the contest for moral authority also played a prominent role, but Sweetnam's discussion suggests a somewhat different ground, having more to do with instilling the work ethic of pure-spirited Protestant pioneers than with gentlemanly behavior *per se*. Rowland's training, like that of the two other great names of American physics at the time, Gibbs and Michelson, tied physics directly to engineering. And that association has continued in the U.S. until very recently, with the basic physics courses at most universities designed for students of engineering and physics together. Many Europeans have shared Heisenberg's impression in 1929 that an engineering attitude characterized American physicists, including theorists, who "behave just like the engineer building a new bridge."[10] In this tradition, Rowland's rhetoric of pure science did not separate science from engineering as an intellectual activity so much as from an overtly commercial activity.

Yet even this distinction is subtle. In the picture Sweetnam draws, Rowland organized his manufacture and distribution of diffraction gratings like a specialized international business. Having surveyed the world market and currently available plane gratings, he invented a much-improved concave grating and a manufacturing process, kept the critical engine and techniques of production secret, hired a first-class machinist to carry out the ruling operation, found a subcontractor to manufacture blanks and distribute the end product, and did his best to promote his gratings as the standard for the entire market, by using them to establish the lines of the solar spectrum with unprecedented resolution and intensity. The enterprise was remarkably successful. It differs from Thomson's invention and marketing of electrical instruments largely in that Rowland did not pursue monetary profits. His rewards (and those of Johns Hopkins University, who subsidized production) were international leadership and prestige. This is a crucial difference in that it allowed Rowland to maintain the connection of physics to practical action without violating his own code of "pure" science. But it should not obscure how fully Rowland shaped his precision physics in the form of an engineering enterprise, not as applied physics, but its reverse, perhaps "purified" engineering. And it was through this paradoxical analogue of commercial activity that Rowland launched the network of users that would establish the credibility, and stability, of his gratings and their spectra.

Overall, then, the papers in this section, like those in the volume as a whole, require that we recognize precision as a remarkable cultural achievement, one rooted in the pursuit of unity. Unity has in this respect taken three distinctly recognizable forms: centralized bureaucratic states, international commerce, laws of nature. Each of these forms, however, depends critically on the others. A great mediator between them has been the capacity to produce objects upon whose value many people could agree, objects that would travel, and precision measurement sits at its center. This capacity is often regarded as a purely technical one, the capacity to produce uniformity through strict methods of control. It is certainly that, but such methods are always contested, because they depend on accepting certain values rather than others, values that are represented in the methods used and the objects produced, whether amperes of current, years of life expectancy, or wavelengths of spectral lines. It is not enough to say that these (partially) universalized entities are conventional, meaning that they are the product of more or less arbitrary definition. The point is that conventions are very difficult to achieve; that the means for establishing them are all-important; that they must be simultaneously material and social. What is true of conventions in general is true of those that establish precision. They are simultaneously agents of unity and products of agreement.

Notes to Chapter Thirteen

1. On Joule's practices and their relation to a new culture of "scientific" brewing as well as machine-shop skills, see Otto Sibum, "Reworking the Mechanical Value of Heat: Instruments of Precision and Gestures of Accuracy in Early Victorian England," *Studies in the History and Philosophy of Science* (forthcoming).

2. *Science* 13 (1944), nos. 2560 (Jan. 21), 2564 (Feb. 18), 2577 (May 19). Bausch and Lomb placed at least thirteen full-page ads for their precision instruments on the cover of *Science* during 1944 alone, for spectrographs, microscopes, metallographic equipment, etc., typically with the same association between competitive success in peacetime and victory in wartime.

3. Paul Uselding, "Measuring Techniques and Manufacturing Practice," in *Yankee Enterprise: The Rise of the American System of Manufactures*, ed. Otto Mayr and Robert C. Post (Washington, D.C., 1981), 103–26, esp. 117–21.

4. David A. Hounshell, "The System: Theory and Practice," in Mayr and Post, *Yankee Enterprise*, 127–52. Importantly for the themes of this volume, Hounshell emphasizes the bureaucratic nature of precision attained in interchangeable parts manufacturing.

5. Deborah Jean Warner, "Rowland's Gratings: Contemporary Technology," *Vistas in Astronomy* 29 (1986), 125–30. Spectra from William C. Winlock, "On the Group **'b'** in the Solar Spectrum," *Proceedings of the American Academy of Arts and Sciences* 16 (1881), 398–405. This picture was obtained from a plane grating ruled by Rogers's chief competitor, Lewis M. Rutherfurd (a wealthy lawyer and astronomical photographer) containing 8,640 lines per inch over a two-inch width, but Winlock also compares results using other gratings, including Rogers's, which had one-third the density of lines. The new spectral lines still only doubled Ångström's count.

6. Sibum, "Reworking the Mechanical Value of Heat," compares the traveling microscope with Henry Maudsley's famous comparator, the "companion of the bench" or "Lord Chancellor," based on a similar drive screw of even higher precision. On Maudsley's instrument, see Uselding, "Measuring Techniques," 110–11.

7. D. Graham Burnett, "Mechanical Lens-Making in the Seventeenth Century: Philosophers, Artisans and Machines" (Senior Thesis, Princeton University, 1993), 30–44. Steven Shapin gives a general account of the issue in "The Invisible Technician," *American Scientist* 77 (1989), 554–63.

8. Mario Biagioli, "Galileo the Emblem Maker," *Isis*, 81 (1990), 230–58. Mary G. Winkler and Albert van Helden, "Representing the Heavens: Galileo and Visual Astronomy," *Isis* 83 (1992), 195–217, also show how Galileo, after having attained high status at court, abandoned his early use of naturalistic pictures of the moon and planets, labeling such representations by others as the work of technicians, devoid of reasoning power. J. L. Heilbron, "A Mathematicians' Mutiny, with Morals," in *World Changes: Thomas Kuhn and the Nature of Science*, ed. Paul Horwich (Cambridge, Mass., 1993), 81–129 (discussed in pt. I).

9. Stan Augarten, *BIT by BIT: An Illustrated History of Computers* (New York, 1984), 164.

10. Werner Heisenberg, "Atomic Physics and Pragmatism," *Physics and Beyond: Encounters and Conversations*, trans. A. J. Pomerans (New York, 1971), 93–102, on p. 95, where Heisenberg ascribes the words to Barton Hoag, an American experimentalist in Chicago.

INDEX

CONTRIBUTORS

KEN ALDER is Assistant Professor of History at Northwestern University where he works on the social history of science in enlightenment France. He is preparing for publication his dissertation on *Forging the New Order: The Origins of French Mass Production and the Language of the Machine Age, 1763–1815* (Harvard University, 1991).

JAN GOLINSKI is Associate Professor of History and Humanities at the University of New Hampshire. He has published *Science as Public Culture: Chemistry and Enlightenment in Britain, 1760–1820* (Cambridge: Cambridge University Press, 1992) and is completing a study of the significance of social construction for the history of science.

GRAEME J. N. GOODAY is Lecturer in the Division of History and Philosophy of Science at the University of Leeds. He has written a number of articles on the genesis of scientific laboratories in Victorian Britain. He has also studied the changing moral discourses and metrological practices of physics and electrical engineering and is currently preparing *The Morals of Measurement: Accuracy and Complexity in Late Victorian Electrical Practice.*

FREDERIC L. HOLMES is Avalon Professor and Chair of the Section of the History of Medicine in the Yale University School of Medicine. He has written extensively on the history of biology and chemistry, especially in the biographical mode and with emphasis on "investigative programs." His most recent study, in two volumes, is *Hans Krebs: The Formation of a Scientific Life* and *Hans Krebs: Architect of Intermediary Metabolism* (Oxford: Oxford University Press, 1991–93).

KATHRYN M. OLESKO is Associate Professor of History and Director of the Program in Science, Technology, and International Affairs at Georgetown University. She is the author of *Physics as a Calling: Discipline and Practice in the Königsberg Seminar for Physics* (Ithaca: Cornell University Press, 1991) and is currently writing *The Meaning of Precision*, a study of the social, political, economic, and cultural contexts of precision in the German states before 1870.

THEODORE M. PORTER is Professor of History at the University of California, Los Angeles, where he studies the uses of quantification and of public knowledge. He is the author of *The Rise of Statistical Thinking: 1820–1900* (Princeton: Princeton University Press, 1986) and has just completed *Trust in Numbers: The Pursuit of Objectivity in Science and Public Life* (Princeton University Press, 1995).

ANDREA RUSNOCK is Assistant Professor of History in the Department of Science and Technology Studies at Rensselaer Polytechnic Institute. She is completing an edition of the correspondence of James Jurin, 1684–1750, Secretary to the Royal Society.

SIMON SCHAFFER is Reader in History and Philosophy of Science at the University of Cambridge. He is coauthor with Steven Shapin of *Leviathan and the Air Pump: Hobbes, Boyle, and the Experimental Way of Life* (Princeton: Princeton University Press, 1985) and is at work on a study of metrology and the labor processes of physical science in Victorian Britain.

GEORGE SWEETNAM is completing his Ph.D. degree in the Program in History of Science at Princeton University, with a dissertation on Henry Rowland and the Johns Hopkins school of physics that formed around his diffraction gratings and his studies of the visible spectrum in the late nineteenth century.

ANDREW WARWICK is a Lecturer in History of Science at Imperial College (London University). He studies the development of the physical sciences since 1750 and is currently completing a book on the emergence of the Cambridge school of mathematical physics during the second half of the nineteenth century.

M. NORTON WISE is Professor of History at Princeton University. Coauthor with Crosbie Smith of *Energy and Empire: A Biographical Study of Lord Kelvin* (Cambridge: Cambridge University Press, 1989), he has also published a series on British political economy and natural philosophy. He is preparing a book on the mediating technologies that in particular cultural "moments" have grounded scientific explanation.